96410

HE
8698   Douglas, Susan J.
D685   Inventing American broadcast-
1987   ing, 1899-1922

JUL 1 '88 DATE DUE		
OCT 31 1988		
NOV 21 1989		
MAR 25 1993		
4/14		
MAR 11 1994		
OCT 29 1996		

JOHNS HOPKINS STUDIES IN THE HISTORY OF
TECHNOLOGY
General Editor: Thomas P. Hughes

INVENTING
AMERICAN
BROADCASTING

1899–1922

SUSAN J. DOUGLAS

THE JOHNS HOPKINS UNIVERSITY PRESS
Baltimore and London

© 1987 The Johns Hopkins University Press
All rights reserved
Printed in the United States of America

The Johns Hopkins University Press
701 West 40th Street   Baltimore, Maryland 21211
The Johns Hopkins Press Ltd., London

Library of Congress Cataloging-in-Publication Data
Douglas, Susan J. (Susan Jeanne), 1950–
Inventing American broadcasting, 1899–1922.

(Johns Hopkins studies in the history of technology)
Includes index.
1. Radio broadcasting—United States—History.
2. Radio—United States—History.   I. Title.   II. Series.
HE8698.D685   1987   384.54'0973   87-3717
ISBN 0-8018-3387-6 (alk. paper)

The paper used in this publication meets the minimum requirements of American National Standard for Information Sciences—Permanence of Paper for Printed Library Materials, ANSI Z39.48-1984.

Photo Credits: Photographs appearing on pp. 18, 24, 69, 74, 83, 96, 99, 152, 157, 158, 198, 201, 204, 277, 294, 302, and 310 are courtesy of the Archives Center, National Museum of American History, Smithsonian Institution. Photographs on pp. 188 and 228 are copyright © 1907/12 by the New York Times Company. Reprinted by permission. Photographs on pp. 21 and 49 are courtesy of the New-York Historical Society. Photograph appearing on p. 301 is courtesy of Westinghouse Electric Corporation.

To my father, Colonel Harry V. Douglas,
and to the memory of my mother, Barbara

CONTENTS

Contents

ILLUSTRATIONS

PREFACE AND ACKNOWLEDGMENTS

MY GENERATION grew up in an environment saturated by the images and messages of the mass media. We were the first television generation, and the broadcast media have always exerted a powerful influence on our perceptions of reality, our self-images, and our dreams for our society and ourselves. It is no surprise, then, that when some of us became historians, we wanted to understand better the rise and impact of the communications systems that had so insinuated themselves into American life and thought. We wanted to analyze intellectually as adults what had gripped us emotionally since childhood.

I became interested in origins. How was America's broadcasting system invented in the first place? I thought the answer lay in the 1920s, and that is where I began. I quickly learned that this was hardly the beginning; nor was this era, in the end, the most interesting to me. As I kept going back in time—initially, I thought, to get background material—I was struck by how the basic questions surrounding broadcasting's role in society were raised decades before radio broadcasting as we know it began. I was also struck by how many precedents were set before KDKA ever went on the air. Thus, I chose to examine in detail what has sometimes been dismissed as broadcasting's "pre-history," and to argue that it was during this period between 1899 and 1922 that the basic technological, managerial, and cultural template of American broadcasting was cast.

This book represents my efforts to draw from and intertwine the two areas of study that have most influenced me: American studies and the history of technology. It was in these areas that I encountered teachers who changed forever my way of thinking and who offered me vastly expanded notions about history and culture. The book has its origins in my undergraduate training and interests, and I would like to thank Malcolm Marsden, who introduced me to American Studies and taught me that historical information was not confined to historical texts; Carmine Dandrea, who helped me learn how to think analytically; and Donn Neal, whose love of history was contagious, and whose constant guidance and support led me to pursue the study of history after college. In graduate school, Patrick Malone encouraged women students to tackle the history of technology, and to link that history to the study of culture. Mari Jo Buhle helped me understand the connections between gender roles and the rise of industrial capitalism. And Hunter Dupree, always asking those "big questions" for which he became famous, was the perfect mentor for those students whose work crossed disciplinary boundaries. Although no one at Brown at the time studied the media, Pat, Mari Jo, and Hunter encouraged my research in this area, and thus allowed me to pursue the work I thought important. I am lucky indeed to have had such generous and gifted teachers. My thinking at this time was also enriched enormously by my friendship with John Kelly, whose intellectual creativity provided essential nurturance and stimulation.

My research and writing were actively supported by a range of colleagues and institutions. Bernard Finn and James Brittain arranged for a research grant from the Institute of Electrical and Electronics Engineers, and without this grant I would have been unable to visit important archives. An early sabbatical allowed me to write a major portion of this book, and I am deeply grateful to my colleagues and the administration at Hampshire College who made this possible. I would like to thank the staff at the Columbia Oral History Library and the National Archives for their assistance. Bob and Nancy Merriam, who run the New England Wireless and Steam Museum in Rhode Island, opened their library to me, talked with me about the project, and provided personal demonstrations of old wireless apparatus. Ellen McGrew, an archivist at the North Carolina State Archives, not only expedited my research with the Fessenden Papers, but also offered support and friendship during my entire stay in Raleigh. Before the Archives Center opened at the Museum of American History, the Clark Collection was housed in the museum's Division of Electricity, and here every possible courtesy and kindness was extended to me. Barney Finn offered guidance and advice, Elliot Sivowitch and Ray Hutt tracked down sources for me and helped me locate stores of archival material, and Elliot shared his office with me while I did my research. Anastasia Atsiknoudas was a constant source of encouragement and help and, whether she knows it or not, was probably the person most responsible for my thinking about the effect organizational structure and behavior can have on technological change. Robert Harding at the Smithsonian's new Archives Center helped considerably with photographic research. Keith Geddes at the British Science Museum and Betty Hance at the Marconi Company Archives expedited my research in England. Gioia Marconi Braga, Marconi's daughter, generously invited me into her home and allowed me to read her father's early letters. I wrote nearly all of the manuscript in the Neilson Library at Smith College, and I am indebted to the staff there for their consistently professional assistance.

When I was circulating the manuscript for comments and reactions, I wanted my colleagues to say that it was great and needed no revision. None of them did this, of course: they were much better friends and more dedicated scholars than that. Instead, they offered thoughtful, careful, and often tough criticisms, and I am deeply grateful to them for pushing me to think harder and revise the manuscript yet one more time. Chris Sterling's criticisms of an early draft spurred me to do more research and to rethink the purpose of the book. Suggestions from David Allison, Alex Roland, and Merritt Roe Smith helped refine the material on the U.S. Navy and radio, and Roe was especially helpful and supportive. Frank Couvares and Barry O'Connell grilled me on the amateur chapter, and their questions prompted a major reevaluation of the material I was using. My many discussions with Bernie Carlson about the history of electricity, communications, and entrepreneurship enriched the book enormously. Nancy Fitch, Joan Landes, and Ted Norton constantly opened my eyes to new ways of thinking about cultural history and have been role models in interdisciplinary work. Two readers, whose identity I do not know, gave the manuscript an especially careful reading, and they posed major conceptual questions with which I was compelled to wrestle. The Society for the History of Technology has provided me and many others with a forum in which we could test and exchange ideas, and this has been critical to my scholarly development. My colleagues in Hampshire's Feminist Studies Program continue to open my eyes to new ways of thinking, and I remain indebted to this impressive group of women for their support. I would also like to thank my many energetic, earnest, and thoughtful students at Hampshire, on whom I have tried out ideas, who have taken an interest in the project, and whose questions, feedback, and concern has meant more to me than they know.

Several friends helped above and beyond the call of duty. Danny Czitrom has been a constant source of advice and support, and to have a colleague with his interests, let alone of his caliber, just down the road has enriched my work enormously. Michael Brondoli, although deeply involved in the preparation of a novel, read the entire manuscript and wrote detailed comments that boosted my spirits during often very dispirited times. David Kerr made invaluable comments on writing style and historical themes, and helped enormously with his specialty, journalism history. Hugh Aitken has seen this project through from its most tentative and awkward beginnings, reading draft after draft, and offering highly detailed comments on everything from specific sources to larger conceptual approaches. Without Hugh's intellectual rigor, his insistence on clear and precise writing, and his unflagging support, my development as a historian would have been deeply impoverished. My debt to Joe Corn is also huge. Few people have had someone read their manuscript with as much care as Joe brought to mine. His considerable skills as an editor, and his commitment to linking the history of technology to cultural history, informed his criticisms of the manuscript, and prompted major revisions and reorganization.

At The John Hopkins University Press, Henry Tom has been the perfect editor: he knew when to be patient and when to prod, when to be flexible and when to be firm. His skills and support have been greatly appreciated. Thomas Parke Hughes offered much needed advice and encouragement during the entire process. My copy editor, Jackie Wehmueller, was a professional in every way. And I want to thank my typist, Leni Bowen, who caught mistakes I had missed, who was always on schedule, and who has been a delight to work with.

My largest debts, both intellectual and emotional, are to my husband, Taylor R. Durham. Simply put, I could not have written this book without him. His love for ideas, his broadly interdisciplinary approach to knowledge, and the delight he takes in intellectual discourse made him the perfect listener, and an even better adviser. There were many times when I lost faith in this project, but he did not, and he never failed to remind me of his faith in me. He also went out of his way to make it possible for me to write whenever I had to, and that meant taking on more than his fair share of extra chores and errands, which ate into his own work time. T.R. did not simply pay lip service to egalitarianism; he put it into practice every day.

Finally, I would like to thank my parents. Of course, they helped sustain me financially during college and then through the seemingly endless years of graduate school. But more importantly, my father, Harry V. Douglas, and my mother, Barbara, each in their very different and often indirect ways, taught me a critically important lesson: that intellectual work has no value unless it is enriched by large doses of empathy, enthusiasm, and a sense of justice. I dedicate this book to them.

INTRODUCTION

IN THE SPRING OF 1922, a "radio boom" swept the United States. The craze was cast, in the pages of the press, as a "fever" tearing through the population, inflaming all in its path. The fever seemed to come from nowhere, it had a power and force all its own, and few were immune to its symptoms. "In all the history of inventing," exclaimed a typical editorial, "nothing has approached the rise of radio from obscurity to power."[1] The words *sudden* and *rapid,* as well as *amazing* and *astounding,* appeared frequently in magazines and newspapers, reinforcing the epidemic metaphor.

Radio was portrayed as an autonomous force, capable of revolutionizing American culture. It was a machine that would make history. It was also portrayed as a technology without a history. Rarely, in those heady, breathless articles about the radio boom, was reference made to the twenty-five years of technical, economic, and cultural experimentation that had led to and produced radio broadcasting. Radio was thus presented as an invention not burdened by a past or shackled to the constraining conventions of the established social order, but as an invention free to reshape, on its own terms, the patterns of American life.

But radio did indeed have a past. It had not simply burst from the blue in 1922, as such press accounts suggested. Nor was it an autonomous force somehow outside of or above the existing order of things. On the contrary, this technology was very much embedded in and shaped by a rich web of cultural practices and ideas. Radio broadcasting resulted from more than two decades of scientific and technical research, institutional jockeying for position, and changing conceptions of how the invention should be used, and by whom. During this twenty-three-year period, precedents were set that would determine, irrevocably, how broadcasting would be managed and thought about in the United States. Yet very little attention has been paid to the complicated social and technical processes that culminated in radio broadcasting. Even most historians of broadcasting, after giving a cursory review of radio's "pre-history," focus on the radio boom of the 1920s, when radio already meant broadcasting and the technology was controlled by a few major corporations. The ascendency of advertising as the source of radio's financial support and the emergence and subsequent dominance of the networks seemed the inevitable results of corporate control. This emphasis on the 1920s as broadcasting's formative era prompts concessions to both economic and technological determinism: it makes the communications corporations appear more prescient than they actually were, and it grants the technology a life of its own, seemingly impervious to the culture in which it evolved. This emphasis also fails to challenge the journalistic myth of a sudden, sweeping radio fever, or to confront the connections between myths such as this and the actual history of broadcasting.

This book is about what led up to the "radio boom": it analyzes how individuals, institutions, ideas, and technology interacted to produce radio broadcasting. The story begins in 1899, when Guglielmo Marconi demonstrated his new invention, the wireless telegraph, during the America's Cup yacht races in New York harbor. The invention sent dots and dashes—the Morse code—through the "ether" without wires. Marconi concentrated on selling wireless to major, commercial customers, such as steamship companies and newspapers, with already established needs for such a device. He was offering what was then wishfully yet too narrowly referred to as "point-to-point" communication between specific senders and receivers. He was not marketing the device to individuals, and he did not conceive of transmitting voice or music. Broadcasting was simply not part of his scheme. But during a nearly twenty-five-year process marked by technical improvements, unanticipated applications, economic and organizational transitions, and considerable though intermittent journalistic fanfare, wireless telegraphy, the nineteenth-century invention, became radio broadcasting, one of the fundamental developments of twentieth-century society.

My aim is to examine this transformation and to show that an analysis of radio's early history—the years between 1899 and 1922—is critical to understanding how and why America's broadcasting system assumed the structure and role it ultimately came to possess. Like virtually all emerging technologies, radio did not simply appear one day in its fully realized form, its components complete and its applications and significance apparent. Radio apparatus, and what all that apparatus meant to a particular society at a particular time, had to be elaborately constructed. Just as individuals and institutions worked, over time, to refine the invention, so did these inventors and institutions, as well as the press and the public, all interact to spin a fabric of meanings within which this technology would be wrapped.

This book, then, is about the social construction of radio.[2] It provides a detailed account of the inventors who made radio possible, and explores how their technical contributions and business practices shaped the early industry. It examines how major institutions, such as General Electric, American Telephone and Telegraphe, and the U.S. Navy, influenced the invention's evolution. But the book also analyzes how radio was thought about and represented in the pages of the popular press, and how that representation changed between 1899 and 1922, for I am arguing that, just as individuals and institutions, and the interactions between them, influenced the course of technical adaptation, so did journalistic portrayals of radio's promise and significance.

How do machines come to mean what they do? How is the public significance of inventions constructed and transformed? Certainly the press plays an important role in this process. For most people at the time, their first vision of radio and its predecessor, wireless telegraphy, was provided through newspaper and magazine articles. For inventors trying to promote their inventions and themselves, newspaper coverage was critical to success. This coverage was hardly neutral or objective: it legitimated certain uses of the invention while condemning others, and it favored a very particular, narrow, and romantic style of technical journalism. Certain stories were told again and again; other stories, important stories, were never told at all. The press, then, by presenting and endorsing certain attitudes toward radio, defined a pattern of ideas and beliefs about how radio should be used and who should control it.

Prevailing media definitions reveal a great deal about the role of technology in American culture, and about dominant attitudes toward technology and power. Yet such media definitions have been largely ignored. It is my purpose here to examine which definitions and interpretations came into play and gained preeminence during the years when wireless telegraphy became radio. Through this study, I hope to establish the importance of analyzing the ideological frameworks within which emerging technologies evolve. We can look at old articles about radio fever as fanciful and misguided stories of little consequence, or we can take them seriously, and analyze the connections they reveal between technology and ideology, and between language and legitimation.

I have chosen the latter approach for several reasons. Most importantly, I believe that our understanding of how America's first broadcasting technology was shaped can be significantly enriched if we view it through the prism of recent thought about the media. Scholars have demonstrated how today's media, by repeating and reinforcing certain values while ignoring or denigrating others, help legitimate and perpetuate the established social order. A faith in capitalism as an economic system superior to all others, the insistence that our existing form of government is democratic, the creed that consumerism equals freedom, and the legitimation of economic and political elites as the rightful holders of authority—all these constitute the dominant belief system as broadcast on television or presented in mass magazines.[3] At the same time, these media images are often filled with cultural contradictions, because there is a strong tension between traditional American values that antedate the rise of monopolistic capitalism and the values monopolistic capitalism requires to survive. Industrial capitalism was initially built on such values as hard work, thrift, self-denial, and deferred gratification. For many Americans those values were tied to Christian principles and community solidarity, altruism being the central anchoring value of both. Advanced capitalism, however, which depends on a large and robust market of consumers, can only survive if people bask in narcissism and come to believe in the importance of leisure, instant self-gratification, and spending.[4] Thus are Americans pulled between competing value systems, torn between the need to feel productive, selfless, generous, and noble, on the one hand, and the desire to luxuriate in the private, status-seeking, self-indulgent offerings of consumer capitalism on the other. What role do the media play in resolving this tension? By visually and rhetorically blending the past with the present, selflessness with spending, and tradition with modernity, they help construct new myths and heroes that justify and romanticize the status quo. Thus do the media mine the nearly depleted veins of tradition to produce an ideological alloy that buttresses the structures of capitalism.

These cultural contradictions are not new; neither are the ways in which media images and language serve to mediate between old and new. Yet we know too little about the early history of this process, and we know even less about how the broadcast media, which transmit this ideology, were themselves shaped by the values and beliefs of the established print media. Our ignorance stems from the fact that too few historians of technology have studied the mass media, and too few media scholars have concentrated on media portrayals of technology, especially past portrayals.[5] Yet the invention of new devices and the role of industrialism in American life were dominant themes in the popular press in the late nineteenth and early twentieth centuries. In fact, one might argue that it was during this period that the press's method of covering and interpreting technological change was developed. In other words, what scholars have identified as the functions of the mass media in the late twentieth century were being formulated and refined during the first twenty-three years of radio's history. Radio was hardly the focal point of such formulations, but it does offer an excellent example of how they evolved.

Ultimately, this book seeks to juxtapose the public, journalistic accounts of radio's development with the private, behind-the-scenes struggles to perfect and control the invention for commercial exploitation. I have, so far, emphasized the importance of studying media definitions of this emerging technology; I do not intend to suggest, however, that the press by itself played a determining role in radio's development, or that such analysis should replace technical and organizational history. Rather, all three stories—how the invention was designed and refined, how inventors and organizations either succeeded or failed in exploiting the invention, and which aspects of the story the press covered and which it did not—must be interwoven for us to understand the process of technological assimilation and legitimation.

The inventors and, later, the institutions seeking to profit from wireless telegraphy had to succeed in three different but interconnected arenas: technology, business strategy, and the press. Control over patents, which gave the inventor a complete technological system and which denied entry to competitors, was, of course, key. But without a well-conceived business strategy that accurately identified clients and both promoted and protected the invention in the marketplace, the inventor would not have been able to survive in America's capitalist setting.

Also of importance, however, and most frequently neglected by historians, were publicly articulated ideas and values as represented in the mainstream media. For those technologies not tied to the factory system or designed for and adapted within an organizational setting, public reception mattered. Prevailing stories about a particular technology legitimated some men as "rightful inventors" while denying recognition to others. Stories in the press could help sell stock and attract clients. They could influence demand for a particular invention. And they could inform readers about the technical inadequacies of inventions. Consequently, developments in the journalistic arena could dramatically affect technology as well as business strategy.

This, then, is the synthesis I am attempting. I see two historical

realities affecting the development of radio: the processes of centralization and institutionalization—private, rarely seen, often incremental and
amorphous, and extraordinarily powerful; and the public, communal
mediation of those processes in the press, a mediation which influenced
at the same time it was informed by competing attitudes toward technology, business, and inventors. These parallel and often interdependent
realities would produce both short- and long-term effects. They would
affect the technology's capabilities—which ones would be developed
and which would atrophy—and they would influence the organizational structures that would emerge to manage the invention. Sometimes
they were in phase, and their pull on the technology would be reinforcing; at other times, their tides pulling in quite different directions, they
left the invention and its promoters adrift. By exploring both the institutional and the popular responses to early radio, and the interactions
between them, we can best understand the richness of that process we
call technical change.

What distinguishes this book from other histories of radio and broadcasting is its emphasis on radio's early development and its interpretive
perspective. Erik Barnouw's three-volume *History of Broadcasting in
the United States* and Christopher Sterling and John Kittross's *Stay
Tuned* cover, in highly readable detail, the rise of broadcasting from
radio's earliest beginnings, but the emphasis of both studies is on the
decades following 1919. The standard general account of radio's early
years has been Gleason Archer's *History of Radio to 1926,* published in
1938. Archer's book, which provides overly generous accounts of the
role the U.S. Navy, RCA, and especially David Sarnoff, president of RCA,
played in the rise of radio broadcasting, simply lacks a critical framework
on the interplay among individuals, institutions, and historical trends.
The book's credibility is undermined by the rather friendly interest Sarnoff seems to have taken in its publication. My suspicions about the
book's reliability on certain points were confirmed when I was working
through the extensive manuscript collection on radio's history compiled
over a fifty-year period by George H. Clark, who worked in the radio
field from 1903 to 1946 and knew the history of radio intimately. Clark
tore out pages of Archer's book, stapled them to typing paper, and wrote
comments such as "Lies!" in the margins.

The most recent, and without doubt the finest, books on radio's
early technical history are Hugh G. J. Aitken's *Syntony and Spark* and
The Continuous Wave. Aitken's books recount, among other things, the
contributions of Marconi, De Forest, and Fessenden to radio's development, and thus there is some overlap between his work and mine. Our

approaches, however, are quite different. Aitken has focused on the intellectual history of the scientific and technical ideas that produced broadcasting and also, in *The Continuous Wave,* on the complicated behind-the-scenes negotiations that led to the formation of RCA. My study goes into the technical history and the origins of RCA in less detail, and analyzes certain topics Aitken has not, such as how the press covered radio, the U.S. Navy's early reaction to the invention, and the business history of the fledgling wireless industry. What links our work, and makes our studies complementary, is our shared interest in the emergence of ideas about what radio might be, the process by which these ideas took shape, and how these ideas evolved within the American economic and social setting.

· · ·

BEFORE TURNING to the radio story, it is important to review the dominant institutional trends, individual aspirations, and journalistic practices this new invention would confront in 1899. The way radio was initially used and the manner in which it was portrayed reflected the larger economic and cultural transformations gripping the United States at the turn of the century. America was recovering from twenty-five tumultuous years marked by wild economic fluctuations and severe depressions, an unprecedented concentration of corporate wealth, labor unrest, and political turmoil. Social conflict—between the city and the country, nativists and immigrants, workers and managers, the wealthy and the poor—had heightened many Americans' perception that with the rise of industrial capitalism they had lost control of their economy, traditions, and destiny. They were caught in an unsettling paradox. Entrepreneurs' efforts to conquer the continent's terrain and vast distances, most notably with the railroad and the telegraph, and thus better unify and coordinate the country, had along the way engendered economic chaos and regional strife. By 1894, when the depression and violent strikes were at their peak, marked by the hemorrhaging of both blood and gold, some writers and intellectuals consumed by "fin-de-siècle melancholy" found escape in utopian or dystopian novels or in prolonged trips to Europe. Their malaise was deepened by a recognition that the sphere of their influence had been small when compared to that of the business community.[6] By 1899, the influence of the business community was steadily increasing as large corporate concentration became established as the dominant method of organizing the American economy.

Radio would, as a result, enter an economic milieu in which large corporations, particularly those involved in transportation and communication, were becoming more powerful and more skillful at managing their interests. Business firms, determined to exert more control over market mechanisms, began coordinating their activities and strategies while reorganizing their internal structures along more efficient lines.[7] Corporate consolidation increased; the merger movement reached its peak between 1898 and 1902.[8] The number of managers rose, too, and they worked in bureaucracies that valued a range of technical skills over family connections or regional ties. By hiring professionally trained engineers to fill many of these administrative slots, industrial concerns sought to extend the application of science and technology to managerial activities.[9] Recent historians have argued convincingly that organizational and economic changes in the business community "had a more decisive influence upon our history than any other single factor."[10] Certainly the management and deployment of radio would be profoundly affected by this trend toward corporate centralization.

The course of radio's early development was also influenced by the professional aspirations and leisure activities of a subculture of middle-class men and boys, who found in technical tinkering a way to cope with the pressures of modernization. Success, survival even, for many of these men, required adjusting to the increasing bureaucratization of the workplace and fitting into hierarchical structures that often ignored or suppressed individual initiative. Conforming to these public roles sometimes engendered rebellion, albeit a necessarily circumscribed rebellion, against regimentation and authority, and against loss of autonomy. The social reform movements, the rise of professional and trade organizations, and the emphasis on education, all attested to the determination of middle-class Americans to regain what power they could within the constraints of the new, centralized bureaucratic setting. These people were seeking both more control and new bases of identification.[11] One such basis of identification was familiarity with mechanical and electrical apparatus. For certain upwardly mobile men, a sense of control came from mastering a particular technology rather than succumbing to the routinization and de-skilling of the factory system. Telegraph operators, for example, considered themselves to be members of a very distinctive, cohesive, and exclusive fraternity, because they had gained command of a technology and a code. Wireless telegraphy would also spawn such a subculture among middle-class boys and men seeking both technical mastery and contact with others in an increasingly depersonalized urban-industrial society. These men and boys were called amateur operators, and by 1910 they had established a grass-roots radio network in the United States. Their use of radio, which was oppositional to both corporate and government interests, played a major role in the emergence of broadcasting.

The tensions among independent inventors, the amateur operators, and interested corporations over how radio fit into American society found expression and, at times, resolution in the pages of the popular press. By 1899, the press itself had become a major American institution. The growth of American newspapers and magazines had accompanied and, in some cases, outstripped the rise of other business firms. The formation of press associations and their intensified exploitation of the telegraph during the Civil War extended the "news net" and increased the speed of news gathering from more distant scenes. Competing publishers' determination to provide the most extensive coverage of the war brought about enlarged news organizations whose stories were critical to ordinary people with relatives directly involved in the fighting. As Michael Schudson has observed, "The war pushed the newspaper closer to the center of the national consciousness."[12] The number of newspapers, their circulation, and the competition among them increased dramatically as their role in American society became more central. Americans, surrounded by rapid industrialization and the intemperance of an undisciplined capitalism, turned increasingly to the press for understanding. Between 1870 and 1900, the number of daily newspapers in the United States quadrupled, and the number of copies sold each day increased nearly six times.[13]

These newspapers, the stories they printed, and the way they were managed, represented larger economic and cultural contradictions and embodied the clash between altruistic goals and the selfishness of the marketplace. Joseph Pulitzer, whose newspapers often contained high-minded crusades against the rich and powerful, himself amassed a twenty-million-dollar fortune, one of the largest ever accumulated in journalism. Many of his crusades, and those of other major journalists, advocated labor reform, yet newspapers regularly exploited their own workers. The press also campaigned against the "trusts," yet newspapers devised monopolistic arrangements to control the gathering and reporting of news.[14] In the more dramatic confrontations between capital and labor, such as the railroad strike of 1877, the press invariably sided with capital.[15] Some reporters and editors became cynical over what they saw as the hypocrisy of the press. Others, however, fused idealism and realism in their outlooks and their prose, and the fusion found a receptive audience, especially among the conflicted middle class, who did not always condone vast inequities in wealth or power, but who hardly wanted capitalism overthrown.

The large newspaper's paradoxical role—capitalist firm, built and dependent on technical advances, yet watchdog over the postwar industrial order—produced an ambivalence about industrialism reflected in its articles and headlines. The press straddled both the world of business and the arena of public perceptions. Newspapers and magazines were owned by men who vigorously took advantage of America's economic system; yet to do so, they had to sell their articles and stories to people who often felt threatened by the vagaries and excesses of that system. Resolving this tension Americans felt about the human costs of the free enterprise system was what the press came increasingly to do. In its exposés, the press did not attack capitalism, the free market, or corporate-government cooperation per se; rather, it attacked the excesses the system permitted. By railing against the more flagrant examples of political corruption, industrial exploitation, and mass poverty, newspapers and magazines helped define what constituted the acceptable range of responsible yet remunerative behavior. Its crusades against privilege on the one hand and working-class anarchy on the other cast the press as ally and protector of its middle-class readers.[16] Yet the values on which industrial enterprise had been built—rugged individualism, self-sufficiency, materialism, faith in technology, and belief in progress—were all still very much celebrated in journalistic prose. By reinforcing and recasting such values, which were in harmony with those of the middle class yet also legitimated the goals of the business community, the press adroitly blended tradition with new circumstances as it helped define an emerging corporate culture. It was in the pages of the press, then, that the tortuous process of accommodation between institutions and the individual, and between technology and culture, was played out most visibly and symbolically.

The journalistic mold in which wireless telegraphy would be cast for the public, thus, was largely formed prior to 1899. Established conventions existed, too, for covering stories about technological change. One of the most durable and popular of these conventions was the inventor-hero. The late-nineteenth-century incarnation of the self-made man, the inventor-hero blended the traditional values of individualism, hard work, and self-denial with the newer realities of rapid technical change.[17] He was used to personify, and humanize, the rise of industrialization.

Since the founding of the country, Americans had worked at reconciling the pastoral ideal with the need for technology, and by the 1840s some had become convinced that technological advances and social progress were intertwined in an upward-moving spiral.[18] Others were not convinced. The robber barons confirmed what earlier doubters had predicted: technology would corrupt those who controlled it and exploit those who didn't. Obviously, the relationship between technology and culture was significantly more complicated than either view suggested, but the press disentangled the complexity, showing a gray skein for what it was: distinct threads of black and white. It wasn't that technology itself was either bad or good; what mattered was the kind of technology and who deployed it. Americans recognized that technological progress was fraught with good and evil, and the press made both their hopes and their fears more manageable by personifying this ambivalence. Jay Gould and Jim Fisk represented the worst sort of amorality and avarice that industrialization permitted, and people could castigate these men as villains, thus venting their fears about what technics was doing to civilization.[19] Men such as Samuel Morse, Thomas Edison, and Alexander Graham Bell provided reasons for celebration: they personified what was best about the American cult of invention. Although they certainly hoped their inventions would make them financially comfortable, these men had no plans, according to the press, which required the exploitation of other people.

But there is another critical aspect of these inventors' achievements which contributed to their popular success: they had harnessed electricity. What could be more heroic, or romantic? People could grasp the building of a machine, but channeling this potentially wild and unpredictable force seemed particularly miraculous.[20] In addition, the telegraph, phonograph, light bulb, and telephone brought widespread and visible improvements to the lives of many people, not just the rich and powerful. Electricity held a special place in the public's imagination, and the men who laid claim to capturing it were obliged to translate it into beneficial, practical uses.

The language used to celebrate inventor-heroes and to convey the wonder of electricity was flowery, naive, and inflated. It was, in a word, romantic. The romantic movement in literature, exemplified by the works of Emerson, Whitman, and Melville, had, by the 1890s, given way to realism and naturalism, which confronted directly the toll of industrial capitalism. So, in literature, romanticism was dead. But in the pages of the press, albeit in a bastardized form, it was alive and well. Heroes had free will and destinies; they conquered their surroundings and the doubts of other men. Instinct paid off. Introspection and action commingled in the hero's spirit, serving together as catalysts for his achievement. There was a profound faith in technological and social progress, even in the face of depressing countervailing evidence. And there was a certain pantheism: nature was wondrous and filled with mysteries; it should be harnessed to do man's bidding but not be destroyed.

Romantic prose sold newspapers, especially when it was used in the service of sensationalism. It also helped journalists distinguish themselves and provided one of the few opportunities reporters had to be creative and individualistic. Most importantly, this romantic rhetoric, when applied to stories about industrialism or technical change, mediated between tradition and innovation. It made industrialism seem less like an impersonal, inexorable force and more like a trend guided by well-meaning men in control of history. An invention might be revolutionary and difficult to understand, but if it was described in familiar, even old-fashioned, phrases and embedded in a recognizable framework of values, it was less remote and threatening. The language used to describe new inventions and the men who designed them was critical to the acceptance of technical change. Thus, using romantic prose when describing the scientific and business goals of entrepreneurs helped legitimate the private, corporate control of machines as the only equitable and progressive method of management.

The romantic picture of wireless painted in the popular press will be analyzed in the following pages to determine whose claims on the invention were legitimated and whose were not. It is reasonable to suspect that the press, an established communications business with a stake in any invention that might cheapen and quicken news gathering, might shape, both consciously and inadvertently, a new communications technology in its own image. Many members of the press, at least those not consumed by cynicism, liked to think of themselves as watchdogs over malfeasance, as educators, and as writers who liberated their readers from ignorance, boredom, isolation, and the oppressions of industrialism. Yet newspapers were also businesses, and none of these altruistic goals would be pursued if the economic cost became too great. Nor would the press's goal of supporting social justice be pushed so far as to challenge the basic assumptions and underpinnings of American capitalism. I hope to illustrate the extent to which these twin desires to uplift mankind and yet to enjoy the benefits of capitalism got imprinted, with all their cultural contradictions, on the early development of radio.

There was one other very important feature of the rapidly changing cultural landscape that remained a parallel but seemingly unconnected trend during radio's early development. This was the rise of mass entertainment. The growth and increasing influence of the press was one part of this cultural transformation. The volume of mass-produced and widely distributed books and magazines had skyrocketed since the Civil War.

The *Nation* noted in 1895 that magazines were being born "in numbers to make Malthus stare and gasp."[21] Completely new genres of entertainment, such as the dime novel, the comic strip, and the amusement park, attracted millions of devotees. Advertising became more prominent and bold, relying on widely recognized cultural symbols and linking consumer goods to specific norms and values. At the turn of the century, vaudeville was the nation's most popular form of entertainment in the public sphere. And moving pictures began to insinuate themselves first into urban working-class and immigrant neighborhoods and later into the world of the middle class. Increasingly, it was technology that brought entertainment, leisure, and escape. This new alliance between technology and entertainment was a marriage of profound economic and cultural significance, a marriage that would eventually produce radio as its most widely adopted offspring.

· ■ ·

THESE, THEN, WERE the larger institutional and cultural frameworks within which wireless telegraphy would become radio. All of the institutions that molded radio's development were themselves undergoing increased centralization and consolidation; they were also engaged in fierce struggles to control those inventions which would help them extend the hegemony of their particular technological systems throughout the country. Radio was swept up in these struggles, its corporate and technical fate increasingly intertwined with those of major private and government institutions. But there was also an important and dynamic reciprocal influence between institutional strategies and individual applications. Individual inventors interacted with corporations, the government, and the press; amateur operators constructed their own sets of meanings around radio, meanings with which large institutions had to come to terms. The press often symbolically mediated these interactions, and in doing so gave voice to certain ideas and silenced others.

It is important to remember that despite the weight and force of these processes within which the invention would develop, radio was not completely malleable. It had very particular attributes that made it difficult to control. It sent messages through space in all directions. It was not secret, or even private, and it was subject to interference. Access was at first unrestricted: anyone with inexpensive homemade apparatus could transmit and receive signals. Establishing financial and technical control over this invention proved problematical. And the invention introduced Americans to an unexplored, mysterious new environment, the electromagnetic spectrum, then known as the luminiferous ether.

The ether was invisible, it was everywhere, and it seemed open to all. Its possibilities were still unknown. No known rules governed its use. As an uncharted frontier, it inspired fear, suspicion, and visions of transcendence and escape. Sending messages without wires was one revolution; coming to terms with this electromagnetic environment was another.

• • •

THIS BOOK IS organized as follows. The opening chapter describes the invention's scientific and technical origins, and its highly publicized debut in the American press. The second and third chapters recount the wireless inventors' early struggles to make wireless a viable business, and focus, in turn, on early technical developments and competition, and on the inventors' varying entrepreneurial styles and management techniques. The fourth chapter describes the U.S. Navy's reaction to the invention and international debates about its proper use. Chapter five explores how inventors managed technical change and business strategy between 1906 and 1911. The sixth chapter looks at individuals rather than institutions, and tells the story of the amateur wireless operators and their particular vision of how wireless might be used by Americans. Chapter seven describes the origins and impact of the first radio regulation in America. Chapter eight examines how World War I, which profoundly accelerated consolidation and centralization in both the public and the private sector, affected the management of radio technology in America. The book's last chapter recounts the rise of broadcasting, and analyzes how the radio boom was interpreted in the press.

All of these chapters explore the interactions among technology, business strategy, and the press. The concept of three arenas of thought and activity has helped me to organize and make sense of a large body of overlapping material. But I also hope that a discussion of what such arenas involved and how they interacted will help shed light on the process by which radio, an emerging technology, was socially constructed in America. For if there is one major theme to the book, it is that technology is as much a process as a thing; it is an evolving relationship between people and their environment which is affected by more than one factor.

As the invention evolved, its name changed as well: in 1899, it was the wireless telegraph; by 1920, it was known as radio. The transition occurred between 1906 and 1912, when wireless telegraphy gave way to radiotelegraphy and radiotelephony (transmission of the human voice as opposed to dots and dashes). We do not know who was the first to use the prefix *radio;* we only know it was considered more precise than *wireless* because it indicated that the waves were radiated in all directions. Gradually the prefix *radio* stood on its own, as the words *radiotelegraphy* and *radiotelephony* seemed too long and clumsy. Between 1906 and 1920, the terms *wireless telegraphy* and *radio* were often used interchangeably; the word *radio* was more general, because sometimes it referred to the transmission of dots and dashes and other times to the transmission of the human voice. By 1920, radio broadcasting was popularly understood to be the transmission of voice and music without wires, and the term *wireless telegraphy* became obsolete in the United States. I have used *wireless telegraphy* in the early part of the book, *radio* toward the end, and have used the words interchangeably in the middle, to reflect contemporary usage.

I must add one final comment to this discussion of the social construction of radio. It is important to point out that this was primarily a white, middle-class, male construction, a process from which most women and minorities were excluded. Several women who have been unfairly dismissed or overlooked in other histories of broadcasting made crucial contributions to the invention's development, and where possible I correct the record. As we consider how meanings are constructed in our culture, and how those meanings interact with machines, we must never lose sight of whose meanings they are and of who had no voice in the process.

Telegraphy without wires—how attractive it sounds.
No more unsightly pole lines disfiguring the streets and highways,
ornamented with the dangling skeletons of by-gone kites.
No more perpetual excavation of the streets, to find room beneath
their surfaces for additional circuits that cannot possibly be
crowded on to the staggering lines that darken the sky
with their sooty cobwebs. A little instrument that one can almost
carry in the pocket, certainly in a microscopic grip,
and if your correspondent be likewise equipped, you may
arrest his attention and talk to him almost any time or place,
with no intervening medium but the . . . ether. . . . Possible?
Certainly. But will it pay?
For this is the final criterion with which this utilitarian age tests
all such propositions, and for the present under ordinary circumstances,
the answer must be NO.

Electrical World, June 10, 1899

CHAPTER ONE

MARCONI AND THE AMERICA'S CUP
The Making of an Inventor-Hero

1899

THE ADULATION OF heroes and the excoriation of villains became a dominant feature of American journalism during the late nineteenth century, when many aspects of American life were in flux. As the society navigated, and sometimes drifted, toward new horizons, heroes served as fixed points during an uncharted voyage. White and black images of good and evil stood out against—and helped make sense of—the complicated and subtle processes of industrialization, urbanization, and centralization which began accelerating in the 1870s. America's ability to cope with great complexity has been accompanied, and probably strengthened, by a reassuring simplicity in idea and symbol.

Hero worship and self-affirmation went hand in hand. When the press lionized a hero, it also flattered its readers and celebrated their aspirations and potential. The hero represented not only what was possible in individuals, but also what was best and most promising in American society. F. Scott Fitzgerald put it more eloquently when writing of Lindbergh's achievement: "A young Minnesotan who seemed to have had nothing to do with his generation did a heroic thing, and for a moment people set down their glasses in country clubs and speakeasies and thought of their old best dreams."[1]

The press relied on heroes most when a signal event dramatized and distilled the larger, more amorphous transformations taking place within society. The hero, by personifying the best of both the old and the new orders, served to "mediate polar tensions" in a culture.[2] Thus, heroes have often embodied both where America's been and where it's going. By examining their qualities, or those that have been accentuated, we can learn something from heroes about the values, dreams, and conceits of the past.

No doubt the rise and increasing hegemony of institutions in American life made the traditional myths of individualism, as apotheosized in the hero, all the more compelling. While writers such as Theodore Dreiser, Stephen Crane, and William Dean Howells came to portray Americans as powerless atoms buffeted about by vast impersonal forces beyond their control, the faith in individualism remained strong and vibrant in contemporary journalism and popular literature. Only the context of the myth was changing. In the late nineteenth century, the heroic individual triumphed more frequently in urban, rather than rural, surroundings and was successful because he had either founded or worked his way up through a modern organization. Organizations did not yet suggest the futility of individual effort; they simply provided a new setting and a new set of challenges.[3] The new heroes were not the men who blazed trails westward or fought their own lonely battles. They were men who stood out in a bureaucratic setting that made their courage, decisiveness, or risk taking all the more glittering and exemplary. Or they were men who, having no such organizational framework, had the vision and audacity to create one.

In late September of 1899, two such men, one already a hero, another about to become one, sailed into New York harbor. They could hardly have been more different. Yet they would be characterized in similar ways, often with the same adjectives and descriptive phrases, and would personify the bridge between the past and the future, the individual and the institution. Examining the way Admiral George Dewey and Guglielmo Marconi were treated by the press and the public reveals much about American culture at the turn of the century and suggests how contemporary journalism would respond to and shape a new communications technology.

America had gone for more than thirty years without a war hero, and even longer without one who could be cheered by the country as a whole. Admiral Dewey ended the drought. As commander of the U.S. Navy's Asiatic Squadron in 1898, he destroyed Spain's authority in the Philippines by sinking or incapacitating that country's entire fleet in Manila without losing one American life.[4] The victory, unbelievably swift and decisive, symbolized America's increasing confidence and influence in world affairs. Although the short-lived Spanish-American War produced other heroes, Dewey eclipsed them all. He had shown the world that the U.S. Navy, and the country it represented, was technically advanced, united, and brash, and was rethinking its role and destiny. When Dewey sailed into New York City in September 1899, he was overwhelmed by a spectacular celebration in his honor. Theodore Roosevelt, governor of New York, declared a two-day holiday. Friday night "pyro-technics" were followed by a Saturday parade in which, it was esti-mated, more than thirty thousand people marched, cheered on by nearly a million spectators. The "admiral who had smashed an enemy's fleet as an appetizer for breakfast" was greeted rapturously.[5] The New York newspapers announced Dewey's homecoming in huge headlines com-plemented by large, lavish illustrations. The entire front sections of the papers, from four to eight pages, were devoted exclusively to the "Greatest Popular Demonstration of the Century for a Living Ameri-can."[6] Popular magazines such as *Harper's Weekly, Outlook,* and *Cen-tury Magazine* featured cover stories, illustrations, editorials, and exten-sive articles on the hero's return. They all agreed that the reception was unparalleled.[7]

The decorations in honor of the hero were extravagant. White, blue, and gold flags and buntings adorned much of the city. A series of ionic columns flanked the parade route for a mile and led to the day's ultimate symbol, the Dewey Arch in Madison Square. The arch was one hundred feet high and was surmounted by and embellished with elaborate statues and depictions of military victories. Twenty gold-and-white "victory" statues lined the route north from the arch to Thirty-second Street. The columns, the arch, and the sculptings were modeled after monuments that had been erected for Roman triumphal processions. They were all temporary, designed and erected solely to honor Dewey, and so were made of wood and plaster of paris.[8]

The newspaper and magazine accounts of the parade showed how a press that had fostered the country's entrance into the war would now both justify that participation and provide symbolic closure to the adven-ture. The emphasis was on how unified America was in its goals: no mention was made, on this occasion, of the very real divisions between the jingoists and the anti-expansionists. Embellishment of the facts was not uncommon in journalistic writing of the period, and facts sometimes became subservient to the requirements of drama and myth. At the hands of reporters and editors, Dewey became the stereotypical hero. At the same time that he exhibited greatness, he also "seem[ed] at times exactly like scores of other average men."[9] He was certainly brave, daring, resilient, and decisive; but he was also cast as modest, considerate, egalitarian, devoted to duty, and concealing reservoirs of feeling under a usually reserved exterior.[10] His voice "shook with emotion," reported the *Herald,* as he accepted a gold loving cup from the mayor of New York. Pointing to the men who had fought with him in Manila, Dewey announced "These are the men that did it. Without them I could not have done what I did."[11] The men who served under him reciprocated, telling one reporter, "We would have sailed straight into hell after him."[12] Special attention was paid to his tact, diplomacy, and humanity. His personal humility and "gentleness" captured and reflected the cultural altruism many Americans believed had motivated their "liberation" of Cuba and the Philippines.

In the eyes of the press, Dewey had struck the right balance between caution and boldness, preparation and action; adherence to routine had been complemented by deft improvisation. Dewey had obeyed orders and had given orders. He had inspired confidence in those above and below him in the navy. His success validated the efficiency and necessity of bureaucratic structures and resources while simultaneously affirming the individual's potential to transcend organizational regimentation. His triumph must have given all those anonymous spectators heart: Dewey himself had been virtually unknown to the general public before 1898, and now he was "classed among the truly great Americans" of the nineteenth century.[13] He came from the ranks of the people; he was a follower and a leader; he was what they were. The admiral whose career spanned forty years of the nineteenth century had shown us how to step into the twentieth. He helped Americans understand their "new place in the world of the future."[14] As portrayed in the press, his success validated the relevance of old values to new demands.

The parade reflected, amplified, and consecrated many of the country's aspirations at the turn of the century. It was as much a celebration of how far America had come since the Civil War as it was a welcome for a war hero. Capping decades of political, economic, and ideological divisions, this highly self-serving war with Spain reaffirmed the country's faith in democracy and provided a renewed sense of unity and ideological purpose. The war also demonstrated that the benefits of democracy could be brought about even more quickly and efficiently by a well-meaning application of technological might; in this case, America's "New Navy." The war with Spain seemed nothing less than a holy crusade, bringing technological and political enlightenment to the infidels. The success of the crusade smothered those recurring feelings of self-doubt, often given voice by American intellectuals and haughty foreigners, which asserted that America would never enjoy the respect and status of European countries. The second wave of immigration and the withdrawal of European investments in America during the 1890s contributed to feelings of resentment and helplessness in the face of foreign influence. And the depression, which brought widespread misery and recurring labor unrest, had driven hope into hibernation. The parade, then, was an expression of optimism, idealism, and self-congratulation. It was a "public act of regeneration," a mass ritual in a mass society.[15] Dewey, who had presided over a series of unpredictable, volatile confrontations with Asians and Europeans, and had handled them decisively with confidence and equanimity, personified what many Americans had come to crave: a sense of unity of purpose, evidence that the United States had a special democratic mission in the world which it was prepared to act on, and a renewed faith that that mission could be realized through the application of American technology. Thus were the Dewey Day celebrations framed within the larger ideological nexus of the American faith in progress, a faith that often served as America's secularized religion.

A conviction that technological and social progress were intertwined in an ever-upward-moving spiral was especially prevalent in the press in 1899. One of the year's biggest ongoing stories was the end of one century and the beginning of another, and journalists marvelled over the extent to which Americans had, in the preceding one hundred years, conquered distance, time, and uncertainty. Believers understood progress to be that continuous advance through time of a civilization in a pattern of successively higher stages of development: an intellectual, cultural, technological, and moral ascent to the ever better.[16] *Scientific American* was convinced such an advance had occurred: it referred to the preceding one hundred years as "A Century of Progress in the United States." A writer for *Popular Science Monthly* admitted that the "men of the nineteenth century . . . have not been slow to praise it."[17] What they praised most were what was easiest to see and count, the innumerable advances in technology and science. The remarkable contrast between the slowness of transportation and communication in 1800 and the swiftness of 1900 especially excited writers, who wrote of the revolutions in these areas. "Mechanics, by its space-annihilating power," observed a writer in *Nature,* "has reduced the surface of the planet to such an extent that the human race now possesses the advantage of dwelling, as it were, on a tiny satellite."[18]

The inflated language used to praise American progress revealed an intermingling of hubris and doubt as men celebrated their technical mastery. *Scientific American* gave voice to technological determinism when it observed that "the railroad, the telegraph, and the steam vessel annihilated distance." There was an uneasiness in this observation, however: "Peoples touched elbows across the seas; and the contagion of thought stimulated the ferment of civilization until the whole world broke out into an epidemic of industrial progress."[19] The disease metaphor, intermixed as it was with words like *civilization* and *progress,* reflected an ambivalence about the relationship between men and machines and an uncertainty about man's ability to manage industrialism. Other metaphors, however, decidedly masculine in their attitudes toward technology and nature, assured men that they were very much in control. One passage in *Popular Science Monthly,* for example, described nature as a fecund seductress whom men had to subdue: "It seems impossible that Nature, now that we have discovered the true method of interrogating her, should not go on revealing herself to us with greater and greater fullness." With nature supine, "the scientific movement," the province of men, was "at its maximum of vigor and productiveness."[20] With nature cast as female, and science and technology as male, progress was naturalized as part of the broader scheme of things, in which men and machines conquered, and women and nature acquiesced. In the journalistic elaborations on progress, such explicit masculine conceits commingled with the more implicit fear that technology, like women and nature, might elude male control.

Dewey's conquest affirmed the primacy of male control. The celebrations insisted that noble, selfless heroes could harness technology and, through it, promote political and social justice. The parade, the neoclassical decorations, and the journalistic fanfare cast military conquest and American expansionism as altruistic gestures, designed first and foremost to liberate the oppressed. In October of 1899, the New York newspapers and northeastern magazines held up a most flattering mirror to their readers. Like Dewey, Americans were brave and selfless; their mission was to spread democracy around the world; and technological and social progress were inevitable in the United States. Modern conquests made possible by modern machines in the end reaffirmed and spread traditional values.

This was how the press constructed the significance of Dewey's heroism. For the twenty-five-year-old, half-Italian, half-Irish inventor who had arrived in New York five days before Dewey, this was the prevailing ideological framework within which he would have to operate. If he could convince the press that he was selfless and that his invention would well serve free enterprise, democracy, and altruism, then he, too, would have a victory in the journalistic arena.

Guglielmo Marconi was already assured press coverage. James Gordon Bennett, the notorious publisher of the *New York Herald* and an avid yachtsman, took particular interest in the *Dublin Daily Express*'s coverage of the Kingstown Regatta in 1898. Marconi had reported the progress of the race by his new method of communication, wireless telegraphy.

Early in 1899, Bennett offered and Marconi accepted, five thousand dollars to cover the America's Cup Yacht Races for the *Herald*.[21]

Bennett's talent for making news happen was legendary. It had been his idea, for example, to send Stanley to Africa to find Livingstone. Bennett and his father had been the first to report about and promote the potential of other inventions, including Morse's telegraph and Edison's light bulb. In the late 1830s, James Gordon Bennett, Sr., had pioneered the use of dispatch boats to intercept ships bringing news from Europe and then deliver the news to the *Herald* offices hours before the steamers docked. His son established the Commercial Cable Company to expedite transatlantic communication. Competition over speedy news gathering and dissemination had prompted the *Herald* and many other penny press newspapers to encourage and exploit a series of technical advances in printing and communications.[22]

To Bennett, wireless telegraphy promised to be another such invaluable innovation. For, although a revolution had taken place in transportation and in communications, a technical and commercial gap existed; the two revolutions had not converged. When moving so quickly from one place to another, travelers were incommunicado. Transatlantic passengers heard no news of the world for however long they were at sea; if their ship foundered, they had no modern way to signal for help. While at sea, they were isolated from the communications networks. The supposedly modern navies of the world, which strutted their new ironclad warships, still used antiquated methods such as homing pigeons and flags for communication and tactical signaling. The news of Dewey's victory, for example, had taken a week to travel from the Philippines, via ship and cable, to the United States. Off-shore islands too small to justify a cable, or separated by a channel too shallow to accommodate one, transmitted and received information by mail and messenger. Basically, where lines or cable connections could not go, there was no communication, even at the close of the "Century of Progress." Marconi proposed bridging this gap by sending the Morse Code through the air, without using any tangible connections at all.

The timing of Marconi's demonstration could hardly have been better. Dewey's parade took place on Saturday, September 30, and the festivities overflowed into the following day. The yacht races began on Wednesday, October 3. In New York, the week seemed to be one continuous holiday. The *Herald* summarized the week's events: "Dewey has had his day, or rather week. Now for the Yacht Race."[23] While the admiral's triumphant return may have been a tough act for a young, relatively unknown foreigner to follow, it also helped set the mood and evoke the optimism, awe, and faith in progress necessary for an enthusiastic reception of the new invention. The success in international competitions which Dewey's parade had honored fueled the already keen interest in the competition between America's *Columbia* and Ireland's *Shamrock*. The *Herald* promoted the use of wireless during the races as "a feat unparalleled in the history of journalism." The paper also emphasized the democratic benefits of the invention: "The *Herald* will thus prove a boon not only to science but to millions of persons who await with eagerness the result of a contest that has excited more interest than any in the history of the America's Cup."[24] The setting was dramatic and highly charged, guaranteeing Marconi and his invention excellent exposure.

· · ·

ALTHOUGH THE PAPERS did not explicitly ask the question, reporters and readers alike may have wondered why the country that had produced Morse, Edison, and Bell now watched a foreigner introducing the latest advance in electrical communications. How did Guglielmo Marconi come to be the "inventor" of wireless telegraphy? What scientific and technical sources did he build on, and which ones did he inadvertently, or judiciously, neglect?

When Marconi introduced his invention to the British in 1896, many members of the academic and technical communities dismissed the apparatus as highly derivative: "The present subject is not new. It has occupied the attention of inventors for at least fifty years."[25] While this assessment was too broad and sweeping, Marconi's wireless, like most inventions, was certainly a hybrid, its lineage a combination of theoretical physics, laboratory experimentation, and seat-of-the-pants testing and tinkering.

The concept of transmitting signals through space without wires had intrigued scientists and inventors for decades. The earliest experiments used electrical induction or conduction to transmit signals without wires. The two men who stimulated practical investigations into transmitting signals without wires were Joseph Henry of the United States and Michael Faraday of England. Both were interested in the relationship between electricity and magnetism, and in the late 1820s and early 1830s, each discovered electromagnetic induction. Faraday published his results first, and thus generally received credit for the discovery. What both men observed was that when a magnet was placed near a conducting circuit and was then rotated or moved, the changing magnetic field induced a current in the circuit. If the magnet remained stationary, the magnetic field was constant and there was no effect on the conductor. Faraday then discovered that in an arrangement of two separate and unconnected wires, a change in the current in the first wire would induce a momentary current in the second wire. The only connection between the two wires was space. Again, change was central to obtaining results: a constant current in one wire did not induce a current in the second. Thus, any change in the lines of magnetic force, whether produced by a magnet moving relative to a wire, or by a change in the current passing through a wire, would induce a current in an independent, physically unconnected wire.[26]

In 1838, shortly after Faraday's and Henry's experiments, telegraph engineers observed a related phenomenon: conduction transmission. They discovered that two wires were not necessary to complete a circuit: one wire could be eliminated and the return made through the ground. As news of these discoveries traveled via the scientific and technical journals, experimenters became excited by the practical possibilities. The Morse Code was transmitted by opening and closing a circuit at different intervals. This on/off operation caused a change in current and could be expected to induce messages in similar circuits not connected by wires, thus achieving wireless telegraphy. Also, water was a conductor. Couldn't experimenters exploit its conductive properties to transmit telegraphic signals?

By the 1840s, in both Europe and America, people began exploring the possibilities of using induction and conduction to signal without connecting wires.[27] One standard experiment involved running two parallel wires along the opposite banks of a river or other body of water, with no connection between the two wires except that provided by the water. This method of signaling came to be known as subaqueous telegraphy. When the circuit was closed on one side of the river, the parallel circuit on the opposite side would register the transfer of energy. A similar test consisted of elevating the parallel wires and transmitting the signals by induction. But wireless telegraphy by induction was severely limited by one crucial drawback: the greatest transmission distance achieved was a mile or two. Beyond this distance, induction was too weak to be useful for signaling. So, despite the initial optimism generated when signals were received across a canal or pond, this method of wireless communication hit a technical dead end. Additional scientific knowledge was essential if progress was going to be made. The theoretical and empirical work that led wireless communication out of this cul-de-sac was accomplished by two university-trained European scientists, James Clerk Maxwell and Heinrich Hertz.

In 1865, Maxwell published his "Dynamical Theory of the Electro-Magnetic Field," which became inspirational to those experimenting in wireless transmission. Like other European physicists in the mid- and late nineteenth century, Maxwell was searching for unifying principles that would link and better explain various related phenomena. This trend toward synthesis prompted scientists to consider less how forces were different and more how they were similar. Maxwell also helped to establish the importance of electricity to the study of physics.[28]

In "Dynamical Theory," Maxwell maintained that accelerated changes in electric and magnetic forces sent waves spreading through space at a definite speed. An electric spark could provide such a necessary quick change in current. He asserted that light, rather than consisting of material particles, was one type of electromagnetic wave, and he suggested that other forms of waves could very well exist, even though they were invisible to the human eye. Maxwell determined the speed of the electromagnetic waves to be exactly the speed of light: 186,000 miles per second. What differentiated these waves, then, was not speed but the number of waves radiated per second.

One important aspect of Maxwell's theory that sometimes is not mentioned in radio histories perpetuated the concept of the ether as the environment in which these waves supposedly traveled. For Maxwell, implicit in his notion of the similar properties of different types of waves was the belief in a single medium that transmitted all these forces. How could there be waves—crests and troughs—without an environment, in nothingness? Some earlier theorists had believed that each force—gravity, electricity, light, magnetism—had its own special medium of transmittal. Maxwell brought unity to these concepts and endorsed the notion of the single medium, *the* ether, and when his equations predicting electromagnetic action were demonstrated so conclusively in experiments by Hertz and others, it was assumed that his theory about the ether had to be correct, as well. Implicit in Maxwell's treatise was the possibility of producing, detecting, and timing the waves he described. But Maxwell's revolutionary theory was to remain untested for nearly twenty-five years.

By the time Maxwell's work was published, the British and German scientific communities had become more closely allied, and the resulting dialogue benefited the advance of physics in both countries. This increased exchange occurred simultaneously with the rise of the university system in Germany, in which organized scientific research began to flourish in lavish, newly built laboratories. Hermann von Helmholtz was at this time one of Germany's foremost scientists and an active member of the British-German scientific community. In 1870 he assumed the prestigious chair of physics at Berlin and began challenging his colleagues to study and assess Maxwell's theory.[29] In 1878 he acquired a brilliant student, Heinrich Rudolph Hertz, who for the next five years would serve as Helmholtz's pupil and assistant. Although Helmholtz asked Hertz to test Maxwell's theories in 1879, Hertz did not begin his landmark experiments until seven years later. After two years of work, Hertz succeeded in 1888 in producing and detecting the waves Maxwell had described. He established, as predicted, that these electromagnetic waves moved at the same speed as light.

Hertz had to devise ways both to generate and to detect these waves.[30] Because the waves were the result of a rapid change in electric current, Hertz had to produce electrical oscillations, or high-frequency alternating current. Scientists knew that electrical oscillations could be produced by the discharge of a Leyden jar, and that was where Hertz began. He connected the jar to an induction coil, which consisted of a primary winding with a few turns of wire which induced a higher voltage in a second coil having a larger number of turns and called the secondary. The electricity passed from the induction coil up through rods on each end, to which were attached perpendicular brass rods. On the outside ends of the rods were attached metal plates or spheres, and on the inside ends of each rod was a hollow metal ball a few inches in diameter. The space between these two metal balls, just small enough to create an air gap, was called the spark gap. The Leyden jar stored the electric charge, the induction coil magnified it, and the spark gap and metal plates radiated it out into the ether. This forerunner of the transmitter generated a high-voltage alternating current that surged back and forth between the metal balls and produced "electric waves which go out into space in the form of ever-increasing spheres."[31] Standing several feet away and using a detecting loop of wire, each end of which was capped with small metal balls, Hertz "received" sparks that could be seen in the dark, demonstrating the transfer of energy which Maxwell had predicted.

Hertz's success was widely publicized and prompted many scientists to pursue research on electromagnetic radiation. But in the development of wireless telegraphy, the next important improvement was in the detector. Hertz's small loop of wire was very crude and could only detect the waves at short distances. The receiver that became incorporated in early wireless apparatus was Oliver Lodge's coherer, added to the system in 1894.[32] Lodge, a British physicist at the University of Liverpool, also had been working toward verifying Maxwell's theories, but he was beaten to the proof by Hertz. His coherer was based on Edouard Branly's observation, published in 1890, that no matter how good a conductor a metal might be in bulk, if finely shaved into bits, its resistance was very great, indeed.[33] Branly's coherer was a six-inch-long glass tube filled with iron filings; the filings' resistance decreased remarkably with the impact of radiated electric waves. Their high resistance was restored, however, if the coherer was jarred or tapped. In his 1894 lecture to the Royal Institution titled "The Work of Hertz," Lodge demonstrated his improved coherer and showed how it could be used as a detector of Hertzian waves. He included a "trembler" or decoherer, which mechanically shook the filings to restore their high resistance. His coherer was more sensitive than Hertz's loop and allowed for the detection of wireless waves at greater distances.

Such was the state of the art in 1894. The dramatic scientific progress that had occurred since 1888 had been made possible by the vitality of and communication within the increasingly well-organized European physics community. America had only recently begun to establish significant university resources for the study of physics, and the nation's geographical isolation worked against comparable intercountry sharing of ideas. Europe enjoyed distinct advantages in elaborating the theoretical foundations and extending the demonstrations of transmitting electrical impulses without wires. Interest in electromagnetic radiation and detection of Hertzian waves remained keen, and experimentation in several countries continued. But the insulated university network that had so advanced the art also displayed little interest in transforming Hertz's experiments into a commercial venture. In fact, such an effort would have seemed vulgar to many academics. Thus, at the close of the century, a scientific reservoir was available for tapping, and a gap in the communications network awaited filling. The missing component was an entrepreneur bright enough, shrewd enough, and persistent enough to establish bridges between the realms of science, commerce, and popular imagination. That man was Guglielmo Marconi.

The very formative experiences that might have subverted Marconi's technical and entrepreneurial goals actually helped the inventor achieve them.[34] Although Marconi's father was a landed proprietor who managed a large estate just outside of Bologna, Marconi had shown little interest in or talent for assisting with the daily administrative duties. He and his father were rarely on good terms, so entrepreneurial skills would not be developed through an active apprenticeship. His mother, Annie Jameson, with whom Marconi was very close, was of the well-to-do and well-connected Scotch-Irish Jameson family of brewers and distillers. So, despite his lack of specific, directed training, he came from a family with a decidedly commercial orientation, an orientation that had not, as yet, channeled him into a particular vocation but that would imbue his approach to experimentation.

Marconi's education was informal and haphazard until he entered the Technical Institute in Leghorn at the age of thirteen. Throughout his youth, he had tinkered with machinery and electrical apparatus and conducted his own experiments based on the work of Faraday and Franklin. When he began studying physics at the institute, Marconi was so excited by the subject that his mother hired private tutors to reinforce his classroom studies. While at Leghorn, Marconi met and became friends with a retired telegraphist who taught the teenager the Morse Code. When Marconi returned to Bologna in 1893, Annie Jameson prevailed upon Auguste Righi, the noted professor of physics at the University of Bologna, to allow her son to audit Righi's classes. The fact that the Marconis were Righi's neighbors appears to have been more persuasive than any confidence the professor had in the young man's potential. Righi also arranged for Marconi to obtain library privileges and allowed him to experiment in the laboratories.

According to Marconi's daughter, Righi was an indifferent and sometimes unencouraging mentor. Yet Righi provided a crucial link in Marconi's development. Trained in mathematics and physics, Righi had been devoting much energy to applied research, and his work served as an example of how such experimentation was conducted.[35] Through his lectures and his lab work, he distilled for Marconi the work of Maxwell, von Helmholtz, and Hertz. Thus he made accessible to an otherwise untrained and spottily educated student the major theoretical and empirical foundations of wave propagation. He helped make Marconi privy to knowledge and a tradition that was rarely accessible to those outside the university system. Like most creative people, Marconi needed both skepticism and encouragement to challenge and sustain him as he worked. Righi provided the raised eyebrows, the doubt, that spurred Marconi on.

Encouragement and conviction came from Annie Jameson. The role of this highly determined woman in contributing to the development of wireless has too frequently been underemphasized. While Marconi's father complained about his son's worthless tinkering, his mother nurtured him at every stage of his development, often at the cost of harmonious relations with her husband.[36] She helped Marconi set up his lab at home, provided him with as much scientific and technical education as she could, discussed his progress with him, and witnessed his first demonstrations. She also provided him with another invaluable advantage:

she insisted that he read, write, and speak English fluently. Without this skill, and all the other support from his mother, Marconi would not have achieved the success he did in both England and the United States.

Marconi's vision for wireless was grand and singular. He was not experimenting with the lecture hall in mind, and he was not interested in writing scholarly articles or in dazzling students or other audiences by producing sparks in the dark. He sought to take wireless transmission out of the university lab and employ it in a practicable, commercially successful system of communication: he meant to have the thing pay. To make it pay, the transmission distance had to be measured not in feet, but in miles.

Through a painstaking process of trial and error (which would characterize his life's work), Marconi improved upon the apparatus of Hertz, Righi, and Lodge. He began in his family's attic, duplicating Hertz's transmissions of a few yards, detecting the feeble sparks with a small metal loop. He adopted Righi's spark gap, which consisted of four metal spheres, and increased their size, which gave him greater transmission distance. He then incorporated his own coherer, which was a refinement of Lodge's glass tube. Coupled with the coherer was his decoherer, a small electric hammer for tapping the filings back to their nonconducting condition.[37] After making these refinements, Marconi introduced the two components that would transport the system from the laboratory to the commercial world: the Morse Key/Recorder and the earthed aerial. As legend has it, the key entered by design, the earthed aerial by accident.

It was when Marconi was experimenting with the metal plates that were connected to each outside end of the spark gap that he made one of his most important discoveries. He had hoped that by replacing Hertz's small plates with larger slabs of sheet iron and elevating them above the ground, he would obtain longer waves and greater distance. What happened was that when he temporarily placed one of the slabs on the ground while holding the other slab up in the air, he noticed a considerable increase in the strength of the received signal. This observation led to his famous innovation of including in both the transmitter and the receiver a connection to earth as well as a vertical conductor or aerial.[38]

With his new aerial design, Marconi was able to signal over hills and achieve a transmitting distance of three miles on his father's estate. With the Morse Key, he transmitted dots and dashes. At this stage of development, telegraphy without wires was a rather straightforward operation. When the Morse Key was closed, either by a quick touch for a dot or a slight longer touch for a dash, the current passed from the batteries and through the thick wire wrapped around the sparking coil. This primary winding induced a current in the secondary, and the current rushed to the solid brass spheres. From the spark gap, "bluish tinted sparks" flashed, sending out intermittent electrical oscillations detected by the coherer.[39] Marconi had a Morse inker connected to the coherer: the signals were recorded on tape, not listened for by an operator.

Annie Jameson, convinced of the potential of her son's achievement, began to seek customers. She discussed the invention with representatives of the Italian Ministry of Posts and Telegraphs, who saw in wireless no advantage over the telegraph and offered Marconi no support. Had the family approached the Italian Navy, they might have received more encouragement. But Marconi's mother would not wait for the Italians to recognize the merits of wireless; in February 1896, she took her son to England, where her family connections were excellent and where interest in improving maritime communications would no doubt be keen.

Marconi and the Jameson family spent the next year and a half patenting his apparatus; demonstrating it before government officials, the press, and the public; and exploring various sources of financial support. William Preece, who had himself been experimenting with electrical transfers by induction, was engineer-in-chief of the British Post Office, and he tried in vain to arrange a speedy takeover by the government of wireless; this arrangement would have provided Marconi with facilities and a salary in exchange for his inventions. Preece did offer the inventor £10,000 for his original patent, a sum too small to lure Marconi away from his cousin Jameson-Davis's suggestion that the family form its own company. Jameson-Davis arranged for the financial backing, and on July 20, 1897, Marconi and his associates incorporated the Wireless Telegraph and Signal Company Limited with a capital of £100,000. At first Marconi requested that his name be excluded from the title, but in 1900 the name was changed to Marconi's Wireless Telegraph Company, Ltd. The company was to acquire Marconi patents throughout the world. To the company Marconi transferred all except the Italian rights to his patents, receiving in turn £15,000 in cash plus £60,000 in paid-up shares and a contract employing him as chief engineer at £500 a year.[40] Wireless signaling had made the transfer from the scientific-academic sector to the marketplace.

With the possibility of government support for wireless eliminated, Marconi had to popularize the invention on his own and demonstrate its advantages dramatically and directly to the public. One of his earliest publicity stunts was establishing a wireless link between Queen Victoria's Isle of Wight residence and the Royal Yacht lying nearby, off

Cowes. Wireless provided the queen with regular bulletins on the health of the Prince of Wales, who was recovering from a fall.[41] By 1899, Marconi had established a link between the East Goodwin light vessel and the South Foreland lighthouse, a distance of twelve miles. On March 3, when a ship ran aground on the Goodwin Sands, signals of distress were relayed from the Goodwin light vessel to the lighthouse, which immediately dispatched lifeboats to the area, saving the passengers and the cargo. After this success, many coastal towns began pressing for wireless installations between all lightships and the shore.[42] On March 28, 1899, Marconi successfully linked opposite shores of the English Channel, sending a message in Morse Code over the thirty-two miles

Replica of Marconi's early wireless apparatus, with spark gap at left, sending key at right, and metal plate for "directing" the waves.

separating England and France. The message was reported to be as distinct as a telegram.[43]

With each successive set of tests, Marconi extended his transmission distance and quickly reported the increase to the newspapers. Both early on and throughout his career, Marconi demonstrated an enviable flair for promotion. He always struck the right balance, obtaining maximum publicity without overstating his case. The best evidence of this talent was the way this Irish-Italian, coming to America when chauvinism was at its peak and many Italians were eyed with suspicion, captured the hearts and minds of the American press and, thus, those of the American public.

•   •   •

ON SUNDAY, OCTOBER 1, the *Herald* announced in bold headlines: "Marconi Will Report the Yacht Races by His Wireless System." The story included illustrations of the apparatus, the race course, and the inventor, and assured its readers that wireless telegraphy was "no longer the dream of the scientist but an accomplished fact."[44] Two steamships, the *Ponce* and the *Grande Duchesse,* would be equipped with wireless and would relay the progress of the race to stations set up at the Navesink Highlands and Thirty-fourth Street in New York. Waiting reporters would send the news across the country and the Atlantic by telegraph and cable, and would post up-to-the-minute reports on bulletin boards in the city. During the last yacht races, the *Mackay-Bennett* had transmitted the news to shore through a submarine cable, but this time would be different: the messages would "come rushing through the air with the simplicity of light."[45]

The yacht races were almost as much of a spectacle as Dewey's parade. Thousands of people crowded onto excursion boats to follow the *Columbia* and the *Shamrock.* Thousands more lined the coast, or formed traffic-blocking crowds near the *Herald* building's bulletin board. This first attempt to use radio to transmit news in America had a sizable and eager audience.

On October 4, the first day of the races, the *Herald* boasted: "Marconi's Wireless Telegraph Triumphs" and "Wireless Bulletins Worked Like Magic." "By flashing his despatches from the steamship *Ponce,*" the *Herald* enthused, "Signor Marconi enabled the public to follow every movement of the yachts from the start. . . . As messages came from Signor Marconi . . . and were placed on bulletin board, there were loud hurrahs." Aboard the *Ponce,* "more attention was given to the mysterious chart room and to Signor Marconi than to the yachts."[46] The speed with which the messages were received was particularly thrilling to Americans. Marconi later recalled that "what impressed the public most was the extraordinary rapidity of the system. Wherever the Marconi bulletins, as they were called, were posted all over the city, the public was less than seventy-five seconds behind the yachts and in many cases less than thirty seconds."[47] Even the other newspapers had to acknowledge Bennett's coup, because wireless was a new invention attracting strong public interest and Marconi was clearly a man to be reckoned with. Marconi's success became front-page news across the nation.

Within days the image makers were at work. By October 6, Marconi had been labeled a public benefactor. A rumor had spread the fear that the *Grand Republic,* one of the excursion boats, had sunk. But Marconi sent a wireless report denying the story and thereby saved "thousands from hours of anxiety."[48] His own charm and skill, the nature of his invention, and the press's hunger for the good copy inventors made, all interacted to place Marconi in the pantheon of inventor-heroes.

Whether Dewey's publicity served as a primer for Marconi on how to win over reporters remains unknown, but Marconi equaled the Admiral in his dealings with the press, and developed a public persona newspapers couldn't resist. He, too, was repeatedly described as being modest about his accomplishments and was praised for being quick to enumerate the contributions of his predecessors.[49] He was cautious in his statements about the progress of his work, and was flamboyant only in his achievements, not in his manner. Although his invention was described as a miracle, he did not boast; in fact, it was observed that he was "not a talkative man."[50] He was consistently described as reserved, courteous, even self-effacing. The *New York Times* stated that Marconi's modesty approached diffidence and related that he had concluded one address "by saying that he had only builded upon the discoveries of other scientists, and gave a list of names of the men who had helped him the most." The paper affirmed its belief that Marconi could "subordinate all professional jealousies and rivalries to the truth."[51] His invention, the *Times* declared, was the result of arduous labor and ingenuity, two characteristics long considered essential to the self-made man.[52] Marconi succeeded in coming across as confident yet modest. Despite what he might have said privately or in his patent applications, when speaking to the press or delivering an address, Marconi conscientiously expressed his indebtedness to earlier workers in the field. He flattered his listeners by expressing his admiration for America, which he claimed was much more supportive of inventors than was England.[53]

Marconi remembered the initial discontinuity between what he

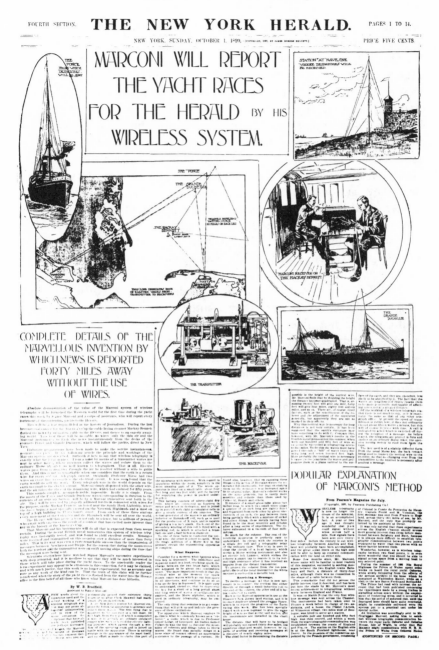

Headlines from the *New York Herald* announcing Marconi's first
American wireless demonstration, 1899.

was and what reporters expected him to be: "For some reason or other it seemed to come rather as a shock to the newspapers that I spoke English fluently, in fact, 'with quite a London accent,' as one paper phrased it and also that I appeared to be very young and did not in the slightest resemble the popular type associated with an 'inventor' in those days in America, that is to say a rather wild-haired and eccentrically costumed person."[54] The *New York Tribune* tried to fit Marconi into this mold anyway. The paper assured its readers that the slight young man had a light complexion and clear blue eyes and, but for a small mustache, was clean shaven. However, the story also noted that Marconi possessed a "nervous temperament" and was somewhat absent-minded, evidently being "more concerned about his scientific studies and inventions than about conventionalities and dress."[55] This characterization did not stick, because Marconi successfully defied it. Marconi was well aware of the handicaps stereotypes exerted, especially those of age and race, and he sought to avoid them through polished manners and elegant dress.[56]

While he may not have looked much like an eccentric scientist, Marconi told reporters what they wanted to hear and print, and thus conformed to the inventor-hero model in other ways. He was not a university-trained theoretician but an experimenter who, by working with his hands—another attribute Americans respected—had accomplished what the professors had not. As *Electrical World* noted in an editorial praising Marconi, "All the world admires a savant, but it will accept a man of only moderate learning if he will create from the remnants of knowledge something for the immediate good of humanity."[57] Scientists were remote to Americans; experimenters were not. Marconi was shrewd enough to play down the theoretical knowledge he had acquired. He emphasized instead how he proposed to integrate wireless quickly into commerce and diplomacy. *Electrical World* stated that Marconi was a "true inventor" because he concerned himself "very little with the theory of wireless telegraphy" and instead "confined his work to experimental changes."[58] Ultimately, the American people did not care who thought of something first; they cared who made it work. And Marconi made wireless work.

Marconi's demonstration of wireless also fit into the journalistic bias toward technological display.[59] Unveilings of inventions that took place during major public events and that involved risk, because the news of failure could not be confined, were much more likely to receive breathless, romantic coverage than private, easily controllable demonstrations. Technological display involved drama; more importantly, it involved the public. Technological display also had to involve a new visual or aural experience for spectators: a street had to be lit for the first time, music played without musicians present, or news bulletins flashed instantly to awaiting crowds. Such demonstrations gave reporters the proper setting for writing about the connections between technology and society: they could speculate about what the new invention would mean to ordinary, everyday people. The bias toward technological display favored the public and the dramatic while tending to overlook the private, the incremental, the small, and the overly theoretical. Marconi understood this journalistic bias well, and he was adept at exploiting it.

While Marconi's style was clearly successful, the fantastic nature of his invention also gripped the journalistic imagination. The *New York Times* captured the excitement in classic nineteenth-century prose: "We of the latter edge of the nineteenth century have become supercilious with regard to novelties in science; yet our languor may be stirred at the prospect of telegraphing through air and wood and stone without so much as a copper wire to carry the message. We are learning to launch our winged words."[60] Sending messages unaided by wires through that strange environment, the ether, was miraculous. Wireless seemed the technical equivalent of telepathy: intelligence could pass between sender and receiver without tangible connection. As one journal put it, wireless stimulated the imagination and made people "think that things greatly hoped for [could] be always reached."[61] Thus, to many, wireless bridged the chasm between science and metaphysics, between the known and the unknown, between actual achievement and limitless possibility.

Such a wonder offered both mundane and more lofty opportunities. One of its most immediate applications to "everyday affairs" would be establishing communication between ships and the shore and thus reducing the number of disasters at sea.[62] Much was written about potential military applications. The telegraph had "dissolved the unity" that had existed between transportation and communication before 1844, and wireless promised to reconnect these two powerful networks.[63] But while such a service was no doubt a great stride forward, for the press it was not the only potential use for the invention. The possibilities seemed so much grander, seemed to extend far beyond the advantages that might be enjoyed by ships' passengers. To fulfill its promise as embellished by the press, wireless had to provide truly democratic benefits and touch millions instead of hundreds. Like its predecessor the telegraph, wireless was cast as a moral force that would bring the world closer to peace. *Popular Science Monthly* observed that, through wireless, "the nerves of the whole world [were], so to speak, being bound together, so that a touch in one country [was] transmitted instantly to a far-distant one."[64]

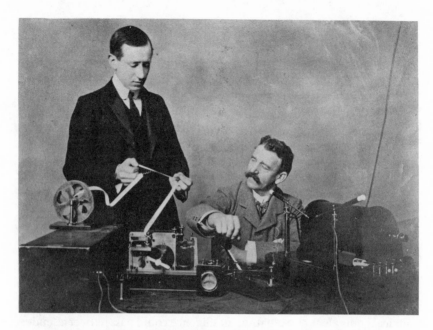

Early publicity photograph of Marconi reading a wireless message while his assistant George Kemp looks on, ca. 1900.

Families and friends separated from one another could now be bound together, too, through this new invention. The *New York Times* foresaw a time when "wireless telegraphy would make a father on the old New England farm and his son in Seattle . . . neighbors—perhaps by the use of their own private apparatus."[65] *Century Magazine* offered a more poignant vision,

of a time when, if a person wanted to call to a friend he knew not where, he could call in a very loud electromagnetic voice, heard by him who had the electromagnetic ear, silent to him who had it not. "Where are you?" he would say. A small reply would come "I am at the bottom of a coal mine, or crossing the Andes, or in the middle of the Atlantic." Or perhaps in spite of all the calling, no reply would come, and the person would then know that his friend was dead. Think of what this would mean, of the calling which goes on every day from room to room in a house, and then think of that calling extending from pole to pole, not a noisy babble, but a call audible to him who wants to hear, and absolutely silent to all others. It would be almost like dreamland and ghostland, not the ghostland cultivated by a heated imagination, but a real communication from distance based on true physical laws.[66]

These wistful forecasts, which envisioned the time when anyone who wanted to could use an "electromagnetic voice" to contact distant friends or relatives, played on the emotional discontinuities people experienced as a result of increasing geographical mobility. Such predictions also reflected widespread frustration over corporate control of the existing communications networks. The early quixotic hopes for the telegraph and telephone had deflated as the inventions came to be managed by large-scale organizations more intent on profits than on public service.[67] By the turn of the century, Western Union was one of America's most resented monopolies, and Bell Telephone hardly inspired public affection. But lines and wires were easy to control. Wireless appeared to offer another chance. For it was not at all clear in 1899 how, or even if, corporations could own or manage the airwaves. It seemed that wireless might be the truly democratic, decentralized communication technology people had yearned for, a device each individual would control and use whenever he or she wanted, without tolls, and without operators.

The journalistic motivations for promoting Marconi and his invention so enthusiastically contained no small dose of economic self-interest. Wireless promised to accelerate and cheapen news gathering, an attractive prospect to publishers resentful of the telegraph companies and their high-priced service. The transoceanic cable companies faced no competition, and they all followed the same rate schedule. Newspapers in the late 1890s were paying ten cents a word for press dispatches from London, while private parties and businesses were paying twenty-five cents a word. The press considered these prices extortionate. Referring to Western Union as a "monopolistic serpent," the *New York Times* complained: "The transmission of dispatches by ocean cable is far more speedy and much less costly than it was fifteen years ago, yet we go on paying the old prices. . . . We are living in an age of lower and lower prices—except for cable dispatches."[68] Complaining about the telegraph monopoly, the *Herald* looked forward to the end of a situation in which "American newspapers, American merchants, and shippers [were] entirely at the mercy of this monopoly and wholly dependent upon its lines for the transmission and receipt of news."[69] The *New York Times* sarcastically noted: "Obviously the claim that the Atlantic Cables are now worked to their capacity, and that any lowering of their tariffs would be ruinous, it not sustained by arithmetic."[70] The paper quoted Professor Michael Pupin, who characterized the telegraph companies as technically reactionary: "The Western Union and Postal Companies are both using antiquated methods. The Western Union Company does not spend ten cents a year for experiments, so far as I can learn. A man who offers them an improvement is treated like a book agent. One seems to feel that there is a big sign over the door 'Inventors and Scientists Not Wanted.' "[71]

So as not to seem too self-interested, however, the press couched its complaints in terms of larger humanitarian concerns. The high cable rates were really everyone's concern, editorials warned, because they affected the quality of international diplomacy and understanding. Sermonizing that "nothing so fosters and promotes a mutual understanding and a community of sentiment and interests as cheap, speedy, and convenient communication," the *New York Times* warned that the high cable rates threatened world peace, and reminded readers that "at the time of the Trent Affair [the United States] nearly got into a war with England through the lack of cable communication." The *Herald* echoed these sentiments, warning of "the danger of a community being dependent upon one sole line of telegraphic communication."[72] Thus, it wasn't simply that the cable companies charged too much: journalistic rhetoric cast them as selfish obstacles to the free exchange of ideas, and as amoral business firms that placed greed before world peace. While these monopolies were damned, romantic phrases such as "community of sentiment" veiled, and thus slyly promoted, the press's own economic agenda.

Newspaper editors reasoned that if Marconi competed with the telegraph, cable prices would have to come down. Furthermore, because there were no wires or cables to maintain, it was believed that wireless would be the less expensive system to operate, especially over water. With wireless, each newspaper could conceivably operate and control its own stations. In 1899, the *New York Times* predicted that wireless competition would produce a "wonderful cheapening of telegraphy and an inconceivable extension of its use in common affairs."[73] The press eagerly awaited every advance in the new art, and with each achievement predicted that the telegraph trust was about to be toppled. Marconi quickly realized that he was offering not only an extension to telegraphy, but also the possibility of an alternative to an increasingly inflexible and unpopular communications system. In many ways he was filling a greater need than he initially had thought.

In 1899, the press began constructing the meaning of wireless telegraphy. What emerged was a web of significance containing noteworthy lines of tension. On the one hand, wireless would be the agent of altruism, because it would save lives and promote mutual understanding. It would reduce modern-day loneliness and isolation by providing individuals with a way to contact loved ones far away. On an individual and societal level, then, wireless would restore a sense of community in an increasingly anticommunal world. And wireless might even undermine the seemingly ineluctable march of monopoly capitalism by allowing Americans to circumvent Western Union and Bell Telephone, and to take modern communications into their own hands. At the same time, however, wireless would expedite commerce, bolster the military, and further the economic goals of the press. Wireless would be antimonopoly but pro-business. All these contradictory desires swirled around the invention, defining it, simultaneously, as the restorer of tradition and the harbinger of a new era. Wireless would, at one and the same time, promote capitalism and defy it.

Marconi had received the favorable publicity he so wanted and needed. Although other Americans had tried to exploit wireless telegraphy and to become its "discoverer," Marconi had successfully preempted the field. His success in the journalistic arena resulted from the dynamic interaction between his own skills as a promoter and the established journalistic conventions of the time. The inventor-hero was a stock character in the press; he was used to humanize technology, and his tribulations and successes provided reporters with a narrative structure within which to embed the more abstract aspects of science and inventing that often defied conventional storytelling. To become an inventor-hero, a man had to have a revolutionary invention that he was willing to display in a dramatic, public fashion. But he also had to be able to charm reporters. Thomas Edison had this talent, and so did Marconi. Both men understood their public roles: they knew that modesty, a sense of humor, openness, and a touch of anti-intellectualism were essential to public recognition. What both men understood, instinctively, was the emerging cult of celebrity in American journalism. Marconi's stunning success with the American public and the popular press led *Electrical World* to comment: "Marconi's managers have shown that they have nothing to learn from Yankeedom as to the art of commercial exploitation of an inventor and his inventions."[74]

But the coverage Marconi received was not an unmixed blessing, for while it applauded him and his invention, it also conjured up uses for the invention which Marconi had not sanctioned or proposed. The predictions in the newspapers and magazines about an electromagnetic voice for all no doubt seemed merely fanciful and harmless musings intended to fan interest. But they would eventually prove less peripheral and less innocuous than Marconi may have appreciated.

Marconi had combined scientific discoveries and laboratory apparatus into a commercially practical system of sending messages without wires. He had succeeded at technological preemption, and he had legitimated that preemption through widespread and enthusiastic press coverage. With his public victory achieved, Marconi now faced the more tedious and frustrating process of establishing his invention in commercial spheres. He had to determine how to market the device and how to make it pay, and this process would involve negotiations with governments. It was not the sort of work Marconi enjoyed, but he was developing a clear conception of where wireless fit into the marketplace, a conception reinforced by the news-gathering needs of the newspaper business. What he could not know was how some Americans—other ambitious inventors and corporate clients, as well as an eager press and certain of its readers—would ultimately substitute their vision of the technology for the inventor's own. The first substitution would have to be technological. Even as Marconi savored his triumph at the yacht races, aspiring competitors considered how to circumvent his patents.

CHAPTER TWO

COMPETITION OVER WIRELESS TECHNOLOGY
The Inventors' Struggles for Technical Distinction

1899–1903

IN OCTOBER OF 1899, at the America's Cup races, the social construction of radio had begun. A foreign-born inventor introduced a revolutionary communications technology, and the newness of both were made familiar through journalistic practices, especially typecasting and analogy. The press presented Marconi as a typical inventor-hero and described his invention as resembling the telegraph, only better. Romantic, flowery language helped make comprehensible this latest advance in electromagnetic theory and practice. News-writing practices thus provided conventional and comforting ways of thinking about something quite radically different. Not incidentally, the press managed to legitimate Marconi as the first and true discoverer of wireless telegraphy.

This was just the beginning, however, and only one part of the social construction process. Such public mediations of a technology's applications interacted with inventors' private, individual, often idiosyncratic ideas about how the invention might evolve. These were technical and highly specified ideas, focusing on the discrete components of, in this case, the wireless system. But these ideas were not formed within a vacuum. Private, technical insights were very much affected by the larger public celebrations of technology which praised certain types of inventions and inventors, and ignored others.

Three aspiring inventors read about Marconi's success at the yacht races with particular interest. Reginald Fessenden, Lee De Forest, and John S. Stone had each experimented with wireless telegraphy by October 1899: Fessenden as a professor in a university lab, De Forest for his doctoral dissertation, and Stone for Bell Telephone. At different stages in their careers, with widely disparate training, goals, and personalities,

Fessenden, De Forest, and Stone provided Marconi with his most formidable technical competition in the United States. Despite their differences, these three inventors shared the conviction that Marconi's system was technologically vulnerable and could be displaced. This conviction, and the way each man gave expression to it through his inventions, had lasting social consequences.

All three men were well aware of wireless's technical limitations, and they were inclined to regard the journalistic hoopla about a new wireless "wizard" with skepticism. At the same time, they saw a man whose wireless apparatus was no more advanced than their own gaining enormous public recognition and succeeding as an independent inventor. Faith in their own technical and theoretical superiority intermixed with envy and resentment of Marconi's success. These men were members of a culture whose popular press had lionized Marconi and had cast his device as revolutionary and filled with unlimited potential, but they were also part of the scientific and engineering subcultures that viewed Marconi and his invention from a considerably more critical stance.

To the press and its readers, Marconi brought visions of enhanced societal and individual control through technical mastery. But to American scientists and engineers, a constituency with more of a vested interest in Marconi's work, his success threatened to compromise the professional control and prestige they had spent the past half-century securing. Members of the American scientific and engineering communities had been undergoing self-imposed professionalization in the late nineteenth century, distinguishing themselves from mere "amateurs" and "tinkerers" by their education, research, and institutional affiliations. Specializations became more clearly defined while barriers to entry in both science and engineering became more imposing.[1] According to the new professional codes, men such as Edison and Marconi had not been properly trained, did not embrace the requisite body of information or outlook, and did not pay homage to the recently enshrined leaders or networks of the scientific and engineering professions. Such inventors, even successful ones, were no longer quite legitimate. In fact, the enormous publicity Edison and Marconi received no doubt further undercut their respectability among academics.[2] Edison's disparaging references to the "old German professor" who studied "the fuzz on a bee" instead of working on practical problems were eagerly quoted in the popular press.[3] The lack of emphasis on and, often, animosity toward scientific training in such journalistic accounts reinforced the growing schism between men of science and men of practice. Thus, when men such as

Edison or Marconi garnered bold, front-page headlines, while diligent and rigorously educated scientists remained unrecognized, resentment brewed. To the scientific community, the headlines obscured the decades of experimental and theoretical work from which the invention had evolved. To the engineers, Marconi's cut-and-try approach represented an outdated and increasingly discredited method that for too long had made their work seem unsystematic and unscientific.

American scientists and engineers placed more credence in what technical journals reported about wireless telegraphy, and these journals pointed out the invention's flaws. Criticism clustered around two major objections: that Marconi had invented nothing new, and that, even if he had, his system was impracticable. The *London Electrician* and New York's *Electrical World* commented that Marconi's device was unpatentable and that successful wireless transmission did not depend on anything originated by Marconi. One editorial stated: "It appears questionable if Marconi really owns a master patent on his system, for the very good reason that the principles underlying its operation are well known to electricians. In producing results he uses in combination certain apparatus which were devised and used by others before him, and whatever he may hold cannot cover much more than a combination of apparatus used previously for other purposes."[4] Marconi's claim that he was the first to discover the advantages of the aerial was labeled as absurd.[5]

While it was true that Marconi's basic components were not new, this criticism missed the point. It was the special combination of these components into a system, and a determined application of that system to commercial and naval communications, which made Marconi's contribution special. Thus, there was a reasonable and persuasive retort to this charge. However, more nagging concerns about the versatility and reliability of wireless surfaced, concerns less easily dismissed.

In 1899, wireless reception was still erratic, maximum reliable transmission distance was about thirty-five miles, and there was no means of tuning. Because all of Marconi's apparatus sent and received on the same general frequency, only one transmitter could signal in a given area at a time. What the newspapers categorized as a major technological advance, scientists and engineers viewed as a technical step backward. Critics charged that wireless offered no more secrecy than semaphores, because anyone with a receiver could pick up the messages. As one writer complained, "Waves are distributed in all directions, and cannot be concentrated in one direction, all methods for doing this having apparently failed."[6] Not only did this scattering of waves preclude privacy, it also limited the number of transmitters that could operate in a given neighborhood. *Electrical World* granted that wireless might have some practical value where there was only one transmitter, but cautioned that "when the transmissions [multiplied] in number and direction, the difficulties would probably increase enormously."[7] Professor Michael Pupin at Columbia went so far as to warn that working "more than one set of instruments at any one time between two continents" was impossible "on account of mutual interference."[8] Because there was as yet no tuning mechanism, and the receiver was not sufficiently selective, the receiver responded indiscriminately to all sorts of frequencies, both man-made and natural, lending credence to Pupin's assertion. Paper tapes from the turn of the century, on which the Morse inker recorded the incoming dots and dashes, reveal what happened during reception: instead of the Morse code, the inker printed either a semicontinuous line, occasionally broken up at arbitrary points along the tape, or gibberish, lines that were neither dots nor dashes.

The energy of the sparking coil was as yet unchanneled into specific wavelengths, so a broad spectrum of wave bands was disrupted by one transmitter in an area. The transmitter, in fact, emitted not one but several wavelengths: the "fundamental" wave and a group of harmonics. These waves were referred to as "highly damped." Today, radio transmitters emit a specific and continuous wave frequency, but the crude spark gap transmitters of 1900 sent out intermittent wave trains of a particular length and strength which declined as the energy subsided. An instructive analogy is that of a swinging pendulum. If a pendulum is powered by electricity, it will swing continuously to the same distance on each side. A smaller pendulum also operated by a continuous power source could swing parallel to and equidistant from this larger pendulum and never collide or interfere with it. However, if a pendulum is merely pushed once, the distance it sweeps will decline with each swing: no smaller pendulum could swing inside the sweeps of the larger one without colliding with it. As the power of the spark gap diminished after each dot and dash, many lesser wavelengths were transmitted, and a very broad band of radio frequencies was disrupted with each message. The highly damped wave, then, produced major interference problems.

The coherer was considered unreliable and slow. *Electrical World* complained: "The present rate of speed of Marconi's system is only fifteen words a minute, which fact has been urged against its practicability in competition with the ordinary telegraph, by which a speed of 600 words a minute has been attained."[9] Another drawback was the noise accompanying wireless transmission, which was described by a reporter for the *New York Times* as deafening: "While [the wireless] exchange was in progress there was a noise from the station like that of a rapid-fire gun in action. So loud were the reports that they could be heard half a mile away. . . . Flashes of light were seen crossing the windows in the receiving house. They were bluish white and flashed horizontally. The house . . . shook with the violence of the shocks."[10]

Critics also claimed that wireless would never transmit over significant distances, because there were physical limits on the height of aerials. In 1899, scientists believed that transmissions of sixteen miles would require an aerial 80 feet high; twenty-two miles would require a 140-foot conductor. Extending these calculations led to the inescapable conclusion that transmitting a message across the Atlantic was impossible: the aerials would have to be between 1,100 and 1,500 feet high, several hundred feet higher than the Eiffel Tower.[11] The popular press, eager to believe Marconi's public statements, either overlooked these problems or assured its readers that improvements in these areas were imminent. The *New York Times* was one of Marconi's greatest champions, and if it printed any stories critical of his apparatus, they were buried in the back of the paper. The technical press, with its more specialized and skeptical audience, remained unconvinced.

Had Marconi introduced his invention as a method of broadcasting information to the public, several of these criticisms would have been less troubling. But his device was conceived and presented as a telegraph using no wires. It was meant to send messages in dots and dashes to a specifically designated private receiver. This was how Marconi marketed wireless, and this was the basis on which it was judged. In his initial comments to the press, Marconi maintained that wireless would not compete with or replace line wires. Rather, wireless would be used where telegraphy was impossible, such as between ships or from ship to shore. Thus, the invention, at least with regard to applications, was not offering a revolutionary new service, it was simply extending an older one. Since at the time there was no service between ships and the shore except that provided by dispatch boats, megaphones, and other limited means, any service that provided speedier communications could be seen as an improvement. This was certainly the *New York Times*'s view. "It is unfair," the paper complained, "to set up for Marconi standards never sought in and impossible of attainment by the systems with which his beautiful invention competes."[12] But a society based on and dedicated to cumulative and orderly technical progress expected the latest marvel to possess at least the advantages of its predecessor. Where wireless could not match a one-on-one comparison of features with the telegraph, it was criticized in the technical press as a failure.

In several respects, Marconi had put his invention on a Procrustean bed. Wireless telegraphy was not best suited to do what Marconi wanted it to, and he was intent on eliminating the very aspects of the apparatus which later made radio such a distinctive and revolutionary invention. Wireless sent waves in all directions; Marconi wanted directional transmission. The transmitted messages were public; Marconi wanted them to be private. But Marconi did not view the properties of wave propagation as absolutes. He was a persistent man, determined to shape the technology to conform to his vision of commercial requirements. This vision was manifest in the drawings accompanying the *Herald*'s coverage of the yacht races. The illustrations representing the "path of electric waves from transmitter to receivers" do not show waves at all but, rather, straight lines linking Marconi, on the *Grande Duchesse,* with the *Herald* stations on shore. The drawing makes clear how Marconi conceptualized his service: if the spectrum would not cooperate, then he would make sure that his apparatus compensated for and ultimately overruled natural phenomena. Commercially, this single-mindedness was Marconi's greatest strength; technically, it proved to be his Achilles' heel.

Marconi had demonstrated that wireless was possible. Now he had to show that it was practical and reliable. Like entrepreneurs before and since, he had to make people believe they needed his invention, not just for special occasions, but on a regular basis. Convincing them would be that much more difficult if only one person at a time could transmit or receive. Marconi had to allay the doubts and misgivings of potential buyers of wireless by significantly extending the distance of transmission and by devising a method of tuning. To meet these challenges, Marconi continued to work as technological revisionist par excellence. Rather than trying radically new approaches or devices that might take him too far afield from his immediate commercial goals, Marconi revised what he had already developed and borrowed directly from the scientific work of others.

Marconi's technical work was always shaped by his conception of the marketplace. The clients he hoped to attract—steamship companies, newspapers, and navies—would require transmission over great distances. To meet this demand, Marconi's approach was to make his existing apparatus larger and more powerful. He also developed wireless as a complete technological system: he concentrated on the components of that system and on the important connections between those components.

Marconi's lack of formal scientific training at times liberated and at other times constrained his technical progress. He rarely allowed contemporary scientific warnings about the unattainability of a particular goal to deter him; he treated theory as an afterthought. For example, he tested wireless under a variety of conditions: over long distances, during the day and at night, in the tropics, over saltwater and over freshwater. He was not trying to verify a particular theory, he was simply collecting data. He did not test hypotheses, he generated information that was yet to be explained theoretically. Like many engineers, Marconi was inadvertently involved in a feedback process. He adapted certain technological innovations that had been designed to test scientific theory and sought to make them commercially successful. Then, in his experimentation, he produced phenomena that had yet to be explained scientifically and stimulated others to embark on new theoretical work in physics and electrical engineering.[13] However, Marconi was sometimes handicapped by his lack of training. Because he could not rely on his own scientific background or an academic tradition, he was compelled to improve wireless primarily by an arduous process of trial and error.

His training and approach may have cost him time and money, but Marconi was not convinced, in 1900, that university-trained scientists would necessarily produce quicker or better results. After all, he, not they, had developed wireless telegraphy. They had thought certain things impossible; he had proven otherwise. But Marconi was under a considerable amount of pressure to improve his system as quickly and decisively as possible. He had to decide whether to bring an eminent, university-trained scientist into the company who might expedite the experimental work while simultaneously conferring academic legitimacy. Yet as a very young man, a man younger than his assistants, it was important to him that he assert himself, technically and psychologically, as the final authority in his company. Here was a shrewd man, aware of his own educational limitations, pulled by conflicting needs, and beset by his own ambivalence. In 1900, he retained John Ambrose Fleming, a scientist twice his age who had worked under James Clerk Maxwell and was professor of electrical engineering at University College, London. This selection brought the company additional prestige and eventually added richness and depth to its experimental work.

Although Fleming and Marconi ultimately developed a productive and highly complementary partnership, their early relationship was uneasy. The tensions between the two professions they represented—scientist and inventor/entrepreneur—were exemplified in their initial contest over authority. Marconi complained that Fleming's technical contributions were highly impracticable. He wrote to his managing director,

H. Cuthbert Hall, "Dr. Fleming seems to introduce so many complications which in practice prove useless, that I think it will be well that the details with regard to the changes in the plant here . . . should be discussed and settled between you and Mr. Entwhistle, as I am afraid that no useful purpose would be served by referring them to Dr. Fleming."[14] Marconi also objected to Fleming's efforts to gain final approval on all technical matters, a function Marconi intended to reserve for himself. "This attitude on his part," wrote Marconi, "opens up again the wider question of his general position in the company and I am desirous that this should be clearly defined to him without further delay. It should be explained to him that his function as Consulting Engineer is simply to advise upon points which may be expressly referred to him and in no way places upon the Company any obligation to seek his advice upon any matters in which it is deemed unnecessary."[15] Marconi wanted the scientist's advice, but only on specific points, and within technical parameters he had already defined. Marconi has been praised for having the foresight to retain a man of Fleming's talents and reputation, but this was not a decision the inventor initially embraced wholeheartedly. The alliance between the man of practice and the man of science was not without its early jealousies, which revealed as much about the changing relationships among inventors, scientists, and engineers as they did about the characters of the two men.

In the early development of wireless telegraphy, the personalities of the various inventors profoundly influenced the course of technical and bureaucratic progress.[16] Marconi, who was sometimes described as aloof, humorless, and self-centered, could not have succeeded as he did if he had really been such a man.[17] On the contrary, when technical obstacles confronted the Marconi Company, obstacles that could strain even the most harmonious working relationships, it was Marconi's personality that propelled the work forward. He experimented constantly and with great energy, setting a standard he expected all his assistants to follow—and they did.[18] What remains of his correspondence to members of the company reveals a man possessing charm, loyalty, and sensitivity to the feelings of others. While he had little tolerance for mediocre technical work, he was quick to praise those who did their jobs well.[19] Even as he insisted that Fleming's ambition be checked, Marconi urged the company's manager to handle the matter with tact, writing, "I do not wish to inflict any unnecessary wound on Dr. Fleming's susceptibilities."[20] Marconi also had a dry, worldly, and occasionally self-deprecating sense of humor, which he felt most comfortable unveiling in private, rather than in public, settings. All of these qualities, coupled with his celebrity status and his grand vision, made him a compelling man to work for and sustained his co-workers when progress was slow and the limelight flickered.

With the various public demonstrations behind them, and technical problems not only unsolved but exposed, Marconi and his assistants intensified their work in 1900, confronting the problems of reception and tuning. They hoped to discover solutions that would overcome interference and lack of secrecy.

Marconi had first considered the problem of tuning in 1897. He had initially hoped to "beam" his wireless signals between two specific points, intending, in this way, to direct a particular message to a particular receiver and thus establish a network of invisible, private, noninterfering lines. He had used parabolic reflectors to direct the Hertzian waves from transmitter to receiver, but these reflectors could only aim short waves, and Marconi was moving toward using longer waves to achieve greater distance. However, longer waves precluded directional transmission: they passed right over Marconi's mirrors. To continue with the reflector method of directional transmission, Marconi would have had to make the mirrors massive, and this was clearly impracticable. Marconi's two goals, greater distance and directional transmission, were at odds, and he had to reconcile them if he was going to progress. Marconi had to abandon the optical approach, with its cumbersome mirrors, and find an electrical solution to the problem. He realized that, rather than using the same broad-banded wavelength, which he would physically "beam" between points A and B, he had to send out different wavelengths and develop receivers that would select between them. Because Marconi believed at this time that an infinite number of wavelengths existed, he thought that tuning would open up the airwaves to as many customers as were interested.

For a solution to this problem of tuning the transmissions, Marconi turned to the experimental work done by Oliver Lodge. By the 1890s Lodge was a well-known authority on electricity, electromagnetism, and the ether, and an instructor of physics at University College, Liverpool.[21] Lodge had experimented with tuning in 1889, but he did not resume his work until Marconi's early public demonstrations. In 1897 he patented his method of syntonic wireless telegraphy, a patent Marconi ultimately acquired. For at least ten years the word *syntonic* was synonymous with *tuned*. Lodge studied selective resonance, a phenomenon in which sound waves produce a sympathetic reaction in similar circuits. For example, a tuning fork, when struck, will generate vibrations in an identical tuning fork nearby. Scientists had discovered that similar electrical circuits could also be resonant, having the same natural frequency of oscillation, and this property provided the basis for Lodge's work. He reasoned that if he could match certain aspects of the circuits in wireless transmitters and receivers and make them electrically resonant, then they would respond "sympathetically" to each other but not to apparatus not similarly adjusted. Lodge discovered that by adding matched induction coils to the aerial connections of both transmitter and receiver, he dramatically increased the selectivity of his apparatus and could tune it to a specific frequency.

Marconi studied Lodge's system of syntonic wireless and improved on it. Realizing that resonance in the two aerial circuits was insufficient, Marconi extended Lodge's principle of resonance, making the two closed sending and receiving circuits resonant, as well, to maximize the effects of tuning. Consequently, Marconi incorporated not two, but four, tuned circuits. He designed adjustable induction coils, and he constructed condensers of variable capacity. He added what we know today as the tuning dial, and, by matching the inductance and capacity of the receiving station to that of the transmitting station, he achieved selective reception. Marconi's improvements in tuning were covered under English patent number 7777.[22] The famous "four sevens" of 1900 turned out to be a crucial patent for Marconi and one of the most frequently litigated claims in wireless history.

Marconi had done his most risky borrowing yet: he clearly had made use of the features specified in Lodge's 1897 patent. Why Lodge did not immediately sue for infringement remains unclear; that he did not do so significantly strengthened Marconi's patent position in Europe and America. Armed with this innovation, Marconi demonstrated for the public first duplex, then multiplex transmission. He would connect two or more differently tuned receivers to the same aerial, and the receivers would only respond to the messages intended for them, even when the messages arrived simultaneously. These public demonstrations in 1900 and 1901 encouraged hope that tuning would quickly overcome several of the invention's major drawbacks.[23]

At the turn of the century, then, Marconi's system of wireless telegraphy was conceptually complete, albeit in rudimentary form. All the components of Marconi's system needed refinement, but the weakest feature was the coherer. Marconi's approach from the start had been to improve on, revise, and make more practicable existing apparatus, and in his attempt from 1897 through 1902 to improve the cohering-decohering process, he also followed this approach. He needed a receiver that would be simultaneously rugged and sensitive, and the quest for the one quality often prevented the achievement of the other. Through painstaking ex-perimentation, Marconi remodeled and upgraded the device. Still, the coherer required a large amount of energy. Two other major problems persisted. The coherer was slow and, worse, it was capricious. This unreliability was due in large part to the decohering process. Because it was necessary to tap the filings back to their high-resistance state after every dot and dash, reception was very sluggish, and it became even more retarded if the filings were not tapped back quite right. *Electrical World* complained about "the changeability of the position of the filings, which [were] as varied as the designs of a kaleidoscope." "In some instances," the journal reported, "the filings are arranged so that they would respond to the feeblest impulse, and upon the very next stroke the marvelous sensitiveness is destroyed and the tube becomes inoperative." One experimenter described the frustrations that accompanied the ap-pliance: "It might go along very nicely, and then, without warning and for no apparent reason, go dead or fail to respond. Then it might remain dead until it got good and ready to come to life. No amount of coaxing would have any effect." Also, "the noise occasioned by the concussion" of decohering was maddening, especially when this was registered on the Morse sounder as a signal.[24] The coherer was also limited by its own internal point of diminishing returns. One could increase its sensitivity, but only up to a point, after which the device would respond to all sorts of static, including its own sparks.[25]

Efforts to upgrade the coherer demonstrated that the device could not be improved any further—it had to be replaced. In 1902, Marconi patented a new, less fickle receiver, the magnetic detector. Like Mar-coni's other innovations, Maggie, as the detector was affectionately called, was based on the experimental work of another scientist. Lord Rutherford in 1895 had discovered that electromagnetic waves could serve as a demagnetizing agent, and he had demonstrated that magne-tized needles could be used as detectors of Hertzian waves.[26] After experimentation, Marconi assembled a box on top of which two wooden disks lay flat at each end. An endless band of fine iron wires moved slowly, by clockwork, around the two disks and past the poles of a pair of stationary magnets. When the iron band passed underneath the two permanent magnets, it underwent a change in magnetism and became sensitive to incoming electromagnetic waves. With each train of oscilla-tions, the band suddenly lost its magnetism and induced a current that produced a signal heard through headphones as an audible note. As the band revolved, a new portion of it came into position under the magnets, making the receiver self-restoring.[27]

Messages were not recorded on tape as they were with the coherer. Now reliance on the discriminating capabilities of the human ear became an important and welcome ingredient in the receiving process, helping to reduce error by selecting real signals from false. But because the magnetic detector, like other receivers of its time, did not yet include a loudspeaker, the operator had to keep the headphones to his ears at all times, or he would lose messages in part or in whole. As a result, transmitting at a predetermined time quickly came to be important. Different wavelengths had different frequencies that were heard as different pitches. Thus, the magnetic detector allowed the operator to identify the tones of different transmitters. Because there was no decohering process, Marconi was able to increase reception to thirty words a minute immediately, twice the speed of reception with the coherer. The magnetic detector was also more stable for shipboard use, and it became very popular, especially in Europe, until it was finally displaced by the vacuum tube during World War I. Marconi wrote enthusiastically to his manager in June 1902 that the new detector worked very well on board ship and that transmission across great distances was possible even by an operator who was not highly skilled.[28]

Through a process of determined adaptation and revision, Marconi retooled the insights and apparatus of others to make his system more reliable, efficient, and elegant. He hired highly competent scientists and engineers to assist him. But they all viewed wireless through the lens he provided, and that view was framed by very particular commercial, technical, and cultural parameters. It was a view that never lost sight of the telegraph, seen as the technological and commercial model to emulate. Marconi wanted wireless to possess all the advantages of wire telegraphy and yet to be free of telegraphy's limitations, such as high cost and lack of mobility. As a result, he concentrated on increasing the distance and speed of transmission and on perfecting tuning, which he hoped would finally ensure secrecy.

Fessenden, De Forest, and Stone each in his own way sought to push wireless beyond this framework. While Marconi considered wireless telegraphy over great distances to be his final product, the Americans gradually came to regard wireless as a necessary steppingstone to their eventual goal: transmitting the human voice without wires. To do this, they rejected Marconi's reliance on intermittent, highly damped waves and instead devised apparatus that would transmit and receive continuous waves.

With the advantage of hindsight, we can see that the work of all three men, when taken together, eventually represented a coherent challenge to Marconi's conception of wireless. But when each man first started out, his approach was quite distinctive. Socialization—family background, education, and work—affected how each of these men would approach wireless telegraphy technically, and also determined the extent to which public visions of radio would influence private tinkering with discrete components. Past experiences and individual aspirations also shaped how each man would respond to the social context of inventing as it existed in 1899.

Fessenden, De Forest, and Stone were surrounded by mixed messages about how a gifted and ambitious engineer might make his way through the world. The process of inventing was portrayed one way in the press, but it was evolving in very different ways within institutions. Men wanting to work as inventors were torn between the compelling if unrealistic image of the autonomous inventor-hero and the very real but less glamorous institutionalization of inventing in the corporate sector. Journalistic renditions of inventor-heroes such as Edison and Marconi suggested that with persistence, patience, and hard work, any technically talented young man could achieve fame and fortune. The newspapers emphasized that the highly individualistic man could, through inventing, establish his own intellectual and financial independence.

The reality, of course, was quite different. Invention increasingly occurred within a large corporation's laboratory. Edison's own work contributed to constricting the range of opportunities: historians agree that Edison's most revolutionary and far-reaching invention was not the light bulb or the phonograph; it was the organized process of invention—the "invention factory"—as embodied in the industrial research lab. Menlo Park, which Edison established in 1876, was the prototype. Ironically, America's archetypal independent inventor designed a communal and hierarchical system of inventing which in many ways would be antagonistic to the next generation of freelance tinkerers. By the turn of the century, General Electric and a few other electrical and chemical concerns had established such labs; by the 1920s, these labs would be a common feature of the corporate structure.[29] Men who preferred security to autonomy and who could accommodate themselves to working in an institutional setting would find in the research lab a haven previously unavailable to scientists and engineers. Such men were able to experiment and invent for both the company and themselves.[30] But for those who were loners, who could never reconcile their personal ambitions with the goals of a corporation, the industrial research lab was either a prison or a powerful competitor.

For Fessenden, De Forest, and Stone, the dream of becoming an independent inventor was tinged with the knowledge that it was institutions that provided electrical engineers with financial remuneration and professional affiliation. All three men worked in research labs sometime in their careers, and all were deeply ambivalent about their own relationship to the institutional setting. They wanted their inventions to be their own, they wanted to set their own technical agenda, and they wanted to break out of the anonymous middle tiers of the corporate, engineering hierarchy. Yet they all required the technical and financial resources institutions provided. Each man, as he entered the field of wireless telegraphy, carried within him these contradictory needs, needs shaped by individual biography and by the larger cultural milieu.

■   ■   ■

REGINALD FESSENDEN was born in Canada in 1866.[31] His father was a minister, his mother the daughter of a farmer and inventor. Family life was warm and supportive, and Fessenden, who quickly proved to be an excellent student, received steady encouragement from his parents. He was given little reason to doubt himself. He attended a military academy for two years, and then went away to the Trinity College School, where he competed fervently, and successfully, for top honors. While trying to decide on college, he was invited by his father's alma mater, Bishop's College, to fill the position of "mathematical mastership." At Bishop's he taught math, Greek, and French, and he began reading the scientific journals in the school's library. Although his wife has written that he completed "all necessary work for the college degree" during this period, there is no evidence that Fessenden ever graduated or received such a degree.[32]

In 1883, he moved to Bermuda, where he became the principal and teaching staff (of one) at the Whitney Institute. The technical journals, which Fessenden followed, were by this time filled with articles about Thomas Edison's achievements, especially with the Pearl Street Station, which was providing a fifty-square-block section of downtown New York with incandescent lighting. These articles were published in America and conveyed the sense that the Northeast was the hub of inventive activity. They apparently helped Fessenden crystallize his goals. In 1885 he left his teaching job and sailed for New York, determined to work for Edison. After repeated rebuffs (after all, Fessenden knew little about electricity at the time), he finally got a position with the Edison Machine Works, which was laying mains along Madison and Fifth avenues in New York. Proving himself to be a quick study and efficient worker, Fessenden moved up to the position of inspecting engineer, and when the laying of mains was successfully completed in December 1886, he was offered the chance to work with Edison in his lab. He eagerly accepted.

Fessenden was twenty-one years old, ripe for a mentor, and ready for lasting professional impressions. His three years with Edison, judging by his subsequent career, shaped Fessenden's attitudes toward the process of invention. He was awed by Edison's facilities, especially the new West Orange lab, which included a complete technical library and extensive, first-rate equipment. He noted Edison's cavalier disregard for expenses as long as his backers kept paying the bills. He witnessed Edison's relentless empiricism, which was often characterized by a stubborn and sometimes counterproductive adherence to ad hoc technical theories and approaches.[33] He participated in the camaraderie among the men, nurtured by long hours, common pursuits, and the sharing of highly specialized information. He became Edison's chief chemist and worked primarily on developing insulation material.

In 1890, Fessenden was about to begin experiments with Hertzian waves when corporate reorganization brought about retrenchment. The Edison General Electric Company was in the process of negotiating a merger with Thomson-Houston, one of its major competitors. (The talks resulted in the formation of General Electric in 1892.) Edison's financial backers, eager to curtail some of his experimental extravagances, insisted that costs be reduced. Whether Fessenden was a casuality of this policy or simply resigned remains unclear. When he left the Edison company, Fessenden became the assistant to J. D. Kelley, the electrician for the United States Company, the Newark, New Jersey, branch of Westinghouse. He worked on dynamos and began experimenting with alternating current. After approximately one year, he moved to Massachusetts to work for the Stanley Company, a small electrical firm based in Pittsfield. These two jobs—the one with United, the other with Stanley—are worth mentioning less because of what Fessenden accomplished in them than because of the contacts he made. While working for United, Fessenden patented, for Westinghouse, a method of sealing incandescent lamps. When these patents proved valuable to the company two years later, George Westinghouse took note of who had authored them. The Stanley Company sent him to England to study British lighting and power systems, and during the trip he stopped at Cambridge. There he met the British physicist Joseph John Thomson, with whom he discussed electromagnetic theory; he also toured Maxwell's enshrined laboratory. In two short years, then, Fessenden had established critical intellectual and commercial links.

After seven years of working in various electrical labs, Fessenden returned to academics. In 1892 he became professor of electrical engineering at Purdue. His department benefited from a generous appropriation, and thus the new professor did not have to scrimp on equipment. Fessenden lectured on and recreated Hertz's experiments. The following year he received a letter from the chancellor of the University of Pittsburgh, who told Fessenden that Westinghouse had developed "a particular regard" for the engineer and wanted him to assume the newly created chair of electrical engineering at the university. Westinghouse wanted Fessenden nearby; Fessenden accepted. Although Fessenden was to have pursued research on the incandescent lamp for Westinghouse while in Pittsburgh, no long-term alliance between him and the company seems to have materalized. Fessenden spent the next seven years, from 1893 to 1900, in Pittsburgh. As the course of study was new, and undoubtedly received some financial support from Westinghouse, Fessenden once again had the freedom to select the equipment he needed and to shape the curriculum. He continued to explore wireless telegraphy, experimented with X-rays, began to receive local publicity for his technical achievements, and established a consulting firm. He had been, both simultaneously and alternately, a scientist and an engineer, a man of theory and a man of practice.[34]

In 1899, at the age of thirty-three, Fessenden had seven years of laboratory and eight years of college teaching experience. Both of these environments had been supportive, promoting the ethos of sharing information, and providing the reassurance that his work was valued and his knowledge expanding. While each setting was attuned to its own practical considerations, both settings valued the pursuit and acquisition of knowledge. These positions had brought Fessenden progressively more prestige and autonomy, which fanned his confidence and enthusiasm. But in the lab, and in the classroom, Fessenden had been insulated from the marketplace. His positions as chemist and electrical engineer did not require selling, either to financial backers or to the public, who would be interested in results, not technical principles. Whether with Edison or at his own consulting firm, problems were brought to him for solving. He had, as yet, no experience in exploiting or developing potential markets. And, while he had supervised the work of others, he had yet to find himself situated directly between financiers who understood little about his work and clients who needed to be wooed.

While he had no reason to doubt himself in 1899, Fessenden also had little reason to doubt others. He was, in fact, still idealistic, unable to "conceive of anything but honest, willing cooperation" from colleagues, backers, and customers.[35] His knowledge of and work in mathematics, chemistry, and electrical engineering, and specifically in dynamos, gave him a decided technical advantage over Marconi and any other competitors. Unfortunately, he thought this knowledge was all he needed to ensure success. Fessenden knew well the realms of theory and experimentation, and he had successfully integrated them intellectually and professionally. He was about to apply that accumulated knowledge, and the fresh perspective it brought, to wireless telegraphy. But the same experience that brought such distinction to his experimental work had ill prepared him for the role of entrepreneur.

In 1900, Fessenden was approached by the U.S. Weather Bureau to experiment with wireless telegraphy on behalf of the Department of Agriculture. The Weather Bureau had been concentrating on improving its ability to predict floods and storms, especially hurricanes, and wireless appeared to be a promising new tool.[36] Cleveland Abbe, acting as representative for Willis L. Moore, chief of the bureau, suggested that Fessenden begin his work in the spring of 1900. Fessenden thought the position would provide him with complete freedom to experiment with wireless, and the contract clearly stated that Fessenden would retain the patent rights to all inventions developed during his tenure with the bureau. He could select whatever apparatus he needed for the work, and he could bring an assistant of his own choosing. The offer seemed ideal to Fessenden. Work began on Cobb Island, Maryland, sixty miles southeast of Washington, D.C.

Fessenden began his experimentation by developing an alternative to the coherer. But to suggest that Fessenden was simply a revisionist like Marconi, or that he focused primarily on improving the components of wireless, would be inaccurate. Fessenden already believed that Marconi's entire system "was based on the wrong principle."[37] Highly damped waves, intermittent transmission, and intermittent reception— all had to go. Fessenden was convinced that the spark gap had to be replaced by a transmitter that sent out a continuous, sustained wave train, and that the receiver had to be constantly receptive to detect these waves. This insight was no insignificant breakthrough: it would ultimately redefine the field, transforming wireless telegraphy into radio. Marconi's intermittent waves, which surged and then ebbed, could carry discrete signals such as dots and dashes. But speech and music are sustained sound and require continuous waves to be transmitted through space. Fessenden was the first inventor to emphasize the importance of striving for the generation and detection of continuous waves. As he nurtured this goal, he sought to develop wireless components that would both outclass Marconi's and serve as steppingstones to continuous wave transmission.

In 1900, he tackled reception first. Drawing on his knowledge of both chemistry and electricity, Fessenden developed what he called the liquid barretter, or electrolytic detector, which consisted of a very fine platinum wire dipped into a small cup of dilute nitric acid. A platinum electrode was sealed in the bottom of the cup, providing an electrical connection to a local battery. When a slight current passed through the circuit, minute bubbles formed around the wire, insulating it from the liquid and thus shutting off the battery current from the headphones. High-frequency oscillations, however, eliminated the bubbles clustering around the wire and permitted the current to flow. When the oscillations stopped, the bubbles began forming again, cutting off the current until the next signals arrived.[38] Fessenden eliminated the tape used by Marconi's coherer and substituted headphones so the operator could hear the incoming signals.

The electrolytic detector possessed clear advantages over the coherer. It provided faster reception, was more reliable, required a fraction of the energy consumed by the coherer, and allowed the operator to distinguish between the different pitches of different transmitters. However, this receiver was not popular with some of the operators, because the acid often spilled or leaked and the end of the platinum wire deteriorated quickly and had to be melted down frequently to maintain maximum sensitivity. Despite these drawbacks, the detector became widely used in the United States, particularly in the U.S. Navy. To Fessenden, the electrolytic detector was important because of its superiority over the coherer. More significantly, because the receiver had the potential of being "constantly receptive," to use Fessenden's words, it would be capable of receiving not just dots and dashes, but the human voice, as well.

Fessenden's most revolutionary contribution to wireless, and the idea that distinguished his from Marconi's approach to wireless, was the radical idea of using a dynamo not merely to power a spark gap or another form of transmitter, but as the transmitter itself. This dynamo would have to generate very high frequency alternating current. In theory, such an alternator would consist of a magnetic field rotating about a fixed coil of wire or a coil rotating through the magnetic field, thereby inducing alternating current. The alternator's source of power came from a direct current dynamo, a battery of storage cells, or reciprocating steam engines. The frequency of the alternator was measured in cycles per second. Sixty cycles per second was considered a feasible frequency in the early 1900s, but Fessenden had something more extravagant in mind: 100,000 cycles per second, a speed considered utterly unattainable.[39] The enormous speed required made the mechanical obstacles to devising such a generator seem insurmountable.

On June 1, 1900, Fessenden wrote one of the most naive and yet important letters of his life. It was addressed to the great electrical engineer at General Electric, Charles Steinmetz. "You would confer a very great favor on me," wrote Fessenden, "if you would induce your company to bid on the apparatus specified and to guarantee delivery within a few months. Possibly the design might be of interest to you personally, as an experiment, though probably you have designed such machines before." He then added optimistically, "We will probably need a number of such sets if these work satisfactorily, possibly 40 to 50."[40] But Steinmetz had not designed such machines before. Nonetheless, he accepted the challenge, suggesting Fessenden try 10,000 cycles first, which Fessenden initially thought would "do very well."[41] By July 1901, Steinmetz reported that he was successfully operating a 10,000-cycle alternator, but Fessenden remained unsatisfied; he insisted on the higher frequencies.[42] He wrote to General Electric placing an order for a 100,000-cycle alternator. A General Electric official responded: "It is with very much reluctance that we accept an order of this kind, and were it not for the fact that we have a very high regard for your experimental work, and desire to aid you in every possible way, we should feel obliged to refuse to undertake the work which is so special in its character and so different from anything which we have heretofore attempted."[43] Despite misgivings, General Electric accepted the order, and the ensuing collaboration between Fessenden and General Electric produced major technical and institutional upheavals.

Confident of the correctness of his vision, and drawing on a rich and diverse background, Fessenden quickly conceived of additional technical alternatives to Marconi's system. By July of 1901, he had devised a wireless telephone. While he admitted that the device was "still a toy . . . only capable of working over short distances," he had great hopes for its potential.[44] The Weather Bureau was so pleased with his new receiver that it authorized the erection of three wireless stations along the mid-Atlantic seaboard, at Cape Hatteras, Roanoke Island, and Cape Henry. The seacoast along the Outer Banks of North Carolina was notorious for its shipwrecks and as the point where hurricanes often intensified as they headed toward the Northeast. If the three new stations helped save lives and property, the bureau planned to extend its wireless chain even farther up and down the coast. Unfortunately for Fessenden, the bureaucratic niche he thought would provide him with both autonomy and security eventually expected him to sacrifice the former to preserve the latter. Because Fessenden was unwilling to make this sacrifice, early government support of wireless telegraphy experimentation was, as we shall see, short-lived.

Fessenden's major American rival, Lee De Forest, had received a more advanced formal education at a prestigious, and some might say elite, institution. Yet De Forest's childhood experiences as a cultural outcast exerted a profound influence on him, making him the wireless inventor most attuned to the aspirations and frustrations of masses of Americans. When De Forest was six, his father, a stern Congregationalist minister, became president of Talladega College, a school for blacks in Alabama.[45] The De Forests, living on a black campus in a black neighborhood on a meager salary, were treated like pariahs by the other members of the white community. For De Forest, there was little comfort at home: his father was a rigid disciplinarian, constantly emphasizing the merits of obedience, humility, denial, and thrift.

De Forest found escape in the realms of fantasy, especially those most far removed from his father's world. He learned to live in the future more than in the present, a trait that would both help and hinder his later work. The myth of the inventor-hero, with its suggestions of environmental mastery, autonomy, agnosticism, and, of course, fame and fortune, firmly gripped the boy's imagination. Edison, whom he learned about through newspapers and magazines, became his idol, the man he most wanted to emulate. In the late 1870s and early 1880s, Edison was cast as a wizard who held seances with nature's most mysterious forces.[46] Thus, for De Forest, Edison embodied the materialism he craved and the spirituality he could not yet escape.

When De Forest was in his mid-teens, he went to the Mt. Hermon Prep School in Massachusetts to prepare for enrollment at Yale. But he would not be attending Yale College, as his father had wished. Instead, he attended Yale's Sheffield Scientific School, which he hoped would prepare him to become an inventor. His classmates remembered him as brash and loud, and he was named the "homeliest" and "nerviest" person in his class. He reportedly also received one vote for "brightest" and sixteen for "thinks he is."[47] A rural Southerner in an elite northern school, a poor man among the well-to-do, and one with no assured future among those sanguine with the confidence money and connections bring, he was even more than before an outcast.

He was, however, undeterred from his goal of becoming an in-

Lee De Forest.

ventor. In fact, it was through his inventions, whatever they might be, that he hoped to gain the trinity comprising the American dream: fame, fortune, and love. He would show them all; he would have the last laugh. His diaries are filled more with purple prose poetry and eager, wistful dreams than with technical discussions. His inventions would be a means, not an end. He wanted to be a celebrity. It is fitting that such a man, excluded from a culture of which he desperately wanted to be a part, and more obsessed with money and fame than with knowledge, would be the inventor who was most responsible for transforming wireless telegraphy into radio. De Forest did not have Fessenden's conceptual clarity about the technology, and he came to appreciate the importance of continuous waves after Fessenden. Yet through his applications of wireless, especially after 1906, he, more than any other figure in the radio community, pioneered in using wireless for broadcasting entertainment to the American public.

After completing the undergraduate course, De Forest stayed on to finish his doctorate and wrote what is described by many as one of the first dissertations relating to wireless. He graduated in the spring of 1899.

He had just moved to Chicago to begin work in the dynamo department of Western Electric when Marconi brought his wireless apparatus to America. De Forest wrote to Marconi, citing his work at Yale and asking for a job. Marconi, preoccupied with his own work, apparently did not respond.[48] Not unmindful of the extensive publicity Marconi received that fall, De Forest resumed his work on wireless, looking first for a substitute for the coherer.

Unlike Fessenden, who concentrated on transforming wireless transmission, De Forest focused on reception. His early technical work, to put it kindly, was highly derivative. In 1900, he and an associate from Western Electric, Edward Smythe, developed the responder or electrolytic anticoherer, which provided increased speed and more sensitivity. A not particularly innovative receiver, the responder resembled the coherer; it consisted of a tube fitted with two metal plugs separated by a space of about 1/100 of an inch. De Forest and Smythe immersed the filings in a glycerine and water electrolyte, reasoning that the electrolytic action of the receiver lowered the electrical resistance of the gap more quickly, thus speeding up reception. Their technical term for this electrolyzable paste was *goo*. The signals were heard through a telephone receiver because, like Fessenden, De Forest wanted to get away from the unreliable recorder method.[49] The similarities between De Forest's and Fessenden's alternatives to the coherer are striking, and they would quickly become less coincidental.

In 1903, De Forest and his new assistant Clifford Babcock developed a receiver they called the spade electrode or electro, which, like Fessenden's detector, consisted of a fine platinum wire sealed in a glass tube and suspended over a cup of acid. The genesis of this receiver is suspect, as Babcock had recently left Fessenden's employ to work for De Forest, and Fessenden won several patent suits against De Forest in 1905 and 1906, establishing the priority of the electrolytic detector. By the time Fessenden was able to prevent De Forest from marketing the electro, De Forest had already profitted from his rival's invention. But his dissatisfaction with the responder, and the court decisions in favor of Fessenden, prompted De Forest to continue looking for a reliable, sensitive, and distinctive receiver. His breakthrough was the receiver he called the audion, the forerunner of the vacuum tube.

At the same time that Marconi was working out his system of tuning in England, an American with little previous practical experience in wireless was also tackling the interference problem. John Stone Stone—both his mother's maiden name and his father's name was Stone—is rarely mentioned in radio histories. His achievements were not as dramatic as Fessenden's or De Forest's, and he did little work on transmitters or receivers. It was the connections within the system that fascinated him. And his work was important because, intellectually and technically, it contributed to the American quest for continuous, undamped, tuned wavelengths.[50] Stone's solution to tuning was an outgrowth of the work being done on automatic switching in the telephone industry; he successfully transposed the principle of selectivity based on resonance from one communications system to another. Stone was a mathematician who studied at Columbia and the Johns Hopkins University in the late 1880s. In 1890, Stone went to work in the experimental department of the American Bell Telephone Laboratory in Boston. At Bell, Stone learned the fundamentals of telephone engineering and began exploring the phenomenon of resonance in electrical circuits. His studies were not unlike Lodge's, except that Stone was working in telephony with relatively low frequencies.

In 1894 the Bell patents expired, prompting the formation of independent telephone companies competing for customers. The low prices that resulted brought more people into the telephone "network," and the central switchboards became congested as operators tried to connect all the calls manually. Gardiner Hubbard, Alexander Graham Bell's father-in-law and partner in the telephone enterprise, approached Stone about developing an automatic switchboard. The phenomenon of resonance Stone had been studying seemed to provide a way to implement Hubbard's plan. Stone proposed equipping telephone sets with resonant circuits, each of which would correspond to a circuit in a friend's telephone. Then, "by a simple operation, as by pressing a button," a caller would be able to connect one of these resonant circuits to his line. Unlike Marconi's, Stone's approach to the selectivity problem was mathematical and theoretical. Stone made drawings of his scheme but did not construct a model of it, as Hubbard had suggested.[51]

Simultaneously, Stone was becoming interested in wireless telegraphy and the phenomenon of high-frequency resonance. In 1892, at the request of his superior Hammond V. Hayes, Stone attempted to transmit speech, without wires, from the shore to a ship at sea. Although his experiments failed, Stone became extremely interested in wireless, and by 1899 he believed that his experimental interests so conflicted with his assigned duties at Bell that he resigned. He then established a consulting firm and continued to explore wireless and the problem of selectivity.

Stone was working toward what he called one-wavedness, his alternative to damped waves, as well as on tuning, and he saw the two goals as inseparable. Building on his previous work on resonance and selectivity for Bell Telephone, Stone applied these principles to wireless. If each transmitter, and the receiver it was intended to activate, contained resonant circuits, then that particular receiver should respond only to its matched transmitter and to no others. Stone's tuning required that the closed and open circuits be resonant, and thus his patent was similar to Marconi's. But Stone introduced a refinement Marconi had not: loose coupling. Loose coupling referred to the separation of the two windings of the transformer to reduce the effects of mutual inductance. In a transformer closely coupled, mutual induction reduced the efficiency of the coil. After the energy was transferred from the spark gap circuit to the open circuit, the aerial was to radiate the waves. However, some of the energy was transferred back, by induction, from the open to the closed circuit, depriving the wave trains of needed energy. These "losses" contributed to the "highly damped" wave whose strength and length diminished as the energy subsided, causing the transmission of more than one wavelength. The closer the circuits were coupled, the greater the damping effect.

Stone found that if the coupling between the circuits was loose, the effects of mutual inductance were reduced and most of the energy was radiated at one principal wavelength. His early inductive coupler consisted of wire wound on X-shaped frames. His later transformers were made of metal tubing in the form of two circular coils. The coils were separated by space, and the secondary slid up and down on a central rod and was locked into the desired spot with a set screw. Because Stone's waves were more defined, interference between them was reduced and tuning was made more precise. Stone began his experiments on the top floor of a warehouse in Boston. He set up two transmitters and two receivers and sent signals simultaneously over a distance of several hundred feet to demonstrate selectivity. He then demonstrated his selective signaling between Cambridge and Lynn, Massachusetts, a distance of about fifteen miles.[52] Stone applied for a patent on his system of tuning in February 1900, months before Marconi's American patent application.

Marconi and Stone had come to their solutions differently, Marconi relying on Lodge's previous work plus trial and error, Stone drawing on theory and his work at Bell Telephone. Marconi had begun with a set of components and bound them together into a system with his "four sevens" patent. Stone, on the other hand, had the linkage elegantly mapped out first, but his components were not as clearly conceived, which gave Marconi an advantage. What Stone called one-wavedness represented Marconi's principal challenge in America.

Stone and Fessenden were driven by a determination to overthrow the concepts and the technology on which Marconi's version of wireless was based. They were quickly dissatisfied with spark transmission and highly damped waves; they were aiming for nothing less than the propagation and reception of continuous waves. Both men's backgrounds, which combined academic work with extensive lab experience, allowed them to synthesize theory and experimentation in novel ways. Fessenden linked wireless to work previously done in electrical power transmission and chemistry, and Stone linked wireless to telephony. They were motivated by intellectual ambition and engineering pride. They had the calm confidence that a supportive upbringing and years of advancement within institutional settings can bring. Although by 1899 each man wanted more time to experiment, more technical autonomy, and professional recognition, neither man was driven primarily by a desire to become a media hero.

De Forest's motivation was different. Gripped by the overriding desire for personal celebrity, De Forest hoped to use inventing to attain widespread public acclaim. Unlike Fessenden and Stone, he did not have his own technological or conceptual alternative which scientific ambition compelled him to refine. His initial technical work was the least original of the work done by the three inventors. Yet precisely because De Forest was so susceptible to the myth of the inventor-hero, and personally felt the connections between wireless telegraphy and individual aspirations, he would soon become the inventor most responsible for transforming wireless telegraphy into radio broadcasting.

While Fessenden, De Forest, and Stone began to feel their way into the wireless business, and started questioning the technical and conceptual foundations on which Marconi's system rested, Marconi continued to concentrate on extending the range of his apparatus. Having patented his method of tuning and begun his work on improved reception, Marconi began considering sites in England and North America for high-power wireless stations. His goal: to establish a regular, commercial, transatlantic wireless service.

The faith and audacity such a goal represented are apparent even to the present-day visitor to Poldhu, Cornwall, the site he selected for the English high-power station. Poldhu Bay is a small inlet bordered by rocky cliffs several hundred feet high. A one-lane country road winds down to the beach and then up a steep hill to the Poldhu Hotel, where Marconi stayed and conducted his experiments at the turn of the century. Every piece of equipment, from the dynamos to the masts for the aerials, had to be painstakingly imported to this remote, isolated, and exposed location. Marconi had to look out over the sea, seeing nothing in the distance, and convince his assistants and his board of directors that they were going to succeed in sending wireless signals to North America.

Work on the English station began in October 1900 and was completed, except for the aerials, in January 1901. In addition to the obvious physical impediments, construction was undertaken in the face of scientific doubt concerning whether such long-distance signaling would ever be feasible. Marconi had to ignore, and ultimately to refute, the assertion that the curvature of the earth posed an insurmountable obstacle to wireless transmission. The general consensus in scientific circles that electromagnetic waves, like light, traveled in a straight line suggested that Marconi would be unable to send messages to ships beyond the horizon because the waves would not follow the curvature of the earth. Michael Pupin of Columbia recounted "many heated discussions" on how severely "the curvature of the earth limited the system."[53] Scientists did not yet appreciate the role the ionosphere played in reflecting high-frequency waves back to earth and thus allowing them to travel beyond line of sight. Marconi did not participate in these arguments, he simply persisted in his experiments. By the summer of 1901, he had succeeded in receiving signals 180 miles from Poldhu. For him, the debate was over; the scientists were left to explain what had made it possible. To the man who had extended wireless transmission from a few feet to nearly 200 miles, crossing the Atlantic did not seem too outrageous a goal.

Marconi's approach was to make all the components of his system bigger or more powerful. To achieve transatlantic transmission, he had to increase the capacity area of his aerial. Early experimenters estimated that transatlantic transmission would require aerials more than one thousand feet high to surmount the "huge curve 100 miles high" between England and Canada.[54] When Marconi discovered that the curvature of the earth did not impede long-distance work, he recognized that his aerials could be effective without being sky-high. Further tests established that a sort of wirework was more effective than a single-wire vertical conductor.[55] Limiting the height of his antennas while increasing their range was essential: first, taller masts were more vulnerable to storms, and, second, the new ocean liners, whose masts were much shorter than those on sailing ships, would require a maximum range from a minimum height in their aerials.

Experimenting with different types of aerials was an expensive and frustrating process. Timber of sufficient size had to be imported to England by steamer, and then brought as close to Poldhu as possible by train. Laborers and animals brought the masts the rest of the way. *Elec-*

trical World reported: "The number of men and horses required increases the cost of handling to an enormous figure in many cases, as wireless stations are almost without exception erected at exposed points where the roads and the means of transportion are about as bad as they can well be imagined."[56] Guy wires made of rope or metal held the wooden masts in place, but even so, the poles were unstable and a gusty coastal storm could level them. Technicians assembled the aerial and the guys while the apparatus was lying on the ground, and then with the help of horses, the men hoisted the mast upright. This process required a full day and perfect weather. Marconi and his assistants had to predetermine what the appropriate tension of the guys would be, because once the aerial was erected, tension adjustment was difficult if not impossible. Despite numerous guys and sturdy wood, the masts continued to blow down.

For his transatlantic experiments, which he planned to conduct between Cornwall and Cape Cod, Marconi had an enormous circular aerial built at Poldhu, "a ring of twenty wooden masts, each about 200 feet high, arranged in a semicircle 200 feet in diameter, covering about an acre." This aerial consisted of four-hundred wires forming an inverted cone.[57] Marconi had multiplied tremendously the number of vertical wires to obtain a greater radiating and receptive surface. On September 17, 1901, only one month after this elaborate antenna had been erected, a severe coastal storm blew it down. In a few hours the storm destroyed what had taken nearly a year to build. A few weeks later, the sister aerial at the Wellfleet station on Cape Cod blew down also. The double aerial fiasco, aside from demoralizing Marconi and his men and imposing a significant financial loss, totally redefined the scope of Marconi's first transatlantic demonstration. He had hoped to unveil two completed, working stations, capable of two-way, transatlantic communication. This was no longer possible for the immediate future, and Marconi had either to wait or to scale down his technological goals.

To Marconi and his associates, the choice was clear: priority was more important than technical perfection. As his manager, Cuthbert Hall, wrote, "It is of the highest importance that we should even in a temporary fashion be the first to get across the Atlantic."[58] Marconi had to settle for one-way transmission and had to change his location from Cape Cod to Newfoundland, a point much closer to England. While the circular aerial was being rebuilt at Wellfleet for future experiments, Marconi opted for a less elaborate system at Poldhu, an inverted triangle of fifty copper wires. In Newfoundland, he would attach the receiving aerial to a kite.[59]

To span the Atlantic, Marconi also had to increase the power of the sparking coil. He did not alter the design of the induction coil, he simply intensified the power supply. Instead of batteries, a low-frequency alternator driven by an internal combustion engine powered the induction coil. Henry Herbert McClure of *McClure's Magazine* described what such a spark gap emitted: "When the operator pressed the telegraphic key, a spark a foot long and as thick as a man's wrist, the most powerful electric flash yet devised, sprang across the gap; the very ground nearby quivered and crackled with the energy."[60]

By the autumn of 1901, Marconi had patented a system for tuning, had made reception more reliable, and had increased his transmission distance to two hundred miles. Yet many of the old problems persisted. Despite the improvements, wireless was still not secret, the messages still could be intercepted or disrupted by a rival station, and, compared to the capabilities of cable and telegraph, transmission was slow and reception haphazard. Furthermore, at this time Marconi confronted, in different ways, two of the Americans seeking to challenge his preeminence in the wireless field. In the summer of 1901, Cuthbert Hall had written to Marconi, warning him about Fessenden's work under the auspices of the U.S. government. Hall had learned about Fessenden from the general manager of the *Herald,* and wrote to Marconi: "He is working with instruments and theories different from Marconi's."[61] Marconi learned of De Forest's work in a less private and more embarrassing manner.

Marconi was scheduled to cover the 1901 yacht races for the Associated Press. He probably assumed that he would once again have the airwaves to himself. But De Forest, eager to gain the spotlight by challenging Marconi and unveiling his responder, left Chicago for New York in the summer of 1901. Through a Yale classmate, De Forest persuaded Charles Siedler, a former mayor of Jersey City, to advance one thousand dollars to support De Forest's demonstrations. He also secured a contract with the Associated Press's rival, the Publisher's Press Association, to compete with Marconi in reporting the progress of the races. The mood was less festive in New York than in 1899; President McKinley had died of an assassin's bullet on September 14, and when the races began at the end of the month the country was still in shock. Nor did wireless perform as effectively, or evoke as much enthusiasm. Marconi's and De Forest's transmissions interfered with each other, and the two had to work out a time-sharing arrangement for sending their reports. In addition, a third, "unidentified," "malicious," and "very unwelcome" transmitter began broadcasting with apparently "no other purpose in view than to upset the carefully arranged plans of the two press associations." This third party was the American Wireless Telephone and Telegraph Company, a firm established primarily to sell stock to the public. The company's goal was to embarrass Marconi; its operator periodically "leaned on the key," making transmission and reception by others impossible. The unfortunate results renewed skepticism in *Electrical World,* which noted that "the problem of securing immunity from interference remains to be solved."[62]

Marconi, always the shrewd promoter, knew he had to divert attention away from his competitors and from the problems of tuning, interference, and interception. He needed another victory in the journalistic arena, something for the newspapers rather than for the critical technical press. On November 27, 1901, he sailed from England to Newfoundland. He downplayed the trip, claiming he was going to conduct some experiments on ship-to-shore transmission. He was, in fact, about to remind his American rivals that whatever their technical visions or pretensions, his skill in seducing the popular press was still unmatched.

. . .

ON DECEMBER 15, 1901, huge newspaper headlines announced that Marconi had succeeded in transmitting the letter *s,* at prearranged intervals, from Cornwall to Newfoundland. The distance covered was two-thousand miles, a tenfold leap from his previous transmission record. The feat was unsubstantiated and unverifiable.[63] Only Marconi and his assistants heard the signal. Yet no voices of doubt were raised in the popular press. On the contrary, Marconi's heroic stature became even more imposing. Reporters flocked to Newfoundland; newspapers and magazines ran "exclusive" stories on the event. And the press easily explained its ready acceptance of Marconi's announcement: "So extraordinary is the achievement that had it been claimed by any other man than Marconi, doubts might well have been expressed; but the invariable modesty and unusual conservatism of the inventor have satisfied the world at large that no such announcement would have been made by Marconi had he not possessed the most undoubted proofs of his success."[64]

Accounts of Marconi's trip to Newfoundland and his experiments quickly crystallized into an extremely flattering legend of a piece with previous popular and formulaic accounts of inventing. Marconi, after "seven long years—years of many disappointments, vexations, setbacks, as well as unequaled success," arrived, without fanfare, in St. John's, Newfoundland. "He came quietly, gave it out that he intended to try signaling to the ships passing the Banks on their way across, and so sent up his kites and balloons with hardly a single spectator present."[65] Reporters emphasized that winter had already begun, and that Marconi struggled valiantly with the elements. He lost several of the balloons and kites in the wind, and often two men were required to hang onto the lines. To dramatize the suspense of the tests, writers recounted how one of Marconi's assistants at Poldhu had been instructed to send the letter *s* for three hours each day beginning December 11, and that the first day Marconi heard nothing. On December 12, at 12:30 P.M., Marconi, not quite trusting his own perceptions, handed the earpiece to his assistant and asked, "Can you hear anything, Mr. Kemp?" Kemp did indeed hear the three dots representing the letter *s*. But, according to the legend, the "quiet, patient, cautious inventor wished to hold his secret."[66] He did not want to notify the press until he was convinced he had heard the signals and until he had more evidence, but his excited and devoted assistants persuaded the reluctant inventor to go public. Playing on his image of the modest, thorough inventor, he later told one reporter that he was "greatly depressed" because he had used the more reliable headphone to listen to the signals instead of a Morse inker to record them and therefore had no "visible evidence of what he had accomplished, such as the recorder's tape would have furnished."[67] Ray Stannard Baker, writing for *McClure's Magazine,* described the achievement in the breathless and awed tones typical of most articles.

> Think for a moment of sitting here on the edge of North America and listening to communications sent *through space* across nearly 2,000 miles of ocean from the edge of Europe! A cable, marvelous as it is, maintains a tangible and material connection between speaker and hearer; one can grasp its meaning. But here is nothing but space, a pole with a pendant wire on one side of a broad, curving ocean, an uncertain kite struggling in the air on the other—and thought passing between.[68]

We can only infer from these credulous and admiring reports how deftly and thoroughly Marconi had ingratiated himself with reporters. To a culture that both encouraged and was repelled by the braggart, and that might have expected someone who had just signaled across the Atlantic to indulge in a little boasting, Marconi's performance was flawless. He had managed to shout and whisper at the same time. He left little room on the stage for other aspirants. As one reporter observed, "In the public mind, Marconi and wireless telegraphy are one; he is its creator."[69] Edison praised him. He was honored at testimonial dinners. Whereas the yacht races had revealed that he was technically vulnerable, his "epochmaking feat" and the skill with which he staged it applied a patina, with significant powers of deflection, to his inventor-hero image.

While the popular press continued its genuflections, the technical press printed reactions from members of the scientific community who refused to believe Marconi had succeeded and considered the story a publicity stunt. What proof, beyond his word, had Marconi offered? Lee De Forest, in his diary, expressed his own skepticism while acknowledging that, strategically, Marconi had been quite clever: "Signor Marconi has played a shrewd coup d'etat whether or not the three dots he says he heard came from England. . . . He has established his fame and stolen thunder from competitors, who may, in a few years, actually send messages across the ocean. His stock is soaring and will make the achievements of others, however meritorious, look cheap enough in the popular eye." De Forest then summarized the academics' doubts: "Whether he actually received those signals or not, he has certainly offered no real *proof* which scientists can accept; and all this great haloo and adulation with his wild talk of transatlantic messages at one cent a word smacks decidedly of chicanery and the methods of the professional newspaper boomer."[70]

Professor Branly, developer of the coherer, wondered about the conditions under which the work was done and questioned whether it had been conducted in a "rigorously scientific manner." Searching for any explanation other than transmission by Hertzian waves, Branly pointed out that "it should be definitely determined that there was no influence from submarine cables, which might be inductively affected by the transmission waves and thus have a part in transmitting the signals to the receiving station."[71] *Electrical World* ventured: "It is quite probable that the publication of the results achieved was unauthorized, the experiments being merely preliminary ones of a scientific rather than practical nature." In the final assessment, according to the journal, "The details reported up to the present time are altogether too meagre to enable any reliable conclusions to be drawn as to this alleged transmission across the Atlantic Ocean. . . . The sudden increase in distance to 2100 miles represents more than a tenfold increase of radius, which we should expect to be overstepped more gradually."[72] These expressions of doubt in the technical press were drowned out for the general public by the din of praise and expressions of wonder which filled the popular magazines and newspapers.

As dramatic as the transatlantic achievement was, and as gratifying as it must have been to Marconi stockholders, it still represented a brief moment of glory which did not guarantee contracts or regular revenues. Duller, more private cultivation was necessary if Marconi was going to translate the public preeminence he had gained through the yacht races and the letter *s* into preeminence in the marketplace.

The transatlantic achievement intensified the fevered expectations of 1899: wireless would bring world peace, freedom from the cable companies, a democratized communications system, transcendence over space and time. There is a certain irony in the portrait of Marconi as a self-effacing and selfless public benefactor. For him, the transatlantic success was the critical first step in achieving a goal he was initially quiet about in public: establishing a monopoly, if possible worldwide, in wireless telegraphy. To the press and the public, that a man such as Marconi could be thinking in these terms, that he could actually consider trying to monopolize "the air," would have seemed both out of character and technically preposterous. Behind the scenes, however, establishing such a monopoly was Marconi's fervent hope, and one with which the American inventors would have to reckon.

Marconi would himself have to reckon with the inventors' aspirations and technological alternatives. Between 1899 and 1901, Marconi could afford to, and did, ignore the American technological challenge. He was obsessed with long distance Morse code transmission, for which he knew there was a market. In the short run, he made the correct technical choice. He also knew how to manipulate the journalistic arena to enhance his legitimacy and divert criticism. But Fessenden and Stone, with their emphasis on continuous, tuned waves, and De Forest, who hungrily coveted Marconi's celebrity status, continued to work in their labs. They also sought to establish wireless firms that would compete with Marconi. Despite the financial and technological uncertainties surrounding them, and despite Marconi's preemptive displays, these three ambitious men sought to design apparatus that would render Marconi's obsolete. Because of the determination of these men, ignoring the American technological challenge was a decision Marconi would live to regret.

CHAPTER THREE

THE VISIONS AND BUSINESS REALITIES OF THE INVENTORS

1899–1905

AS MARCONI, FESSENDEN, De Forest, and Stone vied with one another for technical preeminence and public recognition at the turn of the century, they competed in three arenas: the technological, the corporate, and the journalistic. Already the Americans were pointing, admittedly still in a tentative manner, toward a significant technical departure from Marconi. But without success in the other two arenas, their innovations would matter little, and without an appreciation of how all three spheres interlocked, success could not be sustained. Marconi's stature in the press seemed invincible, but technically, he was vulnerable. Marconi had capitalized on his public image to compensate for his technical vulnerability. He had also devised the beginnings of a corporate structure that would protect and promote his determined entrepreneurial spirit. Again, the Americans had to catch up with Marconi. He had formed his British company in London in 1897, and just after the yacht races in 1899, he had formed an American subsidiary. When he signaled across the Atlantic in 1901, he still faced no corporate competition in America. By 1901, he had already devised a fledgling wireless network guided by highly competitive and exclusionary company policies.

The Americans, seeing the flaws in Marconi's apparatus, may have believed he was not such a formidable rival. But what Marconi may have lacked in technical creativity, he more than made up for with entrepreneurial flair. He was, in fact, a brilliant and determined competitor who was quite prepared by both talent and inclination to control and dominate the wireless market.

What prompted Fessenden, De Forest, and Stone to try to compete with such a man, already established as an inventor-hero and independent businessman? Certainly personal ambition and technical competitiveness drove them on. But individual aspirations were reinforced by the heady economic climate in America at the turn of the century. The society of which these three men were a part was undergoing an energetic and widely publicized financial boom, and newspapers and magazines, by highlighting rags-to-riches stories, suggested that this was a boom in which all Americans, and not just the rich and established, could participate. The stories celebrated Americans who took entrepreneurial risks and suggested that success awaited any financially daring individual. It seemed more possible than ever to strike out on one's own and make good.

The primary source of the get-rich-quick fever was the explosion in the American stock market which accompanied, and helped make possible, an unprecedented merger movement. In the early 1890s, only railroad and government securities, and just a handful of industrials, were traded on Wall Street. Most industrial concerns in the 1880s had been "small, closely owned, and commonly regarded as unstable," and thus were not particularly available or attractive as investment opportunities.[1] Corporate consolidation, triggered by the desire to reduce competition, and made possible by New Jersey's incorporation act of 1889, began to unite such firms into large national companies with national reputations. Further expansion required additional working capital in amounts only generated by selling securities to the public, and such securities began appearing on the market just before the panic of 1893. When these securities fared well during the depression, investors began regarding them as sound investments. By 1897, more than two hundred industrials were being quoted in financial journals, and the number kept increasing. As family or small partnerships became corporations seeking to take advantage of the benefits of selling stock—access to increased amounts of capital, and diffusion of liability—more middle-class customers, eager for new, promising investment possibilities, were lured into the market.[2]

During this revolution in corporate ownership, the press played a critical proselytizing role. Two types of actors brought drama and glamour to the explosion on Wall Street: the so-called Captains of Industry and the everyday, ordinary person who, through shrewd or lucky speculation, made a killing. Certainly J. P. Morgan's financial legerdemain, whereby scraps of paper, properly arranged and exchanged, produced millions in profits, was the most legendary. He and the other "Napoleons of Finance" became heroes and demigods because of the prodigality of their profits and the rapidity with which those profits were made.[3]

Stories in the press suggested that putting together the right merger could make a man a millionaire overnight. These tycoons, and the trusts they personified, evoked a strong American ambivalence toward the concentration of wealth. Muckrakers and reformers attacked the ruthlessness, exploitation, and constriction of competition which frequently accompanied the consolidation in industries. Yet the cleverness, daring, and strength of will required to make such money, as well as the extraordinary sums involved, were sources of admiration and envy to many.

That Americans hoped to emulate, even in a small way, the successes of these businessmen-heroes is evidenced both in the increasing sale of stock and in the newspaper accounts describing many recent rags-to-riches stories. America experienced one of the greatest bull markets of its history. In 1901 an unprecedented three million shares changed hands in one day. According to Mark Sullivan's classic social history of turn-of-the-century America, "a slogan ran through New York, not only downtown but in shops, on streetcars, on commuters' trains: 'Buy A.O.T.—Any Old Thing.'" The press featured stories about waiters, dressmakers, clerks, and barbers who, by buying certain stocks during their lunch hours, found themselves wealthy in a few days.[4] Following the latest developments on Wall Street had become a new American pastime.

Newspaper stories anticipating wireless telegraphy's limitless commercial potential harmonized well with the more general, exuberant stories describing how to cash in on the new prosperity. Wireless, as a promising new technology, might be an excellent investment for those wanting to get in on the ground floor of a new business. After Marconi's transatlantic success, one reporter predicted that "cables might now be coiled up and sold for junk."[5] The cable companies were lucrative firms; if wireless companies displaced them, then wouldn't these new businesses become extremely valuable? All the eager dreamer had to do was recognize what giants Western Union, Bell Telephone, and General Electric had become to calculate where wireless might be in the future and what fortunes might accrue to those who had had the foresight to invest early.

The bull market was an important feature of the economic environment within which the first American wireless companies were established. The Wall Street boom, the rags-to-riches stories, and the excitement over wireless indicated that the wireless entrepreneur would have no trouble selling stock to the public. But competing successfully in the marketplace would take more than credulous investors, for, despite the prosperous economic times, the entrepreneurs faced important challenges in the arena of business strategy. What the popular images of overnight financial success failed to point out was that much more than luck and delusions of grandeur were required to survive economically in the real world of American business.

Alfred Chandler has emphasized how important attention to company structure and strategy were in the early twentieth century for major corporations such as General Motors and Du Pont which sought to become more efficient and gain a larger share of the market.[6] Structure, the organizational design through which an enterprise was administered, and strategy, the determination of the company's long-term goals, were no less critical to the struggling small businessman seeking to establish a need for his product and ultimately to compete with already entrenched firms.

Between 1899 and 1902, Marconi, Fessenden, De Forest, and Stone each formed his own wireless company in the United States. Competition moved from staging demonstrations and courting the media to building corporate structures and articulating corporate strategies. All of the inventors had to decide how to organize their companies. What would be the hierarchical structure, the chain of command? How would information flow between the board of directors making policy and the lowly wireless operators manning the apparatus? How would marketing decisions be made? Who in the company had the final authority? In addition to establishing the company's structure, no matter how skeletal, the inventors had to determine strategy. Obviously, they had to define the market for wireless. If wireless was to provide a new, relatively unfamiliar service, the inventors had to determine how to convince people to use it. The inventors also had to figure out how they would generate revenue. Day-to-day operations had to be guided by long-term planning and goals. Where did the members of the company hope it would be in five or ten years? And how did they expect to get it there? Equally important was the question of visibility and distinctness: How would the Americans avoid looking like latecomers and mere imitators in the public eye, and distinguish themselves from Marconi? And how would they introduce their wireless systems and persuade Americans to use them? To survive as a new business in the age of mergers and monopoly, the inventors had to find workable solutions to these problems.

· · ·

WHILE MARCONI WAS in America in the fall of 1899, he met John Bottomley, a prominent New York attorney with whom he would shortly found the American subsidiary of Marconi's Wireless Telegraph Company Limited. Bottomley was no stranger to the world of science and invention: his grandfather and brother were both scientists, and he was the nephew of Lord Kelvin, the noted British physicist. Kelvin had become an enthusiastic supporter of Marconi's work, and he provided Marconi with a letter of introduction.[7] Bottomley, successful, fifty-two years old, and interested in wireless, was ready to undertake an exciting if risky new project. He and his partner, E. H. Moeran, helped Marconi establish the Marconi Wireless Telegraph Company of America, which was chartered in New Jersey in November of 1899 with a capitalization of ten million dollars.[8] Bottomley became the new company's general manager, secretary, and treasurer, and Moeran served as general counsel. Sometime between 1901 and 1905 they persuaded one of the company's directors, John W. Griggs, a former governor of New Jersey and attorney general under President McKinley, to become the company's president. Both before and after his political career, Griggs was a well-known corporate attorney. Between 1901 and 1907, he served as a member of the Permanent Court of Arbitration at the Hague. For a foreigner who initially had few contacts, Marconi had done quite well. Although Bottomley did not have extensive managerial or entrepreneurial expertise, his connections in commercial circles were invaluable. Griggs, with his reservoir of political experience and allies, knew how to fight and win important legislative battles.

To appreciate the opportunities and dilemmas facing the American Marconi Company, whose operations would be subservient to the plans and goals worked out across the Atlantic, we must first review the emerging structure and strategy of its parent organization in Britain. For Marconi, designing an efficient and respectable organizational framework had been as important as testing and establishing his technical facilities. Orderly proceedings and a clear chain of command were established, unambiguously, from the start. Formalities were important and observed.

When the Wireless Telegraph and Signal Company Limited was established in London in 1897, Jameson-Davis, Marconi's cousin, became the first managing director.[9] The only other officer was Henry W. Allen, the company's secretary. Marconi's experimental staff consisted of himself and two assistants. As the number of public demonstrations increased, so did the size of the technical staff, and Davis delegated supervision of this staff to a senior engineer. Davis had considered his official role in the company temporary and was eager to resign. In August of 1899 he was succeeded by a new managing director, Major S. Flood-Page. The company acquired a manufacturing plant in Chelmsford, thirty-five miles outside of London, in 1898, which was supervised by a works manager.

The company was expanding quickly and was somewhat difficult to oversee because the experimental staff, including Marconi, traveled so much to demonstrate the apparatus. To assist him with his managerial chores, Flood-Page hired H. Cuthbert Hall in early 1901 as manager. In 1900, reportedly against Marconi's wishes, the board of directors voted to change the company's name to Marconi's Wireless Telegraph Company Limited. Changing the company's name was apt, for one thing was crystal clear: this was *Marconi's* company. He had formed it with his British relatives, and he and his family held a controlling interest in the firm. There was never any doubt about whose will was dominant, whose say was final.

By the turn of the century, then, Marconi's company, though small, had several distinct but interrelated departments and a clear hierarchy. The company recognized the importance of having managers in supervisory but neutral positions. While Marconi was deeply involved in his experimental work, he was the driving force behind the company's entrepreneurial activities, often instigating many of the company's organizational and promotional decisions. Yet he was insulated from his board, and from many of the routine daily chores and decisions, by his managing director. Flood-Page resigned in June 1901 and was replaced by Cuthbert Hall, with whom Marconi had a warm and trusting relationship. Hall served as the critical communications link between Marconi and the rest of the company, and surviving correspondence indicates that Hall was very conscientious and that the flow of information within the firm was excellent.

Certain aspects of the company's early structure and strategy deserve special mention. These features were not particularly noteworthy or exceptional in and of themselves, except that they were in striking contrast to the way business was carried out by Fessenden, De Forest, and Stone. The British Marconi Company had regular board meetings, which occurred at least quarterly. Although Marconi's voice was carefully listened to and rarely, if ever, overruled, the inventor did have to get his board's approval for major capital expenditures. Cuthbert Hall wrote memoranda periodically summarizing and assessing the company's status. Marconi diligently participated in many of the managerial activities, such as drafting and reviewing contracts, writing and editing company circulars and directors' reports, setting strategic priorities, and reorganizing the staff.[10]

Unlike the popular press, which had democratic visions of wireless's potential, Marconi viewed his invention as having only narrow commercial applications. He missed few opportunities, when out of the public eye, to emphasize this point. When commenting on a draft of a directors' report in 1903, Marconi instructed Hall to substitute the phrase "commercial purposes" for the phrases "the general public" and "a public service."[11] Marconi polished his public edifice as democratic benefactor, but, privately, he was a highly competitive businessman whose ultimate goal was to establish a monopoly in wireless telegraphy. This goal was eventually referred to as the Imperial Wireless Scheme; Marconi meant to connect the entire British Empire together by wireless, and he meant to own the only company capable of doing so.

Two entrepreneurial strategies dominated the company, and, while complementary in the long run, in the short run they were sometimes at odds. One strategy involved offering a completely new service; the other, offering a less expensive substitute for a service that already existed. Both strategies concentrated on signaling over water. Marconi already knew he got much better results over water, where there were no obstacles to the signals. Beyond this technical advantage, he also saw maritime signaling as his primary market: for him, this was where wireless fit. There was no service at all between ships and the shore, there was a clear need for such communication, and, as yet, there was no competition. The market seemed ready-made. Marconi also altered his earlier statements about not competing with the international cable companies. By 1901 he was publicly stating his intention to offer an alternative system to that of the cable companies, which he knew were viewed as greedy, arrogant monopolies charging exorbitant prices. The cables were one of the monopolies Americans, as well as the British, loved to hate, and it was clear that Marconi could get customers if he could make his service as speedy and reliable as that of the competition.

The key to overtaking the cable business was successfully spanning the Atlantic, and the company believed it imperative that Marconi be the first to do so. John Bottomley urged: "The establishment of commercial wireless telegraphy across the Atlantic Ocean is absolutely essential to the financial success of the Marconi System."[12] The transatlantic stations would provide two main services: they would compete with the cables for press dispatches and commercial messages, and they would keep ocean liners in touch at all times with both sides of the Atlantic. Revenue would come primarily from the charge per message. While the company was overly optimistic about when such a commercial service would be functioning on a regular basis, it did realize that the service would be slow to show a profit. In other words, the transatlantic service represented a long-term strategy requiring major initial expenditures and considerable patience.

The company also sought to establish a dense and interconnected network of short- and medium-range stations aboard commercial ships and naval vessels and strings of shore stations to service these ships. The Chelmsford works was already manufacturing apparatus, the sale of which, it was initially hoped, would return some money to the company's depleted coffers. This short-term strategy, while less grand than the transatlantic scheme, was critical. The company wanted to equip as many facilities as possible, not only to bring in quicker revenue, but also to preempt the market and accustom customers to being members of the growing Marconi network.

The company's two strategies were interrelated, as the company's overarching goal was to create the most efficient and complete wireless organization in the world. The strategies needed to be achieved simultaneously and often required Marconi's presence at various important demonstrations. Thus, when Marconi concentrated on sending the first transatlantic signal, he was unable personally to supervise wireless demonstrations for the British or Italian navies, whose top brass preferred to be wooed by Marconi himself rather than by a stand-in. The company constantly felt it was fighting against time, trying to beat the competition, catch the market at the most propitious moment, and exploit the latest publicity coup before the memory of achievement faded. A letter from Hall to Marconi written in July 1901 describes one of many such episodes: "I believe there is a great naval scare on and that the Admiralty officials are very anxious to make the fleet as efficient as possible in the shortest possible time. If the scare passes we should probably not get such good terms, but we cannot help that, as we should have to sacrifice the Poldhu experiments on the chance of getting a good Admiralty order."[13]

The problems of dual development strategies, technical uncertainty, and the threat of competition confronting Marconi's fledgling company were exacerbated by the trickiest question of all: how to make wireless pay. Electrical communications systems had already been marketed successfully, but one crucial difference confronted Marconi and his competitors: wireless provided no wires or lines. The service was distributed through "the air," which had always been free to all. Even with tuning, anyone with a receiver could listen to messages free of charge. There existed no physical means of restricting access to the wireless network. Traditional western notions of ownership and property laws were completely inadequate in the face of something invisible, intangible, and inherently communal like the ether.

The Marconi Company struggled for years with this revenue prob-

Operators learning how to transmit, receive, and read code
at the Marconi Company's training institute.

lem and would not show a profit until 1910. When the company was
formed in 1897, Marconi had hoped it would become economically
viable by selling wireless apparatus outright, especially to shipping firms.
But a customer could not buy equipment only, because the client would
also need shore stations with which to communicate, and trained operators to handle the messages. No customer was prepared to make that
large an investment in such a new service, especially one for which the
client might have only an occasional need. The Marconi Company had
gone to the expense of demonstrating its system to various potential
buyers only to receive praise but no contracts.[14] So, in 1899, after a
couple of disappointing years, the Marconi Company changed its policy
and its structure: the company would now sell equipment to no one.
Instead, it formed a subsidiary, the Marconi International Marine Communications Company, an operating company that sold not apparatus,
but service. The client paid for access to a communications network
Marconi established and controlled.[15] The company leased equipment
and a trained operator for a specific period of time. No charge was made
for individual messages.

The company's new policy represented a shrewd shift in financing strategy. Leasing encouraged more firms to give wireless a try, and not charging for the messages allowed the Marconi Company to skirt certain British telegraph monopoly restrictions that prohibited a private company from sending telegraphic messages for monetary gain.[16] Also, by controlling the sending stations and the operators, the company could regulate who would receive messages and who would not. The company then established its nonintercommunication rule, certainly its most controversial policy. Marconi operators, on ship or shore, would only communicate with other Marconi operators. Clients using other apparatus were excluded from the network. Only in the event of emergency was this rule suspended.

By 1900, the Marconi Company had negotiated contracts with several steamship companies, including Cunard, was supplying the Italian navy with wireless, and was about to install apparatus aboard twenty-six vessels and six coast stations for the Royal Navy. The company then secured its most prestigious and potentially most lucrative contract: to equip various offices for Lloyd's of London, the prominent marine insurance agency with offices throughout the empire, in all of the world's major seaports. Agents at these seaports were to keep Lloyd's headquarters informed about the status of insured ships, a task that wireless installations would make much easier. The agreement, signed in September of 1901, provided that Lloyd's would install only Marconi apparatus and use the system exclusively for fourteen years. The contract stipulated that the system would neither transmit nor receive messages to or from any apparatus produced by any other company. Marconi's insistence on leasing rather than selling the apparatus appeared in his written comments on the draft contract. He wrote to Hall, "I note that the limitation of Lloyd's power to buy apparatus does not appear in the precis of the agreement, but I presume the point is duly safeguarded in the original. It is not without importance."[17]

This contract provided Marconi with the link he needed to the established international corporate network. With this contract, he had a worldwide presence, as well as an affiliation with a British company so powerful that the Marconi nonintercommunication policy would have real force. If Lloyd's was using Marconi apparatus, and Marconi apparatus only, and communicating with no others, then shipping firms interested in or compelled to be part of this network would have to lease from Marconi. The agreement represented a major step toward making the Marconi system self-perpetuating.[18]

With the Lloyd's contract secured, the Marconi Company erected more shore stations in Europe and established a wireless service on the Hawaiian Islands. All operators were under strict orders not to exchange messages with operators using rival equipment. To Marconi, this refusal to communicate seemed to be the only effective method to control competition, prevent bankruptcy, and pave the way for monopoly. The Marconi Company had gone to considerable expense in erecting shore stations in Europe and North America. Why should shipowners install any sort of apparatus they pleased and then make use of these shore stations while contributing nothing to their maintenance? The Marconi Company believed it could not afford to allow free intercommunication with any other system. By writing the nonintercommunication policy into contracts such as the one with Lloyd's, the company provided shipowners a powerful incentive to lease Marconi.

The Marconi Company gave the impression that communication with rival apparatus was technically impossible. Through oblique and misleading comments to the press, Marconi suggested that different apparatus, of different design and using different wavelengths, would be unable to send to or receive from Marconi stations. He also asserted that because his apparatus was now specially tuned, "it would . . . be a mere accident if [competitors] happened to strike a tune to which the receivers at my stations were responsive."[19] This alleged technical incompatibility did not, in fact, exist. But Marconi succeeded in temporarily convincing some members of the press of the technical obstacles by insisting that his wireless comprised a "system." The excellence of this system depended not only on the superiority of the individual components, but also on the arrangement of and adjustments between these components. Company statements suggested that a components-oriented approach had been tried and abandoned: "We have not found successful attempts to embody apparatus or material of foreign origin in installations such as our own, where every detail is designed with a view to the efficient working of the whole."[20] Marconi maintained that the Marconi apparatus was far superior to any competing equipment, and that by communicating with inferior systems, the company would impugn the reputation of wireless in general. As Marconi explained, "The policy of the Marconi Company has always been that we cannot afford to recognize other systems. . . . We cannot be expected to injure our own cause, which we would certainly do if we permitted these stations to communicate with vessels and stations using our system."[21]

Marconi was trying through company policy to prevent what he could not through technical or legal means. Competitors could easily send to, receive from, and interfere with Marconi apparatus. Many of these competitors were transmitting with equipment that infringed on Marconi's patents, particularly the "four sevens" tuning patent. The company hoped that its nonintercommunication policy would somewhat compensate for its decision not to sue infringers at this time. Cuthbert Hall summed up the company's position on litigation when writing of a suit against De Forest in 1904: "I would rather not fight. We have nothing to gain commercially or in prestige by a win, and we should lose a little in prestige if we lost. . . . Our position now depends far more on contracts than on patents . . . [and] although the case covers a wide area we shall have overthrown an unimportant adversary. . . . I do not see that there is any commercial necessity to fight at all."[22] Litigation was expensive and consumed considerable time and energy; forgoing it was a decision that again reflected the company's patience and ability to take a longer term view of its business. Marconi was naturally more irritated than others by the infringement, and he sometimes found his patent attorneys over-cautious.[23] But he generally agreed to wait for the time when the company was more financially stable and when a legal strike would be more decisive and debilitating to the opposition. Marconi was willing to defer gratification, to subordinate pride to sound business policy, and thus he exhibited a discipline somewhat unfamiliar to his American counter-parts.

The nonintercommunication policy, the agreements between several navies and steamship companies, and the much-publicized Lloyd's contract made the company appear already prosperous. The British and American Marconi companies faced ongoing problems managing and marketing Marconi's system, however. The ambitious goal of trying to establish an international wireless network resulted in the creation of an organization that was often too dispersed and fragmented for a small company to oversee properly. The Marconi Company provided its clients with wireless operators, and those on board ships were Marconi Company employees and not subject to the authority of the steamship companies. Thus, they often had no direct supervision. Some of these operators were slow or lazy, and at least one was caught embezzling from the American company.[24] Under such an arrangement, establishing loyalty and discipline would take time.

Also, the company had to convince both individuals and companies that wireless was a trustworthy and efficient method of communications. A Marconi Company memo recalled the early days when "there were only an average of five or six ships fitted with wireless going into New York harbour each week and at first, as people had not learned to trust wireless communication, . . . most of the messages from passengers were of the 'love and kisses,' 'see you soon' variety."[25] Bottomley complained, "Under the present conditions, the business market aboard ship is practically limited to first-class passengers for ten or twelve hours before the ship reaches its pier, and the same time after leaving. There appears to be no practical way of increasing to any extent, messages of solicitation and congratulations exchanged between passengers and friends ashore."[26]

The strength of Marconi's reputation, the shrewdness of his strategies, the uncertain revenue prospects, and the distance from the parent company—all affected the fortunes of the Marconi Wireless Telegraph Company of America. Financially, the American company was on its own. While its capitalization was based on the value of Marconi's patents, it could expect little monetary assistance from England. Its day-to-day operations were under Bottomley's supervision, and he was not expected to bother England with details. Major policy and marketing decisions, on the other hand, were out of Bottomley's control. The British company determined how Marconi's system would be established, promoted, and operated, and the American company was instructed to adhere religiously to these rules. Most decisions had to be cleared through Marconi and the British office first, which was time-consuming, inefficient, and irritating to American clients. Bottomley was allowed little creativity in adapting or shaping the company's policies to mesh better with the potentially different American circumstances. The British connection, then, in some respects, hindered more than helped. The American company often felt isolated from and neglected by its distant parent, and Marconi's infrequent and brief visits to New York were not sufficiently consoling.[27]

Bottomley reported regularly to London. Like his British colleagues, he convened quarterly board meetings. The American company's structure was minimal when compared with that of its parent, consisting primarily of Bottomley, Moeran, a man named W. W. Bradfield, who appears to have assisted with the managerial and sales duties, and several technicians and operators.

The Marconi Wireless Telegraph Company of America, in the early years between 1899 and 1904, did little more than give Marconi a corporate presence in the United States. It had no network, and its revenues were paltry. The center of corporate structure and the place where strategy was formulated was in England, not America. With Marconi on the other side of the Atlantic, the amount of promotional work and advertising that could be done in America was minimal.

Although shares in the company were available, the officers did not

Wireless operator aboard ship, ca. 1905.

indulge in the sort of stock promotion that characterized other American firms. Pamphlets extolling Marconi's achievements appeared periodically and served as "a guide to those interested in Marconi stocks as an investment."[28] When Moeran received inquiries about Marconi stock, however, he responded that such a purchase represented a very long-term investment and that no dividends would be paid for years, because all available money would go toward improving the apparatus. He advised a Mrs. Florence Hoyt of Brooklyn not to invest in Marconi, but instead to put her money "into the class of investments which may not be regarded as industrial."[29] It is important to emphasize that the company observed certain proprieties when dealing with an eager public, proprieties ignored by others.

Bottomley and the other officers tried to use their influence to persuade prominent men or companies to lease apparatus. Bradfield succeeded in placing an installation aboard the Gould yacht *Niagara*, and J. P. Morgan and other wealthy clients were approached.[30] But the company needed American stations, and the only one being constructed was the transatlantic facility at Wellfleet. The company did succeed in leasing apparatus to the *New York Herald* for installation on Nantucket and the Nantucket lightship in August 1901. In addition to serving the *Herald,* the Marconi Company also reported the arrival of ships to steamship companies, which paid five dollars per vessel for the service.[31]

The Nantucket stations provided the company with considerable visibility, as Nantucket was often the first point of contact for American-bound ships. Thus it was essential that at this station, especially, the nonintercommunication policy be rigidly enforced, driving home the benefits of leasing Marconi and the costs of not doing so. Moeran advised the technicians at Nantucket "not to communicate with ships not having Marconi's apparatus. I will give you a list of vessels."[32] The *New York Herald* was distinctly displeased with this policy: the newspaper had established the station specifically to gather as much news as soon as possible for its paper. Nonintercommunication was completely antithetical to the *Herald*'s goals. Moeran had to notify London: "Kemp [Marconi's longtime assistant] at Nantucket informs me it will be very difficult to prevent operators at Nantucket from sending news of any ship which may be within range and that the *Herald* has thoroughly drilled into them to send every scrap of ship news they can get hold of."[33] The *Herald*'s attorneys notified Moeran of the paper's refusal to agree to the policy and its opinion that its communication with any vessel it chose did not violate the Marconi-*Herald* contract. The nonintercommunication policy, which was supposed to further the company's corporate goals, was already undermining its relationship with its first promoter and major client.

The only other ongoing business in which the American company participated was the transatlantic news service for steamships. Marconi operators were to send news dispatches to the ocean liners, and passengers would pay a small amount for the shipboard newspapers. Shipboard operators were to report any newsworthy activities to the shore stations. But two problems plagued this scheme. Transmission and reception were not yet reliable throughout a transatlantic voyage and, as Bottomley noted, it was not likely that passengers would pay very much for a news service which was offered "but a few hours before reaching shore."[34] In addition, the operators had been trained to send the Morse code, not to compose snappy prose. As a result, their stories were routine and unimaginative compared to the columns of the *Herald* or the *World.* When, several years later, the company referred to the shipboard service as "a dead letter," failure was attributed to "the old story, lack of reportorial ability on the part of the operators."[35]

Bottomley wanted the American company to promote itself more aggressively, to demonstrate Marconi apparatus more widely, but he was often curtailed by decisions in London. For example, the 1904 World's Fair was to be held in St. Louis, and Governor Francis of Missouri invited Marconi to visit the exposition grounds and select a site for an exhibit on wireless telegraphy. Marconi, appreciating that the American company believed such a display might be beneficial, agreed to go.[36] When the company learned that De Forest and a few other companies would have wireless booths, however, it decided not to mount an exhibit after all. Noting that the potential revenues in such a setting would be split among the various competitors, a company officer reported that the projected meager returns did not warrant such a display: considering that the "stockholders of this Company have purchased their stock upon the express understanding that the money thus obtained should be devoted to commercial development of the Company and that, the financial condition of the Company would not warrant an exhibition of the sort proposed, for purely spectacular purposes, the Directors decided to make no independent exhibit."[37] Marconi was simply unwilling to gamble on such short-term ventures that earned only visibility and lent legitimacy to competitors but did not lay the groundwork for future enterprises. For Marconi, public demonstrations had to have a larger purpose and had to buttress company strategy.

The American company had to content itself with serving as a corporate beachhead and basking in Marconi's glow. Marconi and the British company continued to provide one invaluable and as yet unmatched asset: occasional but superb advertising for the Marconi system. On December 21, 1902, slightly more than a year after he had received the letter *s* from Cornwall, Marconi succeeded in transmitting a full message across the Atlantic. The headlines in the *New York Times* blared: "Marconi's Great Triumph" and "The New World Sends Greetings to the Old." The governor general of Canada had sent a message twenty-three hundred miles to King Edward VII. While Marconi had been preparing the stations in Canada and Poldhu for this feat, his assistants had been repairing the aerial and refining the apparatus at the Wellfleet station, for Marconi wanted an American link in the transatlantic network. From Wellfleet, in January 1903, Marconi succeeded in transmitting "most cordial greetings" from President Roosevelt to King Edward VII.[38] This accomplishment made regular transatlantic wireless signaling, a service Marconi was eager to inaugurate, seem imminent. Once again he had wedded publicity and corporate goals.

Meanwhile, Bottomley diligently courted the American steamship companies and worked to establish shore stations. By 1903, the American company had two stations: the one at Wellfleet, and one at Babylon, Long Island, which was little more than a shack used for experiments.[39] Bottomley urged the British company to establish a station in New York City, but to no avail. Bottomley's mandate was to build as extensive a network as possible on the American side of the Atlantic. As a Marconi man, he was expected to add outlets to the growing organization, but it seemed he was to do this with limited funds and psychic support from London. Thus, he was left to try to graft the Marconi system onto the organizational networks of other firms and hope for a ripple effect of influence and revenues. He succeeded in persuading a group of shipping-line owners to contribute to the building and operation of a station at Sandy Hook, New Jersey, which would be equipped with Marconi apparatus. Writing one year after the Sandy Hook facility was built, Bottomley reported that receipts were 50 percent higher than they had been the previous year.[40] The American company also sought an agreement with Western Union, negotiating terms for linking the Marconi Company's primarily offshore system with the telegraph lines. By 1904, the two companies had come to terms, and Western Union was providing land-line links for Marconi wireless messages.[41] Bottomley was building critical organizational alliances for the American company, alliances with major firms, but he had yet to convince the owners of the hundreds of smaller ships operating along America's coast to lease Marconi apparatus. Other than the major shipping lines, then, much of the American market remained unserved.

Without Marconi's regular presence to motivate the staff, and without a clear sense of entrepreneurial purpose—of specific goals originating on this side of the Atlantic—morale and discipline in the American company wavered. Bottomley wrote to Cuthbert Hall complaining bitterly of staff problems: " The chief fault with all these men is that they seem to consider the business as a sort of plaything whilst I am, unfortunately perhaps, terribly in earnest."[42] Although he realized that telegraph men were known for being "wild," he was still unprepared for the "careless living, 'women and wine' practised by those high up in this company." The men's "inattention to business" so distressed him that he fired several of them, although he was "half inclined to clear out the whole force."[43]

Between 1899 and 1904, the Marconi Wireless Telegraph Company of America showed potential for both success and failure. The company remained something of an afterthought to Marconi, who viewed it as an obedient and pliant follower of British policy and not as a company possessing its own strategic initiative geared expressly for the American market. As a result, the American company, while in no immediate danger of floundering, nonetheless lacked an aggressive approach for defining and capturing the American market. Marconi's major contracts were still with European clients. However, while the American company was small and possessed few assets, it was an agent of the growing Marconi network, which seemed likely to increase in size and influence. The leasing and nonintercommunication policies devised by the parent company represented a determined and coherent strategy for generating revenue, enticing clients, and discouraging competition.

Despite its financial and managerial frailties, the American company was still a Marconi company, headed by an internationally renowned inventor who combined promotional flamboyance with a respect for organizational orderliness. Marconi's ability to devise a successful marketing approach stemmed partly from talent and partly from fate. His British heritage and his coming of age in the 1890s had placed him in a strategic location at a fortuitous time: he was a member of a worldwide empire built on maritime dominance which was still at the peak of its power and influence. The necessity of communicating over large bodies of water had stimulated cable construction since the mid-nineteenth century. This was a vast empire that had holdings on nearly every continent and that was sustained by a burgeoning navy and merchant marine—and major gaps existed in its communications network. By interlocking his system with those of other British companies that already possessed an international presence, Marconi could gain a worldwide market much more easily than his American counterparts.

Marconi's success as an entrepreneur was also based on his keen appreciation of how his public performances and his private negotiations had to complement each other. He knew he had two different but related realms to conquer: public perceptions as shaped by the press, and important sectors of the commercial world where the clients were. Paying customers would be less credulous than newspaper reporters and would want assurances about the invention's reliability and range. They would have to be convinced by representatives of the company that they needed wireless. But Marconi understood that well-timed newspaper coverage played a critical supporting role: headlines and contracts often went hand in hand. They were directed toward the same corporate goals, and each one reinforced the other.

Marconi recognized the most important role the press played in the wireless scenario: it was his major advertiser. He had learned by 1899 that the press was eager for his success and would accept his claims, if presented properly, enthusiastically and without question. In other words, it would provide free advertising while conferring a credibility and legitimacy a promotional pamphlet or paid advertisement never could. The press had told the public what an advertisement, had Marconi taken one out, might have: that he was the first and the best, and that America needed his product. The press had also described how the product worked, why America would be better off because of it, and what its applications and uses were. Press coverage, like the best advertising, was both informative and persuasive. Headlines, serving as a supposedly disinterested voice while praising Marconi and his apparatus, reinforced and elaborated on the Marconi Company's claims, providing valuable reassurance to potential clients.

Finally, Marconi understood the role that charm, courtesy, and social station played in building a business. Marconi's upper-class background provided entrepreneurial advantages. His family had influential friends and relations who in turn knew other well-placed people, and this web of contacts provided Marconi with access to social and economic circles closed to his American middle-class rivals. Also, his upbringing had given him a confidence and ease when mixing with the rich and famous which would be very much to his benefit. Wherever he went, from Italy to Newfoundland, he called on the local head of state, both as a courtesy and to further the aims of the company. He appreciated the business value of establishing friendly, informal relations with important people. He occasionally joked about this practice, as when he wrote that he had been meeting "rather often" with Alice Roosevelt, the president's daughter, "always in the interest of the company, of course."[44] Meeting and disarming influential potential allies was not, for Marconi, the tortuous or anxiety-producing process it may have been for men of a different class or background.

Marconi's considerable personal talents and his company's exclusionary business policies posed major challenges for Fessenden, De Forest, and Stone. Their technical alternatives had to be complemented by entrepreneurial innovations, as well. Could the Americans lure potential customers away from Marconi? What business policies would the Americans establish which would generate revenues and sustain their fledgling companies? Would they be able to rival Marconi's exploitation of the press? While the Marconi network as a whole was establishing its hegemony, its American subsidiary was small. Given the major structural faults plaguing the Marconi Company of America during its early years, it is amazing that this company found itself in a preeminent position by 1912. That it did so is, in part, a testimony to the power of the parent company's strategy, which was sustained by the entrepreneurial skills of Marconi and his managers. But the ultimate success of the American

Marconi Company is also evidence of the truly maladroit handling of business affairs by Marconi's American competitors.

· · ·

WHEN REGINALD FESSENDEN began experimenting with wireless telegraphy for the Weather Bureau in 1900, Bureau Chief Willis Moore suggested that if his wireless experimentation was successful, Fessenden should "permanently join the Weather Bureau and superintend the daily work of manufacturing and using the apparatus, especially in case no company is formed for its commercial development."[45] Early in 1902, however, Fessenden initiated negotiations with several independent investors to form his own wireless company, and his extracurricular business dealings piqued government officials. The secretary of agriculture officially stated that Fessenden had been suspended for "disobedience of orders and insubordination." Fessenden claimed that he had good reasons for wanting to leave the bureau. He charged that Moore had tried to pressure him into turning over a half-share in all his patents. Fessenden wrote a letter to President Roosevelt complaining that Moore had wanted to go to the Patent Office and swear out some of the patents in his own name, and that Moore had warned Fessenden that lack of cooperation would compel him to recommend that the Weather Bureau switch to the Marconi system.[46] On Fessenden's refusal to comply, Moore dismissed all but one of Fessenden's assistants and began discrediting Fessenden and his system. Fessenden left the Weather Bureau in August 1902.[47]

Meanwhile, Fessenden's patent attorney, Darwin S. Wolcott, arranged a meeting between the inventor and two wealthy men from Pittsburgh, Thomas H. Given and Hay Walker, Jr. Given had begun his career as an errand boy in the Farmers Deposit National Bank and had worked his way up to president of the bank. Walker was president of his own manufacturing concern that produced soap and candles. The three men organized the National Electric Signalling Company (NESCO) in November 1902. Unfortunately, we know little about what motivated Given and Walker or why they were willing to invest so heavily in Fessenden's inventions. All we know is that they came to believe that wireless, and Fessenden's system in particular, was going to make money quickly. They provided Fessenden with a salary and living expenses, and they retained the option to buy a 55 percent share of his patents. They refused to sell stock to the public, apparently meaning to retain as much control and share in potential profits as possible.

From the beginning, the relationship between Fessenden and his backers was flawed, but early enthusiasm and visions of quick profits obscured the inherent problems. Fessenden, convinced of his technical talents and regarding each invention with a self-satisfied vanity, was a strong-willed and proud man struggling to defer to the business decisions of his backers. His inventions provided the raison d'être for the company, yet there would be no company without Given's and Walker's money. Walker and Given, successful businessmen and equally strong-willed, fully expected to be in charge of managing and promoting Fessenden's inventions. Yet they knew little about electrical communications, and they supervised NESCO on the side, in addition to performing their regular duties with their respective firms. As a result, they were unable to give the company the constant attention it needed, and they failed to understand fully many of the basic economic and technical dilemmas facing the fledgling firm. Their authority over the company's strategy, however, was final. These different orientations, and each man's sense that without his contributions the others would be lost, led to division and resentment instead of a healthy partnership. Throughout the company's history, one basic question was never resolved to everyone's satisfaction: Whose company was it, the investors' or the inventor's? Ultimately, this lack of clarity over whether money or technical expertise entitled one to corporate leadership brought NESCO to an internal impasse. Pride on both sides repeatedly played too large, unnecessary, and debilitating a role in NESCO's business proceedings.

NESCO's organizational structure was informal and spare. Fessenden served both as general manager and technical adviser. Thus, he was in charge of experimentation, promotion, and management, and he did not have the assistance of a trusted and experienced manager to whom he could delegate tasks and who could help neutralize exchanges between the inventor and his backers. Nor did Given and Walker hire a salesman they trusted, someone smooth, persistent, and patient, to promote the company. Fessenden was enthusiastic and believed completely in his system, but he was no manager and no salesman.

Although Fessenden was a prodigious memo writer and correspondent, written assessments of the company's progress were not regularly exchanged between backers and inventor. Fessenden wrote extensive position papers when he was especially agitated about a specific contract or piece of legislation, but no one was in charge of standing back from the company's day-to-day operation and rendering a summary and evaluation of successes and failures (as Cuthbert Hall did for Marconi). Nor were there the quarterly company meetings that provided Marconi's company with a sense of orderliness and continuity.

NESCO began its operations guided by several conflicting strategies, all of them short term and ad hoc, and several of them self-destructive. Because the three men never agreed on a long-term strategy or on how revenue would be generated, they spent their money and energy on competing schemes. Fessenden wanted to sell apparatus outright, and he devoted considerable time and energy to trying to win contracts with foreign governments, the U.S. Signal Corps, and the navy. Given and Walker were more interested in establishing overland operating companies, although who the customers for such a service would be remained unclear.

As soon as NESCO was formed, Given and Walker agreed to advance thirty thousand dollars to erect, equip, and operate three stations, two on the southeastern coast of Virginia, at the mouth of the Chesapeake Bay, and one in Bermuda. Fessenden hired two assistants, Pannill and Roberts, and in the spring of 1903 the three men began work on the Old Point Comfort, Virginia, station, which was just outside Hampton, where the James River and the Chesapeake converge. The company started out, then, focusing on signaling over water. This was what Fessenden had done with the Weather Bureau and what he knew best. They erected shacks and aerials and installed the apparatus. The company then learned that the British government's provisions granting the cable companies a monopoly over all telegraphic communication prevented NESCO from establishing the Bermuda station. Just who was responsible for this critical oversight remains unclear, but Fessenden, who had worked in Bermuda and whose wife was from the island, may have overestimated the influence of his contacts there and assured Given and Walker that the license would pose no problem. This was an unpromising start for the company, but, undeterred, NESCO decided on a new tactic. It would establish headquarters in Washington, D.C., and erect a station in town capable of demonstrating Fessenden's system to various government attachés and people influential in American government circles. The staff was increased again, with H. J. Glaubitz as construction engineer and seven additional technical assistants, some of whom were also operators. The station was to be the first of three—in Washington, Philadelphia, and New York, with the Philadelphia station serving as the intermediate relay.

NESCO, with this plan, had significantly shifted its technical and marketing goals and was tackling an extremely ambitious project: overland transmission between three heavily populated areas in direct competition with the telegraph and telephone. By November 1903, Fessenden was reporting successful signaling between the New York station (which was located in Jersey City) and the Philadelphia station (in Col-

Early NESCO wireless station with aerials, ca. 1903.

lingswood, New Jersey).[48] Later that winter, communication was extended to Washington. NESCO would both sell apparatus and establish this first overland American wireless service. Company correspondence gives no indication of when or how NESCO expected to promote this service, or how it planned to make it more attractive than the telegraph or telephone.

Fessenden, it turns out, was opposed to trying to inaugurate service between New York and Philadelphia. He thought the company should prove itself in less congested and risky areas, such as Baltimore and Norfolk.[49] While Fessenden regarded the New York–Washington tests as experiments assessing the strengths and weaknesses of his apparatus, Given and Walker believed they had the foundation for a regular commercial service between the three metropolitan areas. Given and Walker, who had invested in four wireless stations and were eager to see a return, insisted that the New York–Washington "line" be opened for commercial business as soon as possible. This Fessenden agreed to in August of 1904. In October, Fessenden reported to his backers that he had sent a message from Washington to New York and back in seventeen minutes.[50] After this, reference to these stations disappears from company correspondence, apparently eclipsed by grander and often unrelated projects. We do not know whether satisfactory overland communication remained unattainable, whether there were no clients, or whether Given and Walker simply abandoned the project in favor of a more attractive one. We do know that by 1904, Given and Walker had spent at least one hundred thousand dollars on stations and apparatus that were not bringing in revenue.

One reason work on these overland stations may have stopped was the receipt of an attractive offer from General Electric. In December 1903, Ernst Berg of General Electric approached Fessenden and suggested that NESCO contract with G.E. for two wireless stations, one in Lynn, Massachusetts, and the other in Schenectady, New York, a distance of approximately four hundred miles. There is insufficient evidence to explain how or why G.E. decided to experiment with the invention at this time. Earlier in the year, De Forest had submitted his system for trial at the Lynn works, but he had been unable to get signals through to Schenectady.[51] General Electric officials already knew of Fessenden's work from his early orders for high-frequency alternators. Fessenden responded that he was interested in such a proposition, and, with characteristic presumption, said that while NESCO's standard price for two such stations was sixteen thousand dollars, G.E. would receive a 25 percent discount and pay only ninety-eight hundred dollars. Berg wrote back saying that G.E. officials had "decided to drop the matter at the present time, in view of the rather high price and also the uncertainty of the system." Fessenden promptly halved his estimates on aerial and labor costs.[52]

In March 1904, E. W. Rice, third vice-president of G.E., made Fessenden a generous offer that nearly matched Fessenden's original proposal. G.E. would provide the buildings for the apparatus, the transformers, and the power source, and would pay NESCO ten thousand dollars for the sending and receiving apparatus and the aerials. G.E. would pay NESCO nothing, however, until the stations were operating regularly and successfully during business hours for thirty consecutive business days. Fessenden was to guarantee thirty-five words per minute. G.E. also included a stiff interference clause in the contract which provided: "If the apparatus should fail, through interference, repayment of the purchase price shall be made exactly as if the use of the apparatus were prohibited or interfered with by patent litigation, i.e., if interference should occur within two years . . . you will return 80% of the purchase price, three years 60%, four years 40%, five years 20%, no refund to be made after five years' use."[53] The contract was signed in July 1904, and Fessenden assured G.E. that the stations would be operating by September 15. The G.E. contract initially heartened Given and Walker, who saw the large and well-established electrical firm as the perfect sort of client for a company like NESCO. But the G.E. connection also scared them. Walker wrote to Fessenden that he and Given were "more afraid of the G.E. Company than almost any other interest. . . . We think you should be very careful not to allow them around getting any information on [our] work. It would be untold injury to us if, when we are ready to do business, they would announce that they had a system of their own."[54] Dreams of a profitable contract or even merger were interrupted by the nightmare of industrial espionage and theft.

Because G.E. was slow in providing NESCO with the buildings and transformers, experimentation did not begin until January 1905, when A. A. Isbell, the NESCO technician, worked in $-15°$ temperatures to ready the Schenectady station. Although by February, Isbell and G.E. officials had listened to Marconi's Cape Cod station, the Schenectady and Lynn stations never established regular communication with each other. Isbell complained of interference caused by others testing in the area, and also commented that there were "a lot of strange noises in the works." "Every now and then," he said, "I get a screech in the 'phone like the wail of a dying soul." The technicians found that transmitting during the day was impossible and that at night the signals would be alternately weak and strong. Isbell complained: "At times I can hear him with the 'phone two inches from my head, then out he goes." He began to suspect that the current G.E. was providing was erratic, and Fessenden decided that NESCO should install its own rotary transformers independent of the line voltage. But the problem had been discovered too late, and by August of 1905, G.E. had deemed the eight-month experiment a failure and canceled the contract. Isbell reported: "I have been told by a prominent official here that some strong Western Union influence has been exerted to get us out of here."[55] No evidence has been found to confirm or deny Isbell's report, but it was not an unreasonable suspicion.

Fessenden protested the cancellation of the contract, and M. F. Westover, secretary of G.E., granted Fessenden six more months, until June 15, 1906, to have the stations operating. Fessenden again failed to establish regular communication, and G.E. again canceled the contract. Fessenden suggested that NESCO continue experimenting at the Lynn and Schenectady stations, and when they were improved, G.E. could buy them. Westover turned down this offer for various reasons and said, cryptically, "We would rather state these to you verbally than in writing." Westover tried to mollify Fessenden by promising, "If any time in the future the art shall have reached such a stage that commercially practical operation is assured we shall be ready to consider the question of a contract between us."[56] As with the proposed New York–Washington line, NESCO had invested considerable time and money in a speculative venture that failed to produce revenue, visibility, or prestige. Once again, Fessenden had overestimated what his apparatus could do.

At the same time that NESCO was struggling to establish overland wireless communications service, Fessenden was hustling to sell apparatus to anyone who might be interested. He advertised his apparatus in *Electrical World,* and succeeded in having the journal publish an article about him and the company. He advised Walker, "I expect to have something of this sort published every few weeks."[57] If he read of a foreign government having difficulty with the cable companies, he contacted the country's embassy and arranged an interview to suggest wireless as an alternative.[58] He urged Given and Walker to hire a salesman who used to work for Edison: "As the man seems a hustler, and it would appear to be absolutely necessary for us to have someone to get out on the road to see people, since it is impossible to handle business of this kind from a desk, I am engaging him."[59] This salesman tried to sell Fessenden's apparatus to corporate clients such as B. F. Goodrich and Swift and Company on the basis that they could use it to keep their regional plants in constant communication.[60] Fessenden, his visions of expansion fired by a recent visit to the Columbian embassy, warned his backers "We can never handle big business successfully until we have representatives all over."[61]

By October 1904, Fessenden was bragging to Walker, "Business seems coming in thick and heavy. I am getting results from all of the parties I wrote to in the spring of the year."[62] He had approached clients from Australia, Japan, and Russia, and had received a contract to install apparatus along the Amazon in Brazil.[63] There is no evidence in the company correspondence suggesting how, or if, NESCO expected to oversee such farflung stations once they were installed.

Two of NESCO's most counterproductive policies were its method of setting prices and its stance on patents. In making decisions in these two critical areas, Given, Walker, and Fessenden relied less on good business sense and more on egocentrism and personal pride. Walker and Given, in their calculations of how much to charge a client for overland service, came up with the figure of two hundred dollars per mile. They wanted to recoup their research and development costs quickly, and to show a profit. A colleague in New York delicately told Walker that the price was far too high and would scare off potential clients.[64] The price quoted the U.S. Navy for Fessenden apparatus was ten thousand dollars per set, a preposterously high figure. A strong sense of entitlement—that NESCO should get such prices simply because it deserved to—runs throughout Walker's letters. Fessenden also believed the apparatus was worth that much, but he was willing to be more flexible. He reminded his backers that De Forest was offering similar apparatus, some of which infringed on Fessenden's patents, for fifteen hundred dollars, and that to compete, NESCO was going to have to reduce its prices. Once the navy had become committed to NESCO apparatus, Fessenden reasoned, then the company could raise the price, citing recent improvements as the cause. Fessenden wanted navy contracts and was loath to "undo all the good work [the company had] done with the Navy up to date"—meaning the many hours of demonstrations, explanations, and persuasion.[65] But he was overruled by Given and Walker, who insisted that they and they alone would "furnish all quotations."[66]

Fessenden was willing to swallow his pride in the marketplace, an arena in which decisions were not always final and in which the stakes, for him, were less important. But in the courtroom, where priority of invention would be established, where the considered opinion of a judge, not the vagaries of the market, would establish the "truth," Fessenden's pride preened itself. It was where whatever business sense he possessed deserted him. De Forest had copied Fessenden's electrolytic detector and was offering it to the navy at cut-rate prices. Fessenden was outraged by such piracy and in 1904 determined to sue. The theft was flagrant and Fessenden's response understandable. But NESCO was only two years old, was not yet bringing in any revenue, and suing was expensive and time-consuming. Nonetheless, Fessenden took De Forest to court three times over this infringement, and in 1905 won an injunction against De Forest's production of the detector. The company spent between fifty thousand and one hundred thousand dollars defeating De Forest in the courts; it would have cost far less to beat him in the marketplace. But Fessenden sued even when a victory would bring only psychological returns. He claimed priority over Marconi's magnetic detector and took the case to court in 1904. Fessenden was not even using Marconi's receiver, and NESCO had little to gain from this action. Eventually, in 1907, Fessenden lost the case.[67] Such an expenditure of time and money fit into no long- or short-term strategy; it only sapped a company that had yet to win a major contract.

By June of 1904, Given and Walker had latched onto a new scheme. "It seems to us," Walker wrote, "that there is no way that it would pay so well or that would be so easy to promote as to follow out the plan of small companies and an exchange of stock with a parent company. . . . Suppose, for instance, that we formed companies by states."[68] Walker and Given were not sure how tolls or jurisdiction would be determined or how the "internal business" of such a company would be handled, but they thought their franchising idea excellent. Fessenden was less enthusiastic. In addition to the overwhelming coordination problems, his biggest concern was finding "good men" to oversee the regional companies. He warned that incompetent men would "make any amount of trouble" for NESCO, damaging the reputation of the company and wireless, as well.[69] Once again, Fessenden and his backers had very different instincts about how to manage and promote wireless.

Walker and Given then scaled down the scope of their plan and determined to make NESCO an operating company so attractive and, they hoped, so threatening that one of the major communications companies such as Western Union or Bell Telephone would buy them out. But to make their company sufficiently alluring, they needed a dramatic achievement. Transatlantic service would provide the appropriate bait. Walker and Given decided late in 1904 not to sell any more equipment; all energy and money was now to be devoted to "getting to the other side."[70] This shift in strategy trapped Fessenden between frustrating disappointment and visions of glory, and confounded his technical vanity. He wanted to sell his apparatus and have as many clients as possible using it. He was convinced his wireless system was the best and would ultimately displace all others. To forgo all sales, especially those he had worked so hard to realize, struck him as both unfair and entrepreneurially absurd. On the other hand, he now had a chance to succeed where Marconi had failed. Despite his transatlantic signal and messages, Marconi had yet to establish regular, reliable service across the ocean. The challenge tapped Fessenden's reservoirs of competitive energy and technical pride. Fessenden later testified that he was completely opposed to the endeavor from the start: "The attempt to work across the Atlantic, a distance of 3000 miles, at a time when we had never worked more than 120 miles and had great opportunities for selling sets at a large profit for naval and other uses, was also decided upon against my advice and carried out in spite of my protests."[71] But the zeal with which Fessenden threw himself and those around him into the project belies the fervency of that disavowal.

The site selected for the American terminus of the service was Brant Rock, Massachusetts, just south of Plymouth. The European site was Machrihanish, Scotland. In November 1904, Walker urged Fessenden to complete the stations as quickly as possible; while Fessenden set up the apparatus, Walker would begin applying for licenses. As with the ill-fated Bermuda connection, NESCO sought to install the equipment before determining whether the company would be able to get a license. Fessenden apparently hoped that he would still be able to sell apparatus if major orders came in; he was mistaken. Walker and Given insisted that he not discuss sales with anyone; from now on, they were only interested in selling the whole system. "We do not care to peddle a 20-foot lot out of a plan but sell the whole acreage," wrote Walker.[72]

In July of 1905, Fessenden, his wife, and his son moved to the nearly completed Brant Rock installation, which became NESCO's experimental headquarters and main wireless station. Fessenden was that much farther away from his backers, who had, by their insistence on the transatlantic work, inadvertently given Fessenden much more autonomy and latitude. Transatlantic signaling was a real challenge, demanding a level of technical rigor, power, and precision that NESCO had yet to achieve. It was also a major investment, and Fessenden believed that Given and Walker, who had insisted on the goal, should not now begrudge him whatever monies he needed to fulfill their mandate.

Once Fessenden moved to Brant Rock, which was in fact his personal industrial research lab, the legacy of his years with Edison became more apparent. He became even more dogmatic about certain technical designs, and he spared little expense realizing them. His technical perfectionism reached full expression. Fessenden had not had to scrimp at the Weather Bureau or at the universities, and the stakes there had not been very high; he certainly had no intention of scrimping now. Walker and Given might complain, for example, about the already large amounts of money Fessenden was paying to General Electric to design his high-frequency alternator, but Fessenden could respond by saying that if NESCO really wanted to cross the Atlantic, the company had to have the alternator; no other device would do. How could Given and Walker respond? They knew little about the technology and had to trust their inventor. By May 1906, Given and Walker had put $519,100, an extraordinary amount, into NESCO, and they had to give voice to their frustrations.[73] Accordingly, they admonished Fessenden when the extravagance was one they could identify. Wrote Walker, "We notice this pamphlet came to us in a fine large linen envelope, costing ten or twelve cents, with five cents postage, when an ordinary wrapper with one cent postage would have brought it just as well. It is small things like this that give us an impression that your shop is run in an extravagant manner."[74]

By 1905, NESCO was operating in a private, and sometimes secret, fashion, concentrating on the transatlantic goal. In his lab, Fessenden also continued to experiment with his wireless telephone and the transmission of the human voice, making considerable progress. Now there were no sales and no promotion. The hope sustaining Given and Walker was their plan to sell NESCO to the highest bidder and cash in on the merger movement. It is not at all clear why Given and Walker thought a major communications firm would be interested in a transatlantic wireless system. They knew very little about the industry; in 1904, in fact, Walker wrote to Fessenden asking for information on the capitalization and earnings of Bell Telephone and Western Union.[75] These were not men who knew or understood their potential clients very well.

NESCO did not yet offer the Marconi Wireless Telegraph Company of America any competition. After three years in business, each one costing Given and Walker at least $125,000, the company had settled on the riskiest technical venture it could have tackled, with the hopes of selling its system to companies that had not demonstrated a clear or immediate need for their product. Fessenden was by now well known in engineering circles as well as by those who read the technical journals, but he had not conquered the popular press as Marconi had. The man who had worked as one of Edison's assistants, and had picked up several of Edison's bad habits, had missed important lessons on skillful promotion and marketing.

Fessenden possessed little sense of the dual roles, the public and the private, the entrepreneur had to play. He either failed to grasp or deliberately rejected the way personality affected media coverage and sales. Unlike Marconi, Fessenden showed little interest in using charm, humor, and modesty, however false, to promote his apparatus. He did not cultivate a public image. This inability or unwillingness to master the connections between personality and entrepreneurial success came to have a corrosive effect on Fessenden over the years.

In fact, of all the inventors who acted in the early wireless story, Fessenden seems to have been the most changed by the experience. He has been described by historians as choleric and abrasive, but these qualities were not dominant when he entered the field in 1899.[76] It is true that he had always had a bombastic streak, claiming he had invented more than he had and asserting that his apparatus could achieve much more than it could. He wrote to Marconi in 1899 claiming credit for Marconi's *Herald* contract to report the yacht races; it was he, Fessenden asserted, who had suggested that the newspaper contact Marconi.[77] In December of that year he sent a press release to the *New York Times* announcing that he had invented a receiver two thousand times more sensitive than the coherer.[78] As his inventions became more revolutionary, his egotism became even more exaggerated, and it was a characteristic many potential customers found increasingly unappealing. Yet Fessenden began his work with NESCO in 1903 eager to please his backers, promote his system, and, in his words, "give the people what they want."[79] When, instead of the positive reinforcement he was accustomed to receiving in the lab or in the classroom, he confronted skepticism from clients and resistance from his backers (who he believed knew little about the technology), Fessenden became increasingly disillusioned, bitter, and yes, abrasive.

In sum, NESCO was a company headed by a gifted electrical engineer and two successful businessmen, all three honest and earnest, with several ideas on how to make wireless pay. But they never settled on one strategy that would push the Fessenden system, sustain the company's growth, and bring in revenue. Given and Walker were impatient and fickle: if one approach failed to bring in significant revenue within six to twelve months, the two men would devise a new short-term strategy, often at odds with its predecessor. Their impatience was aggravated by Fessenden's tendency to exaggerate the capabilities of his apparatus; their fickleness was exacerbated by his inconsistency. One week he would assert that his equipment was fully operable, the next remind Given and Walker that it was still in the experimental stage. Unlike Marconi, who was painstakingly building a worldwide organization, NESCO's owners seemingly shied away from the prospect of organization building. As Fessenden and his assistants worked on the Brant Rock and Machrihanish stations in 1905, Given and Walker pinned their hopes on the transatlantic coup they expected would bring them the offer of their dreams.

•   •   •

AFTER LEE DE FOREST demonstrated his system during the yacht races of 1901, he established the Wireless Telegraph Company, headquartered in Jersey City. This was a company in name only; except for De Forest, it had no assets. De Forest was nearly broke. According to his diary, he began a long and fruitless campaign of trying to interest members of the business community and financial promoters in his invention. On December 1, 1901, he wrote in the voice of the still hungry but persistent outcast, "I have approached some twenty-five parties to many of whom I trust to be able to point out, someday, the enormous folly of their timorous mistake." He complained about "the strange lethargy and skepticism of the public in wireless telegraphy as an investment."[80] By the public he meant the Wall Street financiers, who believed Marconi had too formidable a lead, who did not share the journalistic optimism over wireless, or who were unconvinced by De Forest's proposals and manner. After all, he was relatively unknown and fresh out of school, and he had no track record on inventing or engineering. But in January 1902, after months of hustling, De Forest found an ally: "Of all the many men approached in all these months, despite the success of repeated demonstrations, but one man of means has been found with faith in the scheme equal to my own."[81] That man, whom he met through a broker, was Abraham White.

White's avarice was whetted by Marconi's transatlantic feat. White became convinced that wireless could help him amass a fortune, and in February 1902 he and De Forest incorporated a new company, the American De Forest Wireless Telegraph Company, of which White became president.[82] White was not a businessman; in fact, he was a somewhat disreputable character. He was already notorious in financial circles as a speculator and con man, and he was eager to take advantage of the prevailing bull market. Contrary to the conservatism in the business community, White recognized that the public was captivated by the notion of wireless and ready to translate that fascination into stock purchases. Wireless, to White, was no different from a patent medicine. White would convince Americans to invest in De Forest by staging flamboyant demonstrations and publicity stunts. The proceeds from the stock sales would augment both men's bank accounts, although White made sure he benefited more than De Forest. Little research and development was envisioned. The company had no structure or formal proceedings, and its only strategy was to sell as much stock as it could, and as quickly as possible. De Forest's title was vice-president and scientific director, and he started out earning twenty dollars a week, double the salary he could expect to earn elsewhere in the electrical industry.[83] The only other employees were stock salesmen and technicians.

White's business strategy, then, was to use wireless as the basis of a huge stock-promotion scam. In the "stock jobbing" that ensued over the next three years, De Forest was a willing accomplice. Of his own noble role in bringing the advantages of wireless to mankind, De Forest wrote in 1902, "Soon, we believe, the suckers will begin to bite. Fine fishing weather, now that the *oil fields* have played out. 'Wireless' is the bait to use at present. May we stock our string before the wind veers and the sucker shoals are swept out to sea."[84] Fame and fortune seemed within reach.

De Forest and White were enterprising publicity men, and if their tactics failed to entice businessmen, they did succeed in selling stock to the public. White built a penthouse laboratory with glass walls and ceiling in downtown Manhattan. A sister station was installed at the Castleton Hotel on Staten Island. Prospective investors watched De Forest in the glass lab signal across the bay. Then White would paint a picture of "worldwide wireless," with every country on the globe paying tribute to the De Forest Company. On February 7, 1903, White and De Forest parked an automobile equipped with a small wireless station in the Wall Street area, center of the stock market frenzy. From the car, De Forest transmitted stock market quotations to a Dow Jones office.[85] The newspapers featured the story, and White in turn cited the newspaper stories in his stock advertisements. This demonstration suggested that wireless might have a place in Wall Street's communications network, which implied that the invention represented a promising investment. The company also built stations in locations where its salesmen determined that stock could be sold. As these stations neared completion, salesmen and brokers sold the stock for between five and fifty dollars a share. One such station was built in Atlanta. It cost three thousand dollars to build and brought in fifty thousand dollars in stock sales. The Atlanta station, and many others, never transmitted a single word.[86]

White's specialty was the press release, which he sent out with abandon. Few of them were true, but White knew that having his claims printed in the paper imbued them with legitimacy. Like Marconi, White recognized that the press was his primary advertiser. These releases were in fact fraudulent, and had nothing to do with building an organization. The purpose of all press stories was short-term financial gain. One release claimed that De Forest Wireless had absorbed American Marconi. Another declared that the De Forest system was the official system of the U.S. Signal Corps and the U.S. Navy. The newspapers did not investigate White's claims and printed these releases uncritically. White's stock advertisements would include a reproduction of the newspaper article accompanied by the opening line "Did you read yesterday's papers?"[87]

To add further credibility to his promotional material, White reminded readers about Bell Telephone's rags-to-riches story. In 1902 he had persuaded Bell to serve as a consultant to the company. In pamphlets titled "The History of an Opportunity," the De Forest company advised: "Bell Telephone stock, when first offered, went begging at fifty cents a share, and those same shares today are worth $4,000. . . . With the Bell Telephone stock in memory . . . thoughtful persons are buying up wireless stock with avidity." The pamphlet suggested that those who bought stock were being altruistic and financially shrewd at the same time: "All great discoveries which have brought civilized communities into close touch have made millions for those who obtained an interest in them during the early stages of development." The prospectus also continued, on a more personal note, "There is not enough stock to go around. Consider the matter carefully. You have the opportunity. Will you grasp it 'at the flood tide' (now) and ride onto the shore of plenty, high and dry above the adversities which often beset old age . . . or will you hesitate and doubt, and let the chance go by, to remain in senile dependency upon the bounty of others? . . . A few hundred dollars invested now and given to your children should make them independent."[88] White also sold stock on the installment plan, two dollars down, two dollars a month.

The promotional material was highly nationalistic and patriotic. Seeking to emphasize the difference between De Forest and Marconi, whom De Forest and his technicians privately referred to as the Dago, White emphasized in his press releases: "It is the policy of our company to develop its own system with American brains and American capital."[89]

By July of 1902, De Forest stock was selling, and soon White was handling hundreds of thousands of stock promotion dollars. Very little of this cash was allocated for research, development, legal services, or patent applications. It was spent instead on lavish offices, advertising, and showcase wireless stations. White determined the company's policy, and that policy was dazzle, advertise, and keep selling stock. White dispatched De Forest around America to lecture, demonstrate the wireless, and "talk glowing prospects." De Forest was allowed little time and no technical help for experimentation. He did, however, continue to work on his wireless telephone, which he was convinced would be even more successful than wireless telegraphy.

In the autumn of 1903, De Forest demonstrated his "new" system during the yacht races. Although interference was reportedly considerable, De Forest did impress Sir Thomas Lipton, owner of the challenger to the Cup, *Shamrock III*. Sir Thomas invited De Forest to Great Britain to promote the American's system, but De Forest was unable to sell any apparatus. However, he managed to negotiate a contract with the *Times* of London to transmit news of the Russo-Japanese War from China by wireless. White immediately began publishing a newsletter called *Wireless News*. Filled with news items describing the progress of the De Forest system, *Wireless News* also contained a full-page testimonial by Sir Thomas extolling the virtues of the De Forest system. The newsletter notified readers that the *Times*, "the most conservative and influential newspaper in Europe, [was] utilizing the De Forest system in its war correspondence service from Korean waters."[90]

Another early and important client was United Fruit, a company completely dependent on rapid and reliable transportation and communications systems. Incorporated in 1899, and one of the first transnational enterprises operating in Latin America, United Fruit established its hegemony in the region through a combination of wealth, technological prowess, and favorable deals with pliant local governments. The distribution of highly perishable products, primarily bananas, required precise organizational coordination, which was often hampered by the crude communications services then existing in Latin America. *Radio Broadcast* later reported that in 1904, "the entire eastern coast of Central America and the northern coast of Columbia, South America, were without any direct means of communication with the United States, with the single exception of a cable station at Colon, Panama." The journal then described how important business information was transmitted. Messages from the United States arrived via cable and traveled over government land lines in Nicaragua and Costa Rica for delivery at United Fruit's office at Port Limon. These messages were then "entrusted to natives, who would make the trip in a canoe on the open sea between Port Limon and Bocas del Toro in from 30 to 60 hours, depending on weather conditions." Obviously, such an arrangement was unsatisfactory to United Fruit's executives. The company was reportedly so desperate for an improvement on this situation that it was "prepared to go to almost any expense to insure against undue delays to its messages."[91]

In 1904, Mack Musgrave, who supervised United Fruit's telegraph and telephone service in Costa Rica, came to the United States to investigate wireless. De Forest was at the peak of his fame, backed by White's continuous press releases and demonstrating his apparatus at as many well-attended public events as possible. Musgrave bought two complete sets of wireless from De Forest for stations at Port Limon, Costa Rica, and Bocas del Toro, Panama. By 1907, United Fruit had erected and equipped two more stations, and the next year it added four more, giving it stations in Nicaragua, Guatemala, Cuba, Louisiana, and Swan Island in the Caribbean. It had also begun installing wireless aboard its "Great White Fleet," which consisted of approximately twenty-five ships specially designed for the transportation of produce.[92]

United Fruit's wireless operators were plagued by static, which was especially bad in the tropics for nine months out of the year. With De Forest's early spark gap apparatus, which emitted a low-pitched sound not easily distinguishable from static, United Fruit's early work with wireless was extremely discouraging. But for this company, wireless was the only alternative to either rudimentary or nonexistent communications channels, and thus the company made a commitment to con-

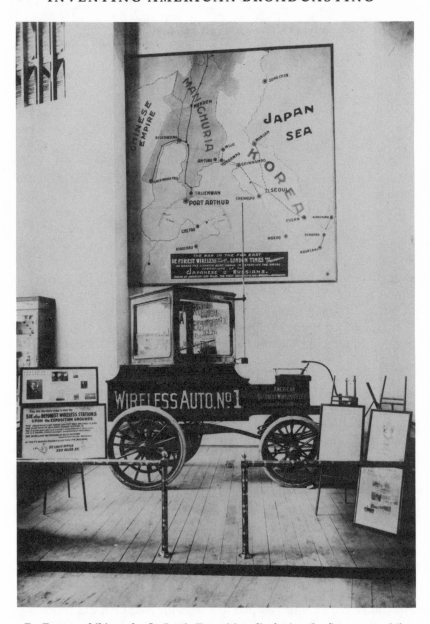

De Forest exhibit at the St. Louis Exposition displaying the first automobile
equipped with wireless, 1904.

tinue trying out and developing the invention. This commitment made United Fruit the only major American corporation willing to invest money and manpower in wireless telegraphy. As a result, the fruit company later became an unlikely partner in corporate negotiations over who would control which sectors of American broadcasting.

Selling a few wireless sets to United Fruit, however, was merely a sideline for White. He was always concerned, first and foremost, with promotion. Although the Marconi Company initially had exclusive rights to display wireless at the St. Louis Exposition, White succeeded in convincing the fair authorities that an American wireless company should also be represented. When the Marconi Company withdrew, White was left in a highly visible position that he was quick to exploit. A sightseeing tower previously used at Niagara Falls was purchased for ten thousand dollars and reerected at St. Louis.[93] De Forest's name, spelled out in lights, illuminated the tower, which was one of the highest structures at the fair. The station participated in legitimate experimentation and in showmanship. Whenever the crowds around the tower thinned, the operators emitted as loud a spark as possible, which reportedly could be heard a quartermile away.[94]

De Forest was also eager to establish whether overland communication was feasible. Professor Michael Pupin, who was frequently quoted in the press on wireless matters, and just as frequently was wrong, had asserted "Wireless messages can never be sent over great distances on land."[95] But on September 5, 1904, a De Forest operator sent a message 105 miles, from St. Louis to Springfield. Six days later, the company set a new overland record of 300 miles.[96] These distances represented not regular communication practice, but fluke successes. Nonetheless, news of them appearing in publications did much to sell De Forest stock and to give the impression that De Forest was outdistancing all other American competition. The overland success also brought the De Forest display the fair's gold medal and grand prize for best wireless system. Such awards provided excellent grist for White's publicity mill.

De Forest reveled in his success at the World's Fair. The newspaper headlines, his name in lights, the money rolling in, a fleeting romance with a senator's daughter: this was all he had dreamed of at Talladega and Yale. But the spotlight blinded him to the precariousness of his position. He was White's marionette, not his partner. De Forest did not have controlling interest in the company that bore his name. His corporate vulnerability was exacerbated by his technical dependence: he was achieving success with a receiver copied from Fessenden. De Forest appears during these years to have been a self-deluded, crass, and entrepreneurially shortsighted inventor who knew and learned little about legitimate long-term corporate strategies. But he did learn one important thing: the American people were enthralled with wireless. They weren't quite sure how they themselves might use it, but the press accounts, displays, and promotional material had convinced many of them that wireless was a significant invention with the potential to make lots of money. There was much De Forest failed to learn, but the American fascination with radio he did not forget.

▪   ▪   ▪

JOHN STONE WAS respected and admired by his fellow experimenters. His university training and work at Bell Telephone had established him as a gifted electrical engineer with an unusually strong background in mathematics. But he was described as being "not much use as a businessman."[97] After leaving Bell Telephone, Stone continued experimenting with loose coupling and selectivity and applied for a patent on his method of tuning in February 1900. Stone's patent attorney, A. P. Browne, then approached potential investors to raise capital for continued experimentation. After canvassing several sources without results, Browne succeeded in interesting B. T. Judkins, a Boston businessman, in forming a small syndicate to support Stone's investigations. Judkins enlisted the aid of ten associates, each of whom subscribed one thousand dollars, and the Stone Wireless Telegraph Syndicate was formed on December 31, 1900. With this backing, Stone rented space in a warehouse and worked to refine his system. He concentrated on the coupling between the open and closed circuits in his attempts to refine tuning and reduce the damping of the waves. He used the spark gap transmitter and the coherer, which continuously gave him trouble. Stone's apparatus emitted a sharply defined wave that the coherer often failed to detect. Stone attempted to develop his own receiver and to refine the coherer. He developed a self-decohering device that used carbon granules instead of metal filings, but this coherer was not very sensitive. The coherer did not work well in his selective system, but Stone struggled along with it until he began using the electrolytic detector.[98]

After conducting his own tests on the simultaneous transmissions of different wavelengths between two differently tuned transmitters and receivers, Stone asked scientists from Johns Hopkins and the Massachusetts Institute of Technology to run independent tests of the system. They verified that Stone's method of tuning was successful: his receivers responded to their intended wavelengths and ignored other signals, including static. Persuaded that Stone's system was marketable, the syndi-

**John S. Stone standing in the doorway of his wireless station
in Cambridge, Massachusetts, ca. 1904.**

cate formed the Stone Telegraph and Telephone Company in July 1902.

Stone erected wireless "huts" on both sides of the Charles River, in Boston and Cambridge, and in 1903 he set up a receiving station in Lynn, Massachusetts. These stations were experimental, not commercial. In 1904, at the request of the Signal Corps, Stone tried to operate his system between Cambridge and New London, Connecticut, a distance of about one hundred miles. Because his receiver was too weak and his transmitting power insufficient, Stone failed in these attempts. However, a year later, he demonstrated his apparatus, at his own expense, at three navy yards on the North Atlantic coast. In November 1905, after a three-month trial period, the navy purchased these three installations. Stone also instituted, in 1904, wireless service between the Isles of Shoals and Portsmouth, New Hampshire. By 1905, Stone's company had begun to sell its apparatus and increase its visibility.

Stone's company was supported by a number of investors, and no stock in it was sold to the general public. The company sold apparatus, it did not provide a communications service. Stone, a gifted mathematician, did not have a skilled entrepreneur to help him run the company efficiently. While his company's policies did not taint wireless's reputation, as had De Forest's, neither did they help to popularize or advance the innovation, as had those of Marconi and Fessenden.

• • •

TWO EXTREMES ON THE corporate spectrum were represented by the fledgling American wireless firms: on the one end, the tiny research lab, refusing to sell either stock or apparatus; on the other, the wireless medicine show, selling worthless stock to anyone willing or able to buy. The contemporary financial climate encouraged both modes of operation and contributed to the shortsightedness and self-delusion inherent in each. Had De Forest and White been more scrupulous, or Fessenden, Given, and Walker more enterprising, one of the American companies might have posed a major corporate threat to Marconi's American subsidiary. In 1905 all these companies were still surviving, and in fact were optimistic, but the fault lines in each are, in retrospect, quite apparent.

Only Marconi had grasped and exploited the interdependence among technology, business strategy, and the press. All of these were interlocked in his mind as he pursued his major goal of building an international monopoly. He settled on a leasing policy in order to attract customers and retain control of the apparatus and the system. Whereas all the inventors referred to their devices as a system, only Marconi constructed a policy that backed up the notion of the complete system and ensured that it would not be compromised or invaded by foreign components. He instituted the nonintercommunication policy to discourage competition. He exploited the advertising potential of the press. Marketing ploys and corporate goals were clearly and inextricably linked: his public demonstrations were designed to legitimate his business strategies. He knew where his clients were and where he wanted his company to be in the future. He had developed a mechanism, albeit imperfect and controversial, for making wireless pay. Although his company was still small, it was governed by the orderly proceedings characteristic of larger, more established corporations.

Fessenden, De Forest, and Stone failed to follow these steps or to offer a solid alternative to Marconi's corporate strategies. They had only short-term responses that obscured the longer view. NESCO, by 1905, was neither selling apparatus nor providing a communications service. The firm had failed to use the pages of the press effectively to promote Fessenden and his inventions and gain wider legitimacy. In the minds of the men running NESCO, and in the everyday operations of the company, the three arenas of technology, business, and the press were not strategically linked. Abraham White, however, appreciated all too well how to use the press to create and manipulate a market. He knew that celebrity would bring legitimacy sufficient to sell stock; why waste time on technology or genuine business strategy? De Forest, with fevered impatience barking at his heels, endorsed the shortcuts taken. Business structure and strategy were the last things on his mind. Stone, meanwhile, devoted nearly all of his attention to refining his apparatus. All three men thought in terms of competition rather than monopoly.

The Marconi Company's most far-reaching strategic breakthrough was, as we can see, conceptual. All of Marconi's strategies, and the structure he hoped would promote them, rested on a revolutionary way of thinking about the ether. As early as 1900, Marconi had regarded the ether as territory he could preempt and privatize. To profit by his inventions, he would stake his claim to the spectrum and then try to deny competitors access to it. He sought to limit who would send and who would receive. In retrospect it is clear that this assumption—that a monopoly, through its control of critical technology and its policies about how the technology should be used, would oversee access to the spectrum—was perhaps Marconi's most historically significant legacy. This assumption extended the basic tenets of monopolistic capitalism to something invisible, ubiquitous, and seemingly communal. Because it was an assumption the Marconi Company jealously promoted and others vigorously opposed, it produced national and international struggles over who had territorial rights in the spectrum. Here was an idea that made broadcasting history.

CHAPTER FOUR

WIRELESS TELEGRAPHY IN THE NEW NAVY

1899–1906

TO WIRELESS INVENTORS eagerly looking for customers, none seemed more promising than the U.S. Navy. Fresh from its victories in the Philippines and Cuba, and in the midst of successful renovation and modernization, the U.S. Navy still lacked a critical tool: a reliable and versatile method of communications which would keep ships in touch with one another and with the shore. Other navies were acquiring such a tool, and the United States Navy could not afford to fall behind. By 1899, Marconi was already supplying the British and Italian navies with wireless, and Kaiser Wilhelm II was providing government support for the development of a German wireless system designed specifically for military purposes. These international competitive pressures were an added incentive for acquiring wireless. If commercial markets for wireless still seemed somewhat uncertain in the United States, the military market seemed assured.

A detailed account of the navy's role in the development of American wireless is essential for several reasons. As the inventors' major potential client and as the part of the government most concerned with the invention's deployment, the U.S. Navy Department between 1899 and 1912 had a determining effect on the fortunes of the fledgling American companies. Later, during World War I, in the interests of national security, the U.S. Navy controlled and operated America's radio communications network of shipboard and shore stations. When the war ended, navy officials assumed a central role in the negotiations leading to the formation of the Radio Corporation of America (RCA). But a retelling of the navy story is also important because several historians have emphasied the salutary effects of what has been called navy patronage while neglecting to examine the often negative repercussions of the naval response to wireless.[1] In fact, the navy offered the wireless inventors a less enthusiastic reception than the inventors expected and historians have recorded. World affairs and the favorable publicity wireless had received prompted the navy to give the invention a trial in 1899. But the navy was a tradition-bound and insular bureaucracy whose organizational structure exacerbated the technical conservatism of many of its officers. By 1915, the secretary of the navy would write, "Wireless telegraphy . . . has come to be regarded as an indispensable adjunct of naval communication."[2] The route to this realization, however, was long and tortuous indeed.

At the turn of the century, the navy appeared, from the outside, to provide a ready-made market for wireless. Ideological shifts within the United States and international military and economic competition had launched the navy on a course of modernization and expansion. Between the end of the Civil War and the early 1880s, the condition of the U.S. Navy had deteriorated so markedly that congressmen, the press, and even naval officials ridiculed the fleet. As the press discovered the journalistic power of scare stories on the subject of naval unpreparedness, rumors and articles circulated depicting the various American coasts as helpless, terrorized by enemy ships from Europe and South America.[3] Such exaggerated fears contributed to support for significantly upgrading the navy during the 1880s. In addition, the country that was emerging as an industrial giant, that was relying on its burgeoning civilian technical arsenal to outdistance other industrial nations, could not tolerate the idea that its navy was technically backward.

The rationale for American naval expansion had been articulated by Alfred Thayer Mahan, recently retired as president of the Naval War College in Newport, Rhode Island. In 1890 Mahan published *The Influence of Sea Power upon History, 1660–1783*. His arguments urging the strategic and commercial importance of a strong navy captured and distilled many of the emerging rationales for American international expansionism that had begun to circulate in the 1880s.[4] Growing American support for expansionism derived from commercial, political, and ideological ambitions. Establishing an overseas presence, however, with the European powers already embroiled in intense colonial competition, depended on a strong merchant marine, which in turn had to be backed by a strong navy.

Mahan's book, by placing the discussion of national and naval strength in a historical context, both synthesized the prevailing sentiments for expansionism and gave them intellectual credibility and weight. His treatise, which received careful attention in both America and Europe, was an "instant success."[5] During the 1890s, and particularly during the depression, Mahan gained many adherents, who viewed the establishment of secure foreign markets as the solution to all the country's economic woes.[6] In December 1897, he published a collection of essays titled *The Interest of America in Sea Power, Present and Future.* Shifting his attention from the past to the present, Mahan called on his countrymen to "look outwards," to take their "rightful place among the nations."[7] Specifically, this meant annexing Hawaii, establishing coaling stations in the Pacific and the Caribbean, and beginning construction of a canal across the Isthmus of Panama. It meant acquiring more property and making the navy stronger than ever. And it often meant acquiring property specifically for the navy. Without "foreign establishments," warned Mahan, "the ships of war of the United States, in war, will be like land birds, unable to fly far from their own shores. To provide resting places for them, where they can coal and repair, would be one of the first duties of a government proposing to itself the development of the nation at sea."[8] By 1898, when America eagerly declared war on Spain to "liberate" Cuba and the Philippines, the U.S. Navy had been dramatically renovated, and imperialism prevailed over isolationism. Dewey's victory offered a convincing display of America's new capabilities and intentions. America was no longer an inexperienced backwoods country with an "alphabet of floating washtubs."[9] It aspired to larger international stature, and it would pursue this goal backed by its "New Navy."

Mahan's most enthusiastic and influential apostle was Theodore Roosevelt. A man whose world view incorporated the assertion of a decidedly masculine brand of adventurism and competitiveness, Roosevelt quickly rose into powerful positions that allowed for full expression of his philosophy. Early in 1898 he wrote, "If the United States is to continue to hold on the Pacific the position to which its great sea-front and its wealth and population entitle it, then we must steadily go on building up our navy."[10] By 1900, America controlled the Philippines, Guam, the Hawaiian Islands, Puerto Rico, and Cuba, and thus had gained strategic positions in the Pacific and the Caribbean. Central to these victories, and to the realization of other international ambitions, was the continued and preferably ever-increasing strength and technical sophistication of the new navy. Determination to continue naval expansion was a major goal of Roosevelt's presidency.

The international obsession with and race for the acquisition of property which the United States had joined exacerbated previously held suspicions, and political leaders became increasingly paranoid over any advantages held by their rivals. The contest was over ownership of property, particularly property in Asia and Africa, which the Europeans had come to realize had considerable economic value. Beyond the direct financial benefits, control over property allowed the owners to determine who would have access to the property (and its products) and to decide how the property would be used. Control over such possessions, then, brought far more than economic returns: it extended the owner's cultural, intellectual, and religious influence. The British Empire was geographically the most extensive and influential empire, and the French, the Belgians, and the Germans aggressively sought to challenge its hegemony. Because, ultimately, this hegemony had been established and preserved by a strong navy, the contending rivals, and Kaiser Wilhelm II in particular, initiated a determined military buildup during the late nineteenth century. Other nations, unwilling to fall behind, matched his efforts, contributing to a rapidly accelerating arms race. Nationalism and militarism, then, were closely intertwined, a lesson not lost on the Americans.

Until 1900, however, few national leaders or their military agents had to concern themselves with what went on in "the air." After all, the air was free; intangible; an age-old symbol of unfettered transcendence; an open, unclaimed expanse. Marconi's achievements and his corporate strategies changed this perception. By sending signals, and then messages, more than two thousand miles across the Atlantic in 1901 and 1902, Marconi aroused territorial concerns, especially in Europe. Marconi's breakthrough meant that wireless messages could cross many national boundaries. Electromagnetic waves did not respect these boundaries; countries could not cordon off their portion of the ether. As *Electrical World* observed, "Wireless telegraphy . . . involves all countries in one circumambient ether, and the air about each is permeated by undulations emitted by others. All countries are, therefore, brought into virtual contact in the aerial ether."[11] Such contact, noted the magazine, would be considered most unwelcome.

> What would be the effect on wireless telegraphy in France and Germany of transmitters pounding away on the ether in hurling messages at America? What would the hypersensitive German Government say if the staff messages during its sacred army maneuvers were broken in upon by John Bull vociferously ordering pork in Chicago? The idea certainly involves the possibility of international complications of a highly interesting sort.[12]

The British had demonstrated in the nineteenth century that authority over one type of property, land, rested in part on control of another type of property, effective communications technologies—specifically, the undersea cables. Now the fear emerged that whichever nation excelled in wireless would be in a dominant position, capable of blanketing the ether with messages invading the territorial airspace and consciousness of neighboring countries, and weaving a fabric of its own influence throughout the ether. A conception of the ether as territory, as property, was jelling. But how could countries control the ether, which was nationally important but at the same time invisible, communal, and mysterious? Making the psychological leap necessary to visualize control of activities in the ether would be difficult even in a period of relative harmony. During the internationally tense years at the turn of the century, when feelings of national pride were at their peak, the realization that even the sky could be violated by a foreigner, an interloper, was particularly unsettling. This fear of foreign occupation of one's airspace crystallized early and never vanished. Aggravating these new concerns was Marconi's nonintercommunication policy.

While Marconi's determination to establish his own complete and exclusionary communications network was a shrewd business decision, it was also a political time bomb. For while the policy represented, to Marconi, his only course for economic survival, it also introduced a revolutionary and distressing precedent: the policy implied that the company viewed the ether as a territory over which it intended to establish property rights. As the first man to demonstrate how this newly discovered resource could be commercially exploited for the benefit of many, Marconi believed himself to be entitled to the rights and privileges that often went to the early pioneers, those who were the risk takers. Although Marconi could not "own" the ether, if he controlled how it was used and who gained access to it, then he would be the domain's master. That, in turn, meant that the British, who already controlled a majority of the world's cables, would control the airwaves, as well. Such an outcome was unthinkable to many government leaders and military officials who were carefully calculating the equations of respective territorial control. Consequently, several governments and their military services began devoting more time and attention to developing their own wireless systems, independent of Marconi and the British. If a country's position depended, ultimately, on a strong army and navy, then those services would have to establish their beachheads in the ether before Marconi staked out the entire realm.

The U.S. Navy, caught up as it was in the international competition over the acquisition and disposition of property both visible and intangible, and in need of wireless telegraphy, would have to confront Marconi's exclusionary business policies. As it did so, the service would be pulled in several directions by the often contradictory goals of different officers. Some officers in the navy, backed by their commander in chief, were determined that the United States keep pace with European navies in every way. Maintaining parity would involve acquiring wireless apparatus and establishing a presence in the ether. Many in the service, however, did not embrace these goals. They valued tradition, not newness. From the outside, the navy of 1899 may have seemed quite modernized and technologically progressive, but a view from within reveals an institution filled with organizational and ideological barriers to adopting a new communications technology. Both ideological frameworks—the desire for expansionism and the allegiance to tradition—influenced the military mindset about how wireless should be used.

Wireless made its debut before a navy adjusting to the physical transformation of its fleet. By 1899, the new fleet was nearly complete. Since the 1880s, the navy, prodded by Congress, and in the face of considerable internal resistance, had begun acquiring bigger, faster, more nearly impervious steel ships.[13] The change from canvas and wood to steam and steel profoundly affected the way a ship was run, what its needs were in port, and how officers thought of their duties and command. The reconstruction, though much needed, was unsettling to the men and to the Navy Department, and, thus, any concomitant alteration in the bureaucracy which might have made this new navy more efficient was not readily forthcoming. The metamorphosis from old to new was initially cosmetic: while the hardware was being modernized, changes in naval administration, organization, and tactics lagged behind. Much as the first steam-powered ships retained rigging for sails, the naval organization sought at least to preserve a familiar structure during such major, unsettling changes.[14]

In a navy like this, torn between the old and the new, there would be psychological and organizational impediments to adopting wireless. The navy's internal structure was a major part of the problem. In 1899 the Navy Department was comprised of eight bureaus, each headed by a bureau chief.[15] The chiefs were responsible to the secretary of the navy, a civilian political appointee who usually knew little or nothing about naval affairs and who served at the pleasure of the president.[16] The responsibilities and jurisdiction of the bureaus often overlapped, yet there were no men, committees, or offices to facilitate interbureau cooperation.[17] Jealously guarding their territory and prerogatives, the bureau chiefs were often embroiled in internecine squabbles that generated "friction, circumlocution, and delay."[18] The difficulty of reconciling and coordinating the duties and objectives of the bureaus was a constant cause of frustration to the secretary of the navy.[19] The bureaus, on the other hand, could not count on long-term or informed guidance from their chief executive.

This lack of departmental coordination and direction was exacerbated during the first decade of the twentieth century by the activities of the navy's commander in chief. President Roosevelt took such an active interest in the Navy Department that he became its de facto secretary. Even someone as energetic as Roosevelt could not provide the department with sustained leadership and continuity while serving as president, however, and as a result, the navy's top management and public relations position was compromised. Between 1902 and 1909, there were six secretaries of the navy, none of whom had much power or influence.[20] This leadership vacuum worsened the organizational isolation of the bureau chiefs. Consequently, the chiefs learned to rely on "precedent and routine," and the department was guided by the daily grinding of the bureaucracy, which "ruled with an iron hand, usually ignoring, sometimes penalizing, those who attempted to introduce reforms and innovations."[21]

Tactically, the seagoing navy was equally decentralized until the twentieth century. Before then, the fleet had been divided into "small groups of cruising vessels thousands of miles apart," although each ship in actuality usually cruised by itself. "Even when in company, the ships rarely engaged in group maneuvers," the men accustomed to thinking of each ship more as a "potential solitary raider than as a unit of a fighting fleet."[22] Although the department began mandating periodic exercises and maneuvers in 1894, there was no accompanying "fleet policy," no long-term vision of coordinated activities or strategy within the Bureau of Navigation. Not until 1907 was there a permanent fleet consisting of ships and commanders trained to operate cooperatively.[23]

Thus, at sea and on shore, autonomy and independence at the higher levels of the bureaucracy prevailed. Within each bureau and on each ship, the lines of authority and communication were clear and strong. But between bureaus, and between ships, the lines, if they existed at all, were no more than fragile threads. Once ships were at sea, their lines literally and figuratively cast off, there was no web, either organizational or technical, to connect the ship to shore.

The communications system available to the navy both served and reinforced its decentralized administration. By 1890, telegraphic or cable communication was available in most ports and navy yards, and this somewhat eroded autonomy: when in port, squadron commanders could be closely in touch with Washington. Although by the turn of the century there was a "growing tendency to make naval strategic decisions at Washington instead of [in] the theater of operations," this was still only a tendency and not a practice that could be enforced when ships were incommunicado. Flag signaling by day and newly installed light signaling by night were used for intership communications. During rain or fog, or across long distances, intership communication was impossible. Many ships were not equipped with lights and therefore could not signal at night.[24] And, once out at sea, no ship could communicate with the shore.

Responsibility for providing ships with signaling apparatus fell to the Bureau of Equipment, which furnished the vessels with other supplies, including coal, rigging, navigational instruments, cordage, and hammocks.[25] Thus, the Bureau of Equipment, a procurement and supplies division with no authority over or expertise in engineering, ship construction and design, or maneuvers and fleet tactics, would be responsible for assessing and acquiring wireless telegraphy, which would alter all three.

The navy's response to wireless also reflected its increasingly delicate political position. President Roosevelt, the navy's primary lobbyist, began to encounter well-organized congressional opposition to extending the new navy. Senators and congressmen opposed to Roosevelt's brand of imperialism (and allied with others opposed to his liberal reforms) succeeded in reducing several of Roosevelt's requests for large naval appropriations.[26] These budgetary battles increased financial uncertainty and reinforced departmental caution. Thus, even if a bureau chief was technically sophisticated and sought to sponsor a particular innovation, he would confront obstacles above, below, and lateral to him in the organization. On the other hand, an officer's reluctance to make use of a particular technology was protected by the navy's decentralized structure.

The navy of 1899, then, was not the sort of organization in which technical sponsorship, especially of an invention that threatened autonomy and decentralization, was either desired or possible. This was the bureaucracy that wireless and its inventors would confront. During two separate but overlapping processes—the navy's acquisition of and the navy's implementation of the invention—the wireless inventors wrestled with unanticipated obstacles. The inventors were working at the forefront of electrical engineering, tackling both scientific and technical mysteries. The mysteries of organizational dynamics, however, would be more difficult to crack.

· · ·

IN OCTOBER 1899, the Bureau of Equipment, taking advantage of Marconi's American visit, sought to inspect his apparatus on behalf of the navy. Marconi agreed to allow four officers, all electrical experts, to witness the operation of his equipment throughout the yacht races.[27] In his report to the bureau, Lieutenant J. B. Blish stated that the demonstrations "were most convincing that the system was already excellently adapted for use on board ship." "My investigations since then," Blish added, "have strengthened that conviction."[28]

During these observations, Marconi was persuaded to allow naval testing of the apparatus after the yacht races. Marconi agreed to the tests only after issuing several disclaimers: he had not expected to give such a demonstration, and thus the equipment he had with him was not "sufficient for a government test . . . on a large scale." Nor did he have with him his "devices for preventing interference" from competing transmitters, because these devices were not yet "completely patented."[29] Marconi wanted it understood that this was not his standard demonstration for naval vessels, that he did not have all of his state-of-the-art equipment with him, and that, consequently, he could not guarantee the same success in these tests that he had achieved during the yacht races. Marconi was on the verge of patenting his method of tuning, whereby several wavelengths could be used with a given antenna. Because he was the only one signaling during the yacht races, he had no need for tuning and had not brought the additional apparatus to America. Whereas Marconi was seeking to protect himself from unjust criticism, the navy eventually came to believe that he was trying to cover up a major and unavoidable defect of the system.[30]

Marconi's apparatus was dismantled from the press boats in October of 1899 and installed on the armored cruiser *New York* and the battleship *Massachusetts*, both anchored in the New York harbor. A third set, at the Navesink Lighthouse in New Jersey, served as the shore station. Members of the navy's "Marconi Board" were to assess the equipment's accuracy, establish maximum operating distance, determine the best location for the instruments, and report on interference. After several days of tests, one of the board members, Lieutenant Commander J. T. Newton, advised the bureau that sending accuracy was not always achieved and that Marconi's temporary setup aboard the ships would be inadequate for a permanent installation. Transmission speed averaged twelve words per minute. While the two ships exchanged messages over a distance of 36.5 miles, and the *Massachusetts* received the *New York*'s transmissions from as far away as 46.3 miles, this success was overshadowed by a persistent drawback: interference occurred whenever more than one set was signaling, because only one wavelength was being used for all transmissions. Although Marconi had claimed that he could prevent interference, Newton complained that Marconi "never explained how nor made any attempt to demonstrate that it could be done."[31]

Despite these failings, Newton and the board recommended that the navy give the system a trial. Newton pointed out that the system could be adapted for use on all navy vessels and had the distinct advantage of performing well in "rain, fog, darkness, and motion of ship." "Excessive vibration at high speed apparently produced no bad effect on the instruments," Newton observed. Within the working ranges, accuracy was good. Newton noted that the best location for the instruments would be "below, well protected, in easy communication with the commanding officer."[32] Another board member wrote, "Even in its present state the instruments can be made useful in signaling between ships, and ship and shore."[33]

Admiral R. B. Bradford, chief of the Bureau of Equipment and himself quite knowledgeable about electrical technology, was persuaded by this report and appealed to the secretary of the navy on December 1, 1899: "This system is successful and well adapted for Navy use. The chief objection to it is known as 'interference.' . . . Notwithstanding this fact, the Bureau is of the opinion that the system promises to be very useful in the future for the naval service." Citing Marconi as the recognized inventor and noting that no other "makers of electrical instruments" had been able to duplicate Marconi's apparatus successfully, Bradford recommended acquiring sets from Marconi for continued naval experimentation.[34]

Despite this favorable endorsement from the Bureau of Equipment, the navy did not acquire any of Marconi's apparatus. Why was this admittedly imperfect yet extremely promising invention not adopted by the new navy? The explanation most often cited is that the navy rejected Marconi's contract specifications because they were too expensive and restrictive. The dispute over the terms of purchase reflected misunderstanding on each side about the needs of and constraints on the other party to the contractual negotiations. Because of the company's new leasing policy, Marconi would not sell his apparatus to the navy or to anyone else without royalties. Under his terms, the navy would purchase no fewer than twenty sets at a total cost of ten thousand dollars and agree to pay a ten-thousand-dollar annual royalty. The royalty would be reduced if more than twenty sets were purchased.[35] For Marconi, these terms represented a concession: the navy was allowed to keep the sets, whereas other customers could only lease the apparatus. But the navy found the terms, and the corporate strategies underlying them, completely unacceptable.

The navy's reaction to Marconi's terms and policies was influenced by finances, by precedent and law, by nationalism, and by a continuing suspicion of inventors and business firms. The Bureau of Equipment did not have enough money to pay Marconi's price, and the department was constrained, by law, from obligating funds beyond the current fiscal year.[36] In addition, the navy viewed Marconi's leasing and nonintercommunication policies as unnecessary and as monopolistic ploys designed solely for the purpose of granting yet another British company complete control over international communications. As one official noted, "Such a monopoly will be worse than the English submarine cable monopolies which all Europe is groaning under and I hope the Navy Department of the U.S. will not be caught in its meshes."[37] Anti-Semitism reinforced these fears. Although Marconi was Irish and Italian (and Catholic), and the managers and technicians of his company British (and not Jewish), he was sometimes thought of as part of a Jewish corporate cabal, having reportedly "sold himself to the Jews," as one officer put it.[38] This prejudice made the monopoly seem more sinister and threatening. From the navy's point of view, Marconi was trying to prevent anyone else from gaining access to a resource—"the air"—which traditionally had been free. The navy had no way of understanding the financial difficulties surrounding wireless: the research expenses, the patent and legal fees, and the revenue problem. To the navy, the Marconi financing strategy was not protective but avaricious. Marconi, in turn, had no way of appreciating the financial, legal, and political constraints operating on the navy. John Bottomley, in a report to the British Marconi Company, referred to Bradford's "trite argument that the government could not legally enter into a contract to expend money beyond an appropriation for a given fiscal year." Bottomley then added derisively, "It is hardly necessary to comment on this argument."[39]

During the next twelve years, negotiations between the navy and Marconi rarely transcended this early stalemate. But the Marconi Company was not alone in provoking negative reaction from navy officials over prices and contract terms. Every company trying to do business with the navy encountered an attitude inhospitable to inventors and unappreciative of their technical goals and financial needs. The navy, on the other hand, found the inventors to be overly sanguine about the capabilities of wireless and insensitive to the navy's various organizational and political handicaps.

Naval officers and wireless inventors were, in fact, approaching each other from two strong but opposite cultural traditions, traditions that influenced self-image and behavior, traditions that were laden with prejudice and stereotypes that often affected negotiations. A navy man and an inventor were very different types of people, differently socialized, with contrary and often conflicting orientations. The naval officer was an organization man. He spent his life obeying orders, moving gradually up through the ranks, preserving and identifying with the status quo, honoring tradition, defending the organization that provided him with security and recognition. Except during wartime, success in the naval service involved diligence and diplomacy, keeping a low profile. Organizational stability surrounded and insulated the naval officer, and that was what he came to prize.[40]

The wireless inventor, on the other hand, had no such large organizational affiliation. Often he was a loner, sometimes seeing himself as an outcast who would redeem himself through his invention. Driven by a desire for fame, money, love, or all three, the inventor sought to make his mark on history by making change possible, by disrupting the status quo. Initially, sometimes constantly, plagued by problems of financing or solvency, and determined that their contributions would remain distinctive, these men, among the last in the independent inventor tradition, built their reputations and careers on technical change and improvement. Stability, established ways of doing things, existing schemes—these were what the inventor disrupted, sometimes deliberately, sometimes inadvertently.[41] Because he lived on possibilities, he was of necessity overoptimistic, often given to exaggeration.

Inventors and organization men acquired and used money differently, and this was an additional and powerful source of mutual distrust. To paraphrase Hugh G. J. Aitken, inventors responded to market demands, to "signals" they received from the economy, while military men, not usually subject to outside forces, responded more to "internally generated signals," which were rarely tied to the marketplace.[42] Their contrasting pecuniary orientations, coupled with widely divergent socializations, induced the members of each group to view the members of the other with suspicion and, occasionally, contempt. As the navy continued to investigate wireless over the next ten years, these conflicting traditions, cultures, and attitudes played a salient part in contract disputes.

Not until the autumn of 1901 did the navy conduct further tests of wireless. That fall, the department decided to explore what European inventors other than Marconi had to offer. Why the department resumed its investigations into wireless at this time remains unclear. Although Reginald Fessenden had been experimenting with wireless under the auspices of the Weather Bureau, and Lee De Forest had recently formed his own company, the navy, also for reasons that remain unclear, believed that there was "no American wireless telegraph company ready to furnish apparatus."[43]

Commander Francis M. Barber, USN, retired, an old classmate and friend of Bradford's, was living in Paris at this time. Well connected in diplomatic circles, knowledgeable about electrical engineering, and fluent in French and German, Barber seemed the perfect middleman between the European inventors and the U.S. Navy. From 1901 until 1908, he monitored the European technical press, solicited information from inventors and naval officers, visited the various companies, and sent complete and lively reports on all aspects of wireless to the Bureau of Equipment. His thirty-year tenure with the U.S. Navy had given him a keen appreciation of how to get technical information from foreign military organizations. He ingratiated himself with the senior officers first, and after he had won them over, he felt free to go to the real source of information. As he wrote to Bradford, "it's no use commencing with junior officers anyway. They have all the knowledge; but the old busters have to be coddled first."[44]

Barber's correspondence provides a fascinating view of how the bureau's official representative perceived and dealt with the inventors and the still young wireless industry. During 1901 and 1902, he investigated the apparatus of two French inventors, Rochefort and Ducretet, and of two German firms, Slaby-Arco and Braun-Siemens-Halske. He liked Rochefort because he was a "modest gentlemanly little man and not at all captious and prejudiced as inventors usually are."[45] He also observed that "an inventor is a visionary, a visionary is a genius, and a genius is a lunatic or next door to it."[46] Some inventors he heard about aroused his interest, but he decided against satisfying his curiosity: "One better have the itch than encounter an impecunious inventor. He never lets up once he makes your acquaintance."[47] One day he would visit Ducretet, who would call Rochefort a liar and a thief, and the next day he would hear Rochefort say the same things about Ducretet. Barber found sorting out the wireless situation in Germany particularly frustrating, because he was unable to obtain what he considered to be reliable information. "These manufacturers are such liars," he complained, "that one often wonders with St. Paul 'What is truth?' " Barber's suspicion of inventors was compounded by his attitudes toward many foreigners. He was unimpressed with the German company Slaby-Arco, which he found "too slippery." This opinion seems to have been reinforced by his impression of Count Arco, whom he described as "a weedy little chap with a great big head—he looks like a tadpole."[48] In his assessment of the British, Barber commented, "You can't hint to an Englishman, you must *kick* him. In my long business experience the English are the most dishonest people I know."[49]

Barber reserved his most stinging scorn for Marconi. Any information, whether rumor or fact, which reflected badly on Marconi's apparatus or his business was eagerly reported to the Bureau of Equipment. Barber heard—and believed—that Marconi had "walked off" with others' inventions in developing wireless, and that he was therefore operating with an extremely vulnerable patent structure. Thus, Marconi was sure ultimately to fail, but in the meantime, his "system" deserved to be circumvented, since it was all stolen anyway.[50] Barber doubted the accuracy of press accounts hailing signaling successes, such as Marconi's celebrated transatlantic *s* in December of 1901.[51] He took particular delight in recounting a conversation he had with Colonel Hozier, the secretary of Lloyd's and a director of the Marconi Company: "He thinks Marconi had never yet got a signal across the Atlantic or 2000 miles at sea either. The whole thing was a stock-jobbing operation worked in the interest of 'a lot of Jews.' This from a director of the company is rather good."[52] He continued to hope and expect that the U.S. Navy would "be able to drive the American Marconi Company out of business."[53]

These are the words and attitudes of the man who was the navy's primary source of information regarding the European wireless community. As the Bureau of Equipment's eyes and ears in Europe, he was in a highly influential position. The inventors, no doubt unaware of his true feelings, opened their laboratories and factories to him, advised him, confided in him, boasted to him, and, of course, tried to win him over. While transmitting important technical and business information to the bureau in the United States, Barber was also reflecting, and reinforcing, a particular way of viewing and dealing with inventors. He also articulated what appears to have been the prevailing naval attitude toward patents, wireless systems, and how properly to negotiate with the wireless companies. To Barber and the bureau, inventors were those eccentric and frequently deceptive people the department was forced to do business with in order to get the apparatus it needed. The bureau's subsequent business practices were certainly consonant with the overall spirit and outlook of Barber's correspondence.

During the next several years, the navy experimented with various kinds of wireless, both European and American. These sets were usually tested between the Washington, D.C., Navy Yard and the Naval Acade- my at Annapolis, as well as between Annapolis and one or more ships. The distance between Annapolis and the Washington Yard was only thirty miles, so, as Barber noted, "almost anything ought to work there."[54] In the spring of 1902, Barber arranged for the navy to purchase two sets each from Ducretet, Rochefort, Slaby-Arco, and Braun-Si- emens-Halske.[55] These were tested between August and October 1902. That autumn, De Forest succeeded in having the navy purchase and test two of his sets, but the trials were hindered by a dearth of skilled oper- ators and officers knowledgeable about radio.[56] A Wireless Telegraph Board was established to oversee and report on the tests, but its members had other, conflicting, duties and were unable to continue with the board for long. Three of the five members had to be replaced during the course of the tests. The officers ultimately "went their respective ways," leaving three enlisted men to oversee the tests and then notify their superiors of the results.[57]

Admiral Bradford, chief of the Bureau of Equipment, complained to the secretary of the navy about the lack of departmental commitment to the experiments: "The Bureau desires to express its great regret that these important experiments have been interrupted for the want of ves- sels necessary for the work; also that two members of the Board are under order for sea. It is feared that no important results can be reached unless a Board can give its uninterrupted attention to the subject."[58]

The early negotiations leading to the purchase and testing of these sets indicated how the navy would do business with the wireless com- panies over the next eight years. The navy enjoyed a buyer's market, and Barber seemed well aware of his advantages. To ensure that their appa- ratus performed well, the various companies wanted their own en- gineers to be present at the tests. They expected the navy to subsidize the travel expenses, especially since naval operators and engineers would need the sort of instructions and advice not conveyable in written specifi- cations. From the inventors' perspective, the navy would be getting the best possible results and free training, and should be obliged to cover the travel expenses. The navy, of course, did not see it this way, and declined to support such assistance.[59]

On instructions from Bradford, Barber indicated to the companies that the navy would not employ the services of any private specialists, and while it would be helpful to have experts on hand when the tests occurred, the navy's engineers would probably be able to figure out the apparatus.[60] Of course, the thought of amateurs tinkering with their instruments drove the inventors wild, especially because proper performance could mean a big contract. Barber knew this, and quickly made the inventors see that sending representatives at their own expense was better than sending none at all. By April of 1902 he was able to advise Bradford: "I have them all corralled and they will go at their own expense rather than not at all."[61] Extra expenditures had not been the only consideration: naval pride was operating, as well. Barber acknowledged, "It is rather humiliating to be obliged to have 'square heads' come over and show us how to run things, but after all the main idea is to succeed and to get the best apparatus."[62] The navy spent nearly $12,000 on the eight sets of wireless, whose prices ranged from $2,250 to $3,500 for two sets.[63] When Slaby-Arco, citing recent improvements, tried to raise its prices, Barber notified the company that it could either return to its previous prices or cancel the navy's order.[64]

During the tests, both the Slaby-Arco and the De Forest apparatus outperformed the French.[65] Slaby-Arco, hearing of its success, wrote Barber what he described as a "very cheeky letter": "They wanted to know how soon now they might expect the orders which would repay them for the vast expenditure to which they had been subjected in sending engineers to the U.S. and they wanted me to write and urge that the orders be placed immediately. I replied laconically."[66] Slaby-Arco finally was awarded a contract. Its prices were low, and, more importantly, its apparatus was better suited to the navy's need for easily adjustable instruments. The receiver used by Marconi and Slaby-Arco, a filings coherer, was connected to a recorder that printed signals on a strip of paper. The receiver, which was insensitive and erratic, would sometimes print static as well as signals, but it provided a written record and required little skill to operate. De Forest had substituted headphones for the tape, so the operator could distinguish between true and false signals. As one navy technician later noted, "The De Forest method had the advantage of enabling any speed of reception to be used, depending on the skill of the operator, but the very fact that the Navy did not have even one operator [who] was skilled militated against the De Forest method."[67]

Thus, the needs of the navy and the goals of the inventors, especially those Americans seeking to reconceptualize and improve on Marconi's apparatus, were completely at odds. The inventors were striving for greater distance, greater selectivity, and faster reception. The inventors assumed the navy would welcome all three. The navy, on the other hand, preferred apparatus of moderate range which was easy to adjust and operate, even if it was less sensitive or accurate. Naval officers may have wanted apparatus that supplied a written record if the apparatus were going to be operated by enlisted men. The inventors were thinking in terms of technical improvement, not organizational accommodation, and they assumed the navy would want the most up-to-date apparatus available. The navy, however, needed equipment that would compensate for its organizational idiosyncrasies, a factor the inventors were slow to grasp and reluctant to address.

In March 1903 the navy ordered twenty units of Slaby-Arco apparatus. The purchase prompted *Electrical World* to condemn the navy for the "cold shoulder it had consistently turned to American workers in the field"; the journal referred to the system the navy favored as "that of the German Emperor's court jester."[68] Reginald Fessenden, whose company was now six months old, wrote to the Bureau of Equipment and the secretary of the navy suggesting that before buying foreign equipment, the navy should test his apparatus. Fessenden quoted Bradford a price of four thousand dollars for two sets and offered "to send a couple of men with a pair of sets" to whatever location the navy desired.[69] Bradford agreed, being careful to specify that the entire test would be at Fessenden's expense. The bureau would not, as it had before, purchase two sets.[70] Why should it? Already it had spent approximately six thousand dollars on French apparatus that barely worked. If they wanted the business, the Americans would have to take risks. Fessenden and De Forest continued to brag about their apparatus, particularly to the press. Navy officials, and Barber in particular, believed the claims to be hyperbolic public-relations statements (which they sometimes were).[71] Yet if the inventors were going to boast, the navy was going to hold them to their word in subsequent tests. More frankness on both sides might have better served all concerned. But none of the inventors behaved as if he believed that candor would sell wireless.

Bradford notified Fessenden of two other conditions for the tests: the navy had no room available on board ship for Fessenden's apparatus, so he would have to set it up in a hallway, and the specifications for the apparatus included the filings coherer.[72] Fessenden complained that the Slaby-Arco people had not been relegated to a hall, and said that his company was disinclined to supply the navy with special—and outdated—apparatus at the company's expense.[73] Bradford responded that the navy gave "available space only, not that best suited for the apparatus."[74] As a result of this impasse, no experiments with Fessenden apparatus were made until August and September of 1904, when the department scheduled tests of American apparatus between the Brooklyn Navy Yard and the Navesink Highlands.[75]

By early 1904, four and a half years after Marconi's first demonstrations before the "Marconi Board," the U.S. Navy had barely begun to exploit wireless, and only a few of the instruments in its possession were made by an American company. The apparently dilatory fashion with which the navy was adopting wireless was the subject of several biting editorials in the technical press. *Electrical World* wrote, "If the lack of such apparatus in our army and navy is due to the neglect of the moss-backed bureaucrats who sent our artillery into action at Santiago with black powder, the public ought to know it, that the authors of the negligence may be properly pilloried."[76] Noting that, in the military, it had "too long been the fashion to fore-damn anything and everything devised by civilians," the magazine warned that the navy's "procrastination bureau" was indulging in dangerous dawdling.[77] "As matters stand now," the magazine concluded, "we would be at a great disadvantage in this respect if attacked by any reasonable power."[78]

While Fessenden and the navy squabbled over where to put a receiver on board ship, events were occurring which prompted reconsideration of the strategic importance of wireless. Although the primary actors in them were Europeans, these episodes drew the United States into the controversy surrounding how activity in the ether would be regulated among nations, and awakened some members of the navy to the realization that other navies were much further along in their use of and control over the invention of wireless telegraphy.

·   ■   ·

IN MARCH OF 1902, Prince Henry, the Kaiser's brother, was returning to Germany after a highly publicized visit to the United States. He was sailing aboard the German liner *Deutschland*, which was equipped with Slaby-Arco apparatus manufactured in Germany. According to the prevailing story, none of the Marconi stations on either side of the Atlantic would communicate with the ship because it used rival apparatus. Prince Henry, who had tried to send several messages to both the United States and Germany, was outraged. The ship, with its royal cargo, might as well have had no wireless equipment at all. A writer for *Electrical World*, describing German reaction as "malignant Marconiphobia," reported: "There arose in Germany a chorus of effervescent indignation, which quickly sped along the cables to the *Herald* office in New York, so that the American hardly less soon than the European public was treated to the unedifying spectacle of a learned professor, a Noble Count and various other potent and distinguished personages alike, foaming at the mouth with a species of almost berserk fury."[79]

When the *Deutschland* incident was publicized, Marconi calmly contended that due to technical incompatibility, the Germans could not "communicate with [Marconi] instruments." He elaborated, evoking the special Marconi tuning mechanism as an obstacle to intercommunication: "The instruments at these stations are intended to work with suitably tuned apparatus, and we have no information as to the kind of waves radiated from the installation on the *Deutschland*. . . . The bad working of the so-called Slaby-Arco system must be set down to its own defects."[80] This rather disingenuous response epitomized, for the Germans, British commercial arrogance. The Kaiser's own brother had been incommunicado, not for lack of apparatus, but because of the exclusionary policies of a British firm. Although the *Deutschland* incident appeared at first to be a petty confrontation between two rival companies and their respective countries, it was actually a watershed in the early history of wireless. The emerging problems surrounding the technology and its financing and regulation, and the sanctity of each country's territorial air, were embodied in the Marconi-German clash. Could a private company, whether it had technical priority or not, gain dominance over a resource such as the airwaves and become the arbiter of who could use them and who could not? This was the question the Germans indignantly submitted to the world community, fully expecting the weight of opinion to be with them and against Marconi.

German officials announced a campaign to thwart the Marconi Company's attempt to achieve an international monopoly. The Kaiser ordered that all German military and civilian stations would use only Slaby-Arco equipment, and that German experimentation would be stepped up.[81] In addition, he sought to involve the other major powers in the resolution of the conflict. His indignation over the *Deutschland* incident prompted him to invite seven nations to join Germany in an International Wireless Conference, scheduled for the autumn of 1903. Citing the preservation of world peace and free enterprise as the primary reasons for the meeting, the Kaiser sent invitations to Great Britain, France, Spain, Austria, Russia, Italy, and the United States. A month before the conference, in July of 1903, the two competing German firms, Slaby-Arco and Braun-Siemens-Halske, merged to form Telefunken in order to present a united German commercial front against Marconi. Telefunken, with the full support of the German government behind it, initiated a campaign to compete with Marconi on every continent.

Marconi knew full well why the Germans had called such a conference, and he did not hesitate to question the Kaiser's motivations: "I regard the proposed convention as neither more nor less than another attack by Germany upon British industry. . . . The time is not ripe for international consideration. It is better to let fair competition work out its natural results. You cannot well yoke my system to an inferior and unsuccessful imitation. The attempt is to place inferior imitations on terms of equality with the original system."[82] The British press noted that had Germany been the first country to develop wireless, that country's leader would not have been so eager to call such a meeting. The Germans' urging of such a conference in the name of peace had the ring of false piety. The *Edinburgh Review* described initial British reaction: "Any proposal hailing from the Kaiser affects a certain class of statesman—or, rather, politician—much as presenting a red tag to a bull affects that quadruped."[83] Meanwhile, Marconi asked Cuthbert Hall to begin drafting a statement that the company would provide to both the British and the Italian delegates. "And I hardly need tell you," wrote Marconi, "that [the statement] should be based rather upon what we take to be the interests of the shipping community in general than upon our own special interests."[84]

Although the Kaiser's invitation suggested that a range of issues of mutual concern would be discussed at the conference, the only real issue, as Marconi expected, was the Marconi Company's refusal to communicate with other systems.[85] John I. Waterbury, one of the U.S. delegates to the conference, summarized the arguments on both sides of the issue. If the Marconi Company was forced to receive messages from what it alleged was an inferior system, it would injure its own interests and jeopardize the world's good service. If the Marconi Company had an advantage over its rivals, the advantage was the result of fair competition. Although wireless had been demonstrated in 1896, it had been in commercial operation only for three years, and world leaders were ill advised to try to "fetter a new discovery in its development." On the other hand, the Marconi Company allegedly had declined to compete with anyone else over the same proving ground. The German delegates charged that the company's results were no better than anyone else's, that others could communicate as satisfactorily as Marconi, and that it was never too soon to prevent the company from achieving a worldwide monopoly.

All countries at the conference except Italy and Great Britain favored compelling the Marconi Company to communicate with other systems, because how could an invention with such potential for saving lives and property be monopolized by one company? The question of compensation for the Marconi Company was then raised: Could governments restrain a system without leaving it stripped of reward or protection? The Italian delegates argued that the Marconi Company was entitled to indemnity should the recommendation of the majority become law. English representatives favored a surtax on messages sent by rival concerns. Germany objected. The compromise effort failed, and the final resolution of the conference stated: "Coast stations for wireless telegraphy are obliged to receive and transmit telegrams going to or returning from ships, without distinction as to the system of wireless telegraphy employed by the ships."[86] However, the conference did not restrain Marconi at all, as the resolution did not have the force of law. Both Italy and Britain had government contracts with the Marconi Company for equipping their navies, and they maintained that signing the agreement would place them in violation of their contracts. The British delegates said they had no power to impose regulations and restrictions agreed on by other nations; in fact, they were not about to let Germany move effortlessly into an industry as potentially valuable as wireless.

The other two American delegates to the conference were Brigadier General A. W. Greely of the U.S. Signal Corps, and the navy's European wireless informant, Commander Barber. Barber arrived with his anti-Marconi sentiments and was happy to have them reinforced. The American delegates also got a firsthand view of the importance other governments, and especially other navies, attached to wireless and to ensuring military priority in the ether. The European armed forces were not going to permit Marconi to gain control of the ether; rather, they were going to stake their own claims. Each country was developing its own brand of wireless, often under strict government control. This gave them an advantage, not only over Marconi, but also over the American military services. To the Europeans, the ether was a resource of such importance that its exploitation primarily for commercial purposes seemed shortsighted and risky. Marconi saw the ether as a resource having economic value, whereas the Germans and their allies believed its strategic value outweighed any other claims, especially those of business. This was the perspective that the Americans would take home with them and that would galvanize some members of the navy to become more actively involved in the deployment of wireless.

The American delegates quickly came to believe that their status as representatives was inferior and that they were regarded as relatively powerless because their government had done little to promote or gain jurisdiction over the American wireless situation.[87] Thus, although the conference accomplished little that was formal or enforceable, it impressed on the American delegation the advantages and international importance of a strong military presence in the airwaves. Barber wrote to Bradford at the Bureau of Equipment describing the extent of government control in Europe, and Bradford used this information to argue for an extension and acceleration of the navy's wireless work. But the chief of the Bureau of Equipment was not powerful enough to orchestrate such a program; he would need an organizational ally with greater authority, such as the secretary of the navy, to propose and implement the scheme. Because the naval secretaries of the Roosevelt administration were rendered impotent by the president's naval activism, the bureau would need support from the president himself. Until 1904, however, issues much more important than wireless occupied the president. The invention moved closer to the forefront of his concerns in February 1904, when the Russo-Japanese War broke out and wireless played a role in an armed conflict for the first time.

The use of wireless by the combatants and by observers raised complicated questions. Both the Japanese and the Russians relied on wireless during the war; in addition, many correspondents and observers had converged on the war zone. De Forest, under contract to both the *Times* of London and the *New York Times,* had operators dispatching news of the war from China via wireless. Marconi operators were also in the vicinity. What was the status of such a correspondent and his press boat during hostilities? Could the governments involved prevent the press and civilian observers from usurping the airways during a particular emergency? Russia was strenuously opposed to the use of wireless by neutrals within the zone of hostilities. This opposition stemmed from Russia's embarrassing showing during the war; wireless was only one of many strategic areas in which Japan demonstrated its superiority. Some observers believed that Japan's advanced wireless had provided a clear and decisive advantage.[88] But advanced wireless was not sufficient for successful exploitation of the airwaves during war; those airwaves also had to be available to government stations.

The issues of who had priority in the airwaves, how respective spheres could be delineated, and how commercial stations could be prevented from interfering with government stations became pressing. No guidelines existed for appropriate conduct in the ether during peace, much less during war. As one journal noted, "The ordinary precedents cease to have any direct value."[89] President Roosevelt followed the war closely and was personally involved in the negotiations that led to the Treaty of Portsmouth in 1905. Consequently, he was cognizant of the ways in which Russia's second-rate wireless system had served to undermine that country's position. He was also concerned about the interference caused by the press and private wireless concerns in the war zone when government access to the airwaves was critical. During the war, seven different wireless systems often were operating at once.[90]

Roosevelt began appreciating that the strong navy and the international preeminence he desired required an efficient, well-organized communications system. But within the American government, the navy was not the only organization pursuing a viable wireless system. Several government entities, in fact, had been quietly, though not very successfully, competing with one another for control of wireless. The Weather Bureau of the Department of Agriculture, which had begun sponsoring Fessenden's work early in 1900, argued that it had been first in the field. The Signal Corps of the U.S. Army, under the direction of General A. W. Greely, had already tested and purchased De Forest and Fessenden apparatus. The navy, which was lagging behind in the race, maintained that wireless was best suited for signaling over water and that for national security purposes, the navy should control the government's wireless system.

Roosevelt was determined to end this bureaucratic struggle and to consolidate the government's wireless activities. After six months of monitoring the war in the Orient, he announced, on June 26, 1904, the appointment of the Interdepartmental Board of Wireless Telegraphy, better known as the Roosevelt Board, to report on the question of consolidation and management of wireless for the government and "to quiet the spirit of competition which has sprung up between three departments of the government, each desiring to control the operation of wireless on the coasts." Roosevelt's pro-navy bias was clear: he appointed to the board Rear Admirals Robley D. Evans and Henry N. Manney and Lieutenant Commander Joseph L. Jayne of the navy; General Greely of the army; and Willis L. Moore, chief of the Weather Bureau.[91] In addition to making a recommendation concerning which department should oversee the development of wireless, the board was to determine how private companies and government stations could operate most harmoniously. Roosevelt also charged the board members with considering the rights of inventors and determining the circumstances under which the government could claim a monopoly of the airwaves. A spokesman for the Navy Department assured the press that a "government wireless monopoly [was] not planned" and that the board would not "seize on the art and science of wireless telegraphy."[92]

In August of 1904 the board members submitted their report. The navy, in an effort to keep pace with other navies, was to manage and operate the government's wireless system and begin establishing "a complete coastwise radio telegraphy system, covering the entire coasts of the

United States, its insular possessions and the Canal Zone in Panama." Wireless stations within this system, the board recommended, should transmit to or receive messages from ships at sea free of charge, provided the navy was not competing with a nearby commercial station. The board also suggested that private companies not be allowed to erect stations where they might interfere with naval or military operations.[93] In other words, if the navy erected a station in a particular area first, the navy would gain preemptive rights and be able to prevent private companies from locating in the vicinity. This was no insignificant recommendation, since the various American companies had only recently begun to establish their own stations. The board proposed eliminating the coastal stations of the Weather Bureau because the meteorological data necessary for that department could be collected by stations of the Navy Department. The legislation urged by the report would provide for the licensing of all private stations and the placement of them under the supervision of the Department of Commerce and Labor to prevent "the exploitations of speculative schemes based on a public misconception of the art" as well as "control by monopolies and trusts."[94] Despite all this, the board's report reaffirmed that the government did encourage private enterprise.

The board addressed only some of the questions formally assigned for its consideration. With board membership stacked in favor of the navy, naval control of the government's wireless was inevitable. The board did not tackle the difficult problems of standardization or peaceful coexistence between government and private stations. The report issued was not in the spirit of the board's initial assignment, and the contrast between the stated intent and the final recommendations angered the press and private companies.

The *New York Times* described the plan as nothing less than confiscation. The proposals sounded too much like a government takeover of "an art which is yet only in an embryo state of development."[95] Such government control was not part of the American tradition, proclaimed the *New York Tribune*, which predicted that the proposals would be unacceptable to the American people.[96] *Electrical World*, which had advocated varying degrees of government regulation, preferably through one of the civilian departments, expressed indignation:

> The Navy Department is particularly disqualified at the present time from becoming the custodian of wireless. . . . Such a policy cannot be too strongly condemned, not only because it involves an extension of military authority over what in times of peace is a purely commercial function, but because of the deadening effect on development of the art

that would inevitably result from bureaucratic control. . . . That such development would occur under military domination none, we believe, will seriously assert. . . . As to the probable result of naval control, we need only point to the humiliating rank the Naval Observatory holds among similar establishments in this country and throughout the world.[97]

Suspicions concerning the board's motives abounded. *Electrical World* called the national defense argument a pretense designed to "appeal to the jingo spirit." Its editors hoped that the government could "keep the etheric peace without owning the entire ether as well as the earth beneath."[98] The *New York Times,* in a more moderate tone, endorsed *Electrical World's* position that the government, which might have special needs in time of war, should not "hamper enterprise by unduly restricting the application of wireless in time of peace."[99] In press releases issued after the board's report, Fessenden's company, NESCO, claimed that the Commerce Department would in practice only grant requests for station sites to the government, and would ignore similar commercial requests. "Under these conditions," NESCO asserted, "the wireless companies might just as well go out of business because there will be nothing for them to do."[100] H. J. Glaubitz of NESCO described the government's plan as "a socialistic scheme for stealing property" which would not be approved by Congress.[101]

While Glaubitz correctly anticipated congressional response to the board's proposals, not all of the recommendations required legislation for adoption. The suggestion that commercial stations be licensed by the Department of Commerce and Labor, and the proposed restrictions on the location of private stations, would have to be voted on by Congress. Building naval wireless stations, however, only required additional appropriations or, failing that, skillful use of money already available. So while the various wireless companies could begin lobbying against the proposed legislation, they also had to confront these new naval ambitions. On the one hand, the navy was going to need more apparatus; on the other, it seemed intent on establishing its own network, which threatened the civilians with major and possibly crushing competition.

The board's recommendation that the navy build a nationwide coastal wireless network employing a standardized system brought additional competitive pressures to the navy's wireless trials scheduled for August 1904. The inventors believed that the company that performed the best would gain the lion's share of the government business, and possibly even monopolize sales to the navy. They were eager to please. The navy, now in a rather commanding position as the government's primary and growing wireless customer, enjoyed complete control over the terms governing the tests and any subsequent contracts.

The provisions for these tests were different from those of the 1902 demonstrations. Now all expenses related to the demonstration (except expenses incurred in supplying the necessary current) were to be assumed by the companies. The companies were allowed to send their specialists to help with the tests, but the Bureau of Equipment's operators were to be given every opportunity necessary for the bureau to determine whether its operators could successfully operate the system. In addition, the Americans' apparatus had to conform to the technical standards previously set by Slaby-Arco (a system inferior to and different from their own).[102]

As the date for the tests approached, the navy issued an additional requirement. Reacting to Marconi Company policy, Lieutenant Jayne insisted that the American companies guarantee that their systems could maintain communication with any other type of apparatus. Fessenden's sales representative pointed out that NESCO could not possibly make such a guarantee, as the company could not be responsible for the performance of competitors' apparatus. NESCO could not ensure, for example, that an 1899 Slaby-Arco filings coherer could detect NESCO transmissions. NESCO did guarantee that the company's apparatus would be able to communicate with any comparable apparatus, but the navy adamantly refused to accept such a proviso and insisted it would not do business with any firm that could not meet the guarantee provision.

The American Marconi Company scoffed at the terms set for the tests. The company would participate only if the bureau would guarantee that successful performance would lead to a contract. "In view of the fact that we are working on a commercial basis over greater distances and under varying conditions all over the world," wrote Bottomley, "no outlay for the purpose of demonstration only commends itself to us."[103] The bureau would not consider contingent contracts, and the Marconi Company saw no reason to incur an expense that, in its opinion, "would be out of proportion to the value of the result."[104]

Fessenden also chafed at the navy's approach to the tests. He was anxious about the nearly one thousand dollars he estimated the demonstrations would cost his small company. He tried to arrange for an alternative method of testing, preferably at his own stations, but the navy refused. The bureau wanted to have control over the tests. The navy had its own needs and requirements: only by testing wireless on its ships and at the navy yards could it determine suitability. Its men had to be able to operate the equipment. Continued mistrust of inventors' claims reinforced the navy's desire to test the apparatus on its own turf. As Admiral Manney wrote, the bureau preferred "to conduct the tests in its own way."[105] The conditions the navy imposed during these and subsequent tests were, from the inventors' point of view, niggardly and demoralizing.

The navy demonstrated even less faith in the inventors in negotiations over purchases and contract specifications. The wireless market was still small, and the various inventors competed fiercely against one another. Pride as well as money was at stake, and the mutual hostilities provided the navy with bargaining advantages.

Once the navy had decided to acquire apparatus, its first goal was to drive down the price, and its policy was to buy from the lowest bidder. Barber took great pride in his negotiating skills, reporting that Slaby-Arco lost about seven thousand dollars on the first twenty sets it sold the navy: "The company inferred from my letters that they were competing with other people, especially with Braun-Siemens (I *did* mislead them intentionally in that respect) and the result was an impossibly low bid which I accepted by telegraph before they had time to think it over."[106] One year later he persuaded Telefunken to lower its price for a strategically important station by threatening to buy from the French at lower prices. Barber exulted: "Evidently they are red hot on the subject and the Bureau can name its own figure—It isn't often that you get a German down on his stomach like that."[107] The navy paid nothing in advance; in fact, no payment was sent until the apparatus was installed and operating. If the apparatus arrived late or was damaged in transit, or if the enlisted men mishandled the installation, the payment to the supplier was reduced.[108] While it was clearly in the interest of the navy to obtain the best possible price and not pay until the apparatus was working, its tactics compounded financial uncertainty for the inventors.

If an inventor would not reduce his prices, the navy would have a competitor copy the invention and supply it at lower cost. This was the tactic the inventors found most infuriating. One instance of this tactic involved Fessenden, who introduced the navy to his receiver, the electrolytic detector, during the 1904 demonstrations. Fessenden's assistant wrote that naval officials were "highly pleased with the results," as NESCO had performed "much better than any other system tested by the Navy."[109] Evidence bears this report out: by 1905, the electrolytic detector was the navy's standard receiver. But Fessenden's prices (two thousand to five thousand dollars per set) were considered too high, so the navy arranged for De Forest (who had already copied Fessenden's invention), Telefunken, and Stone to supply imitation receivers at a lower cost.[110] George Clark, one of Stone's technicians, offered his company's rationale for obliging the navy: "We were really making use of the . . . electrolytic detector, but since the U.S. Navy was making free use of it . . . we felt that we were violating no patent."[111] Within a year, the navy itself began assembling the receiver.

Fessenden, who after a year of courting military officials thought he was "on good terms with them," was outraged. He knew his apparatus was more expensive than that of the Germans, who received government support. Yet he was infuriated at the navy's seeming failure to understand research and development costs and to respect patents. For more than two years he wrote bitter letters of complaint to the bureau.[112] Fessenden notified the Navy Department that the navy was buying pirated apparatus and that such transactions were illegal. He advised the officials that "one who does not own property is not allowed to sell it cheaper than the rightful owner."[113] The secretary of the navy, William Moody, found Fessenden's letters so "extreme" in their tone that he began making inquiries about Fessenden's character.[114] Meanwhile, since he had no basis on which to judge the inventor's claims, the secretary suggested that Fessenden prove he was, in fact, the rightful inventor and owner of the detector. If the courts upheld Fessenden, then the navy would consider his protest carefully. Fessenden filed suit against De Forest and Telefunken. After Fessenden's first victory over De Forest in 1905, Moody's successor, Charles Bonaparte, advised Fessenden that the victory was not conclusive, and that the department still felt free to buy from De Forest. Fessenden won three more consecutive decisions against De Forest and considered his victories quite "conclusive." Secretary Bonaparte, however, now dismissed the importance of the patent suits and informed Fessenden that the navy felt "relieved of any moral obligation" to honor Fessenden's claim because his prices were still too high.[115] The navy continued to send orders to De Forest, who continued to fill them despite the decisions of the court. Fessenden's only alternative was to obtain an injunction and contempt of court citation against De Forest and his backer, Abraham White. He did, and their bail was set at ten thousand dollars. In addition, they now owed sixteen thousand dollars in fines and risked going to jail if they continued making and selling the detector.[116] In 1905, Fessenden also won his patent suit against Telefunken.

Fessenden may have expected infringement from a competitor, but he was truly offended and disillusioned to have his inventions appropriated by the government. Throughout 1906, the navy tried to persuade Fessenden's competitors to sell the electrolytic detectors by stating that the attorney general had decided that the government "could use the liquid barretter on account of Fessenden's contract with the Weather Bureau." Fessenden continued to issue complaints to the secretary of the navy, the attorney general, and the president of the United States. He demanded that Secretary Bonaparte be impeached for knowingly buying stolen property. He wrote to the president, saying, "This manufacture of apparatus by the departments is particularly objectionable because the wireless companies threw open their stations to the officers of these departments and gave them the fullest information possible with the distinct understanding that the information so given was to be treated as confidential."[117]

To inventors, patents were central: patents established priority in scientific and technical circles, in history books, and in the courtroom; they could ensure an inventor's prestige and fortunes. With so much riding on them, patents were considered inviolate by their owners. The navy, on the other hand, believed it could not be constrained by patents; in fact, the navy considered itself under no legal obligation to recognize patents. Amid all the press releases, claims, charges, and countercharges, how could the navy tell who the legitimate patent holder was? The navy's policy was to acquire apparatus "independently of patents."[118] The military argued that it was unable to determine priority and could not serve as "a court for the settlement of disputed claims as to inventions."[119] Barber advised the Bureau of Equipment that he doubted whether anyone truly had a defensible patent on a wireless telegraph system. He did not think Fessenden, who was threatening to sue the government for back royalties, should be taken seriously: "I doubt if any of the present owners of wireless telegraph patents will ever do anything more than they have done in serving these preliminary notices."[120] Fessenden's threats against the navy were empty; the government at this time could not be sued for using patents without permission.

Throughout 1903 and 1904, Fessenden and his backers, Given and Walker, had disagreed about how to negotiate with the navy. Despite the navy's testing and contract terms, Fessenden had favored accommodating the navy and offering it NESCO apparatus at reduced prices to initiate what he hoped would be a close and lasting relationship between the two organizations. Walker and Given, who wanted more immediate returns on their investment and who were becoming increasingly exasperated with the navy's procedures, favored maintaining their prices—by early 1905, $12,500 per set.[121] By late 1904 and early 1905, with the navy both buying from competitors and making the electrolytic detector itself, Fessenden, too, had had enough. In fact, NESCO's experience with the navy no doubt contributed to Given and Walker's determination not to sell apparatus to anyone, but instead to sell the entire system. By 1905, NESCO refused to have any further dealings with the navy. "If we do not communicate any more of our inventions to the government," wrote Fessenden, "the government cannot steal them."[122] The entire affair left him angry, bitter, and defensive.

The navy's efforts to circumvent Fessenden's patents and make use of his detector represented, to the inventor, not just a legal outrage, but a technological affront, as well. The navy was indicating that it wanted to buy only components, not entire wireless systems. The debate over whether wireless was a system, and whether different wireless systems existed, provided another source of controversy between several of the inventors and the navy. Since 1900, the Marconi Company's strategy had been to market wireless as a complete system or network. The company would erect the shore stations, equip the ships, and establish channels for communication. Other companies tried to follow suit. Although this systems policy was motivated primarily by business considerations, technical considerations played an important part, as well.

In each competing wireless set, the various components were carefully engineered and adjusted with the efficient operation of the entire system in mind. From the number of turns in the induction coil, to the type and number of condensers, to the aerial arrangement, all the interconnections were designed to meet the system's special needs. Chances were excellent that rival apparatus would not integrate well into a competing system and would cause poor performance. For example, a very sensitive and reliable detector that was connected to incompatible or second-rate headphones would function below its capabilities. No inventor could allow alien and possibly inferior components to discredit his system or the merits of wireless. Inventors were trying to protect their business, but they also took pride in the distinctiveness of their apparatus and recoiled at the thought of it being dismantled and recombined with competitors' devices.

The navy preferred to regard wireless components as individual inventions like telephones or light bulbs. The navy considered the inventors' systems rationale nothing more than a justification for monopoly, but by 1904 it recognized that control over the technical system brought control over the airwaves. Consequently, the navy determined to buy components and establish its own "composite" system instead of buying any of the competing systems being offered by the inventors. As the chief of the bureau advised Barber in 1902: "It is proposed to conduct tests of composite sets, made up of portions supplied by different makers and such a combination may be adopted as standard for the service in case it is found to work better than an entire set supplied by a single maker."[123]

The Bureau of Equipment, which did not think civilian "square-heads" were attuned to the needs of the navy, and which may have wanted, out of pride, to develop its own system, no doubt sought to achieve standardization through the composite route. The bureau was also trying to reduce technical uncertainty. If it mastered the components and designed its own system, it could better anticipate or avoid overly rapid technological turnover. Certainly, if the navy intended to gain hegemony in America's airwaves, it had to believe that it, and not a civilian, controlled the technology that provided access. But acquiring various components and implementing a composite system were two very different processes. As Walker complained in 1905, "The government gets a kind of a hotch-potch of a system that is not the best and is no credit to anyone."[124] Nevertheless, the navy began acquiring apparatus, assembling its "hotchpotch" system, and erecting stations, primarily at navy yards and lighthouses around the country.

One of the areas the navy targeted for wireless was the Caribbean. Mahan and his disciple Roosevelt were convinced of the strategic importance of the Caribbean area, and American expansion in the region had been dramatic in the late 1890s and early 1900s. By 1904, coaling stations were established, Cuba and Puerto Rico were under American control, and plans for the Panama Canal were well underway; the necessity of a communications network linking the various American outposts was clear. The navy, no doubt eager to buttress its bargaining position on the Roosevelt Board, began entertaining bids for four high-power stations in Key West, Puerto Rico, Cuba, and the Canal Zone in the spring of 1904. The contracting company was to guarantee the ability "to maintain at all times communication and under all atmospheric conditions between stations 1000 miles apart." The navy specified that the contractor was to complete the installations within six months of the date of the contract. The navy would provide the power, the aerials, and the buildings, and it was responsible for transporting the equipment to all the locations except Florida.[125]

De Forest, citing his overland successes and gold medals from the St. Louis World's Fair, assured naval officials he could cover the great distances required, and submitted a bid for $65,000. Fessenden bid $324,000. De Forest won the contract, which was signed in late June of 1904, just as the Roosevelt Board was about to convene. In addition to providing various guarantees, De Forest had to put up a bond of more than $16,000 which would be forfeited if he did not complete the stations on time and meet all the other terms of the contract.[126] De Forest would not be paid until all the stations were working satisfactorily. While the navy's desire to ensure good performance was certainly understandable, these were very stringent requirements to impose on a small company erecting radio stations far away from its base of operations and sources of supply. The territory was unknown to De Forest, and there were more and more reports that static was particularly relentless in the tropical regions. No one, anywhere, was maintaining "at all times" communication a distance of one thousand miles. Even Barber questioned the navy's specifications: "When I said some time ago that I did not think that his contract with the Department was legal, I meant that if it came into court, the court would decide against the Department."[127]

De Forest began working on the Caribbean stations in January 1905. His working conditions were extremely unpleasant and difficult; he described the Cuban station as the "hellhole of wireless." Cyclones, lightning, gales, and earthquakes often destroyed the recently completed stations and aerials.[128] In addition, the navy was slow in sending equipment and supplies, and De Forest could not meet the six-month deadline. He complained about the "delays of months" and the "breakdowns of Navy apparatus," and he was demoralized by the "hostility, open or concealed, on the part of officials, from whom [he] had every reason to expect cooperation and interest." He warned his attorney that the navy might charge that De Forest had fallen down on the contract, and said that the company should protest this charge. Revealing his antimilitary prejudices, De Forest wrote: "If the Navy, through their cheap outfits and red tape[,] delay our success, we will not let their still cheaper officers with more gold tape than brains throw the hooks into us."[129] Abraham White convinced the Navy Department that his company should not be held responsible for the delays, and by early 1906, all the stations were completed. They were, however, failures. De Forest had difficulty maintaining a transmission range of two hundred miles at night, and during the day transmission was usually impossible.[130] These stations did not perform according to the contract specifications and could not have improved the navy's opinion of wireless. About one year later, George H. Clark, then an employee of Stone Telegraph and Telephone, went to experiment with and make improvements on the navy's New Orleans and Pensacola stations. Clark was only allowed to experiment during the day, when static was at its worst. He could not test the system at night, because the lighting system of the navy yards were powered by the same mains that fed the wireless transmitters, and if Clark transmitted at night, the lights in the commandant's home flickered and dimmed.[131] Like De Forest, Clark was caught between one faction in the navy that wanted wireless improved, and a much larger group of officers who were not about to be inconvenienced to achieve such a seemingly farfetched goal.

By 1906, then, certain naval officers were determined to establish an extensive wireless network along America's coasts and on its new possessions which would be capable of exchanging messages with the ships of the new navy. The apparatus installed in all these stations, while consisting of components invented by civilians, would be arranged and assembled according to naval needs and specifications. These naval officers—Barber, Bradford, and the handful of other men interested in wireless—were not opposed to a monopoly of wireless, they were only opposed to a civilian monopoly. They came to believe strongly that, like the European governments, the American government was entitled to control its nation's wireless; furthermore, they believed that the navy should be the agent of that control. The U.S. Navy had acquired a significant amount of new property in the past ten years: new ships, coaling stations, and bases. Like any organization, it was proud of its new role and influence and jealous of its latest acquisitions. As the naval officers directly involved in the acquisition and international regulation of wireless came to share the European perception that the ether represented a territory of national importance, they wanted to acquire dominion over this possession, as well. Ironically, they wanted what they condemned Marconi for pursuing: a monopoly of the airwaves.

While certain navy men hoped to thwart Marconi's corporate ambitions and to establish a strong American military presence in the ether, many others remained completely indifferent or hostile to such goals. A review of how the navy first tried to utilize the new invention illustrates how individuals and the organizational structure of which they were part compromised the usefulness of wireless. When the first twenty Slaby-Arco sets were ordered from Germany, there were no naval engineers who knew how to install them properly. In the summer of 1903, there were only eight enlisted men capable of taking charge of a station.[132] There were no wireless operators on board ship. And few commanders welcomed the apparatus. As L. S. Howeth has written, "No serious effort was made by the various commanders to organize, utilize, or supervise radio communication within the fleet."[133] These men, especially out at sea, enjoyed complete control of their ships, and they did not want that authority subverted by wireless. This invention threatened to render their leadership merely titular. As George Clark observed, "The traditional power of a commanding officer to do as he felt best with his ship or command as soon as he got out of sight of land would have been completely wiped out if someone in the Bureau of Navigation or elsewhere could give him orders. So often the instructions to the wireless room were to shut down the wireless and not acknowledge calls from shore at all."[134]

Flag lieutenants were to supervise wireless on board ship, but they knew nothing about the equipment and had no incentive to learn. Wireless was installed below decks to protect the apparatus from the rigors of battle. Did that mean that the flag lieutenant would be consigned to a remote cabin, away from the captain and the action on the bridge? This prospect was hardly appealing and was quite naturally opposed. One flag lieutenant, T. P. Magruder, while inspecting a new installation on his ship, objected to the "unsymmetrical appearance" the antenna wires and guys produced and ordered the lines and wires realigned to parallel the rest of the ship's rigging. The new arrangement significantly reduced the efficiency of the apparatus. When it was suggested that the new arrangement rendered the sets nearly useless, Magruder said he "didn't give a damn about wireless . . . but he did give a damn for the appearance of the ship."[135]

The performance of wireless was also affected by the ability of the enlisted men and the quality of the facilities available for maintenance and repair of the apparatus. Lieutenant J. M. Hudgins, who had helped Barber investigate European apparatus, complained to the secretary in 1904: "We are not getting one-half the service possible out of the apparatus in use, owing to the lack of skilled operators." He warned that few of the men assigned to take charge of the navy's new stations were qualified for such duty, particularly since they had no experience adjusting or making quick repairs to the sets.[136] Strong criticism of the operators' general incompetence came from both civilian and military quarters and persisted for ten years.[137]

The navy's methods of installing and maintaining wireless also undercut the value of the equipment it acquired. Wireless was installed aboard ships while they were docked at the New York Navy Yard or the Washington Navy Yard. The apparatus theoretically could be repaired at any navy yard. The yards were also the sites for navy shore stations. The nature of the work and supervision at the yards did not promise to provide wireless with a favorable environment, however. Administration of the navy yards epitomized the department's decentralized structure and management. Although nominally controlled by the Bureau of Yards and Docks, the yards contained offices and staffs affiliated with and loyal to the other bureaus. Predictably, this led to confusion and waste.

For example, several different engineering departments and machine shops, each working for a different bureau, were dispersed throughout a yard. This arrangement militated against a concentration of effort and a sharing of expertise.[138] The inexperienced operators charged with installing and repairing wireless would have carried out their duties more efficiently had they been part of a unified engineering department at the yard. Under the existing arrangement, they had little supervision and often found themselves caught between conflicting orders, one set from the Bureau of Equipment, another from the commandant of the yard.[139] No technical standardization or uniformity existed between navy yards; disregarding whatever standard plans the bureau may have tried to issue, each navy yard pursued its own method of wireless installation and repair.[140]

Exacerbating this lack of continuity and fragmentation was what the navy called its composite system of wireless. *Composite* did not mean that the navy used only one kind of transmitter or one kind of receiver connected according to standard specifications. The navy concocted these systems from whatever components were available at the time at the lowest price, and left it to the operators to place them together. This encouraged untrained and inexperienced men to tinker with the apparatus and to conduct their own trial-and-error experiments. The use of the composite system also meant that, often, an operator transferred from one yard to another or from one ship to another "had to learn an entirely different run of wiring and placement of apparatus."[141] The composite system and the independence of each navy yard and of each wireless station led to a proliferation of different wireless sets throughout the service. The chief of the Bureau of Equipment in 1907 described the costs resulting from lack of supervision and standardization: "Certain operators when first ordered to a station, and who were perhaps familiar with other systems, would not use that provided but improvised systems of their own. The original instruments would thus fall into disuse and deteriorate, and when these operators were detached they would take away the improvised instruments. The stations would thus remain inefficient for a considerable period and in some cases could hardly be operated at all until new instruments were provided."[142] Inventors were exasperated by the situation, which they believed caused their apparatus to be abused.[143] One company claimed that some apparatus it had loaned to the navy was in such poor condition on its return the company had to discard it "as a lot of junk."[144]

Some navy yards, particularly those on the West Coast, complained of hand-me-down equipment and unsuitable facilities. Once a ship or station was equipped, little effort was made to update its apparatus. The commandant of the Mare Island Navy Yard suggested in 1904 that the yard's wireless station be moved from the deteriorating pigeon coop in which it had first been installed.[145] Six years later, it was too dangerous for operators to work in the wireless building, the building was so decrepit and leaky.[146] Inhospitable conditions existed in wireless stations in navy yards throughout the country, and as the years passed, conditions worsened. By 1909, the Bureau of Equipment was receiving reports from dozens of navy yards criticizing the barely functioning, obsolete, and poorly maintained wireless sets at the shore stations.[147]

Indeed, as late as 1912, wireless was passed down the naval hierarchy until it was housed in the least desirable facilities and used by the men with the least power and responsibility, the enlisted men. Wireless had reached an organizational dead end. Not until 1912 would critical realignments take place which would promote the invention's integration into naval structure. While certain naval officers argued for American military control of wireless similar to that enjoyed by European military organizations, their arguments before Congress were undercut by frequent reports of the poor performance of military operators and apparatus. Several officers continued to lobby for such control, and their effort intensified after the second International Wireless Conference, which took place in Berlin in October 1906.

Because nothing had been solved at the first International Wireless Conference, and because the use of wireless during the Russo-Japanese War had generated new diplomatic problems, the Germans called a second conference. In September, just before the meeting, the Institute of International Law in Ghent, Belgium, adopted rules governing the use of wireless during war. Captured wireless operators were to be treated as prisoners of war rather than as spies. Neutral ships and balloons that had been used to furnish an adversary with information helpful in the conduct of hostilities could be removed from the zone of hostilities and the aggrieved government could seize any wireless apparatus found on board. A neutral state had the right to close or take over the wireless station of a belligerent operating in the neutral state's territory.[148] The adoption of these regulations, which helped set the tone for the conference, provided official recognition that wireless was an important weapon and that certain transmitters, innocent though they might be, could be restrained by the different warring states. Most significantly, the regulations established that each nation had its own territorial radius in the ether and that violation of this invisible realm was as unacceptable as any other incursion.

In Berlin, the twenty-seven countries attending the second conference faced old and new dilemmas. What constituted a satisfactory arrangement between private wireless companies and government stations during war and during peace? Should all seacoast wireless stations be tuned to one or two different wave bands so ships could quickly locate a shore station in an emergency? Should standard and separate international wavelengths for mercantile vessels and navy vessels be established in order to avoid friction and ensure safety? The different nations had to agree on a universal distress call. They also had to decide whether an international wireless bureau such as that proposed by Germany could fairly and equitably arbitrate international disputes.[149]

Yet, as in 1903, one issue dominated: Was the Marconi Company entitled to communicate only with its own stations and with no others? Editorials had been appearing in the American press for months insisting that life and property depended on free intercommunications. Diplomatic pressure to thwart the Marconi Company's monopolistic practices had also been intense since the 1903 conference. The Marconi Company's wireless station on the Nantucket lightship had come under attack with increasing frequency because of the station's refusal to exchange messages with those operating rival systems. After the *Deutschland* incident, the German ambassador in Washington, Baron von Sternberg, registered complaints about the exclusionary practices of this important station and charged that the efforts of the Marconi Company to establish a worldwide monopoly resulted in "most serious injury to the interests of German shipping and commerce."[150] As a result, in November 1904, American diplomatic officials had ordered the Nantucket station to exchange messages with all systems. Attorneys for American Marconi answered that the company refused to comply, insisting that such a proviso was unnecessary because there was "not a single transatlantic liner equipped with apparatus other than Marconi."[151] However, American officials feared a British monopoly, as well, and they directed the Marconi Company to remove its wireless system from the lightship. The navy took over the installation, equipped it with Telefunken apparatus, and announced that the station would be available for commercial use with any other wireless system.

Three of the four American delegates in Berlin were military men, two from the navy and one from the army, and over time the sentiments of these men had become increasingly anti-Marconi. Marconi was expected to have few allies at the conference. Barber, anticipating the tone of the meetings, predicted that most of the delegates would be government officials who had "small consideration for the private businessman."[152] When the American delegates arrived in Berlin, they quickly endorsed the mandatory ship-to-shore intercommunication resolution and submitted a motion for compulsory ship-to-ship intercommunication. Participants and correspondents expected this to be Germany's first motion, and America's preemptive strike reportedly "came as a 'regular bombshell.' "[153] Few had anticipated that the Americans would assume an activist role. Britain offered a compromise, agreeing to accept compulsory intercommunication between ship and ship on matters pertaining to navigation alone. But, as the *New York Times* reported, the "United States delegates declined to agree to the compromise, affirming that they were willing to stand or fall on the principle of intercommunication."[154] They were joined in this resolve by the German delegates and all the others except those from Britain, Italy, and Japan. Delegates from these three countries were often deliberately ignored by the others during the conference to illustrate what might happen on the high seas to countries that refused to abide by the free intercommunication policy. The German representatives used tactics that were less subtle: they threatened that their new technology enabled them to destroy every message in the air and wipe out all their rivals' transmissions, and that German operators might resort to this tactic if necessary.[155]

The compromise that emerged required every public shore station "to exchange wireless communication with each and every wirelessly equipped ship, and vice-versa, without regard to the system of wireless telegraphy used by either."[156] Under this agreement, ships at sea could always communicate with a shore station, and no ship would be rendered incommunicado, as the *Deutschland* had been. To mollify the Marconi Company, a schedule of charges per word, over and above the regular land charges, was established at the convention. Barber reported with obvious satisfaction: "The Marconi monopoly is not dead: but it is mortally wounded."[157]

At this point the 1906 conference began tackling other substantive issues that the 1903 conference had left unresolved. With the intercommunication issue settled in its favor, the German delegation, which represented German military interests, worked to codify other features of wireless communication to bolster military control. The German efforts were supported by the Americans, who had gained influence and respect at the conference in part because they appeared to have no financial or political interest in the outcome. Like the Germans, however, the American delegates wanted to secure, through law, military priority in the ether. To that end they supported the revolutionary German proposal that the ether be divided into regions by wave lengths, with the military getting the largest and the best tracts.

At this stage in wireless communications, in the early 1900s, experimenters used wavelengths of approximately 300 to 1,600 meters. Marconi continued to move toward using longer wavelengths, which he believed traveled farther and encountered less congestion. The Germans recommended a range of 600 to 1,600 meters for naval and government use and 300 meters for merchant ships and commercial stations. Commercial stations could use other wavelengths, but none exceeding 600 meters. Not surprisingly, the Marconi Company strongly opposed the assignment of these shorter wavelengths to private stations. At first reading, the allocation appeared to be a reasonable initial step toward dividing the spectrum into more manageable zones and preventing interference between government and commercial stations. However, the various government stations had use of the longer, more desirable wavelengths, which traveled farther. In England, except for naval stations, most of the ships and shore stations were equipped and operated by the Marconi Company, whereas in Germany, all the stations were government owned and operated. This apparently impartial assignment of 300 to 600 meters to the commercial stations was an attempt by the German military to relegate Marconi and most of the major British stations to the inferior portion of the spectrum. The American delegates supported this allocation, hoping it would help ease the U.S. Navy into a preeminent position in American wireless: the navy hoped to gain through regulation what it had failed to achieve technically.

The delegates worked out other regulations. All ship stations were to be licensed by the country under whose flag they sailed. Shipboard operators were to be licensed after having passed an examination on signaling and apparatus construction and operation. These operators had to be able to transmit at a speed of at least twelve words per minute. Each ship would take a three-letter call number assigned by its government. The delegates affirmed that distress messages had priority over all others, as did certain government messages relating to navigation information and weather conditions at sea. Wireless operators had to take an oath of secrecy which bound them to protect the privacy of wireless messages.[158]

Few of these resolutions were achieved without German-British friction. While the debate over compulsory intercommunication received the most publicity, the delegates from the two rival countries also bickered over the less weighty details, even over the selection of an international distress code. Britain preferred its own CQ (supposedly from "seek you"), but agreed to add a *d* to the end to reduce the possibility of error. The Germans insisted on SOE, their distress call. However, because the letter *e* was only one dot, it could conceivably get lost during transmission. The delegates finally agreed on SOS as the distress code, and by 1908 most of the participating nations had adopted it. In his first transatlantic tests, Marconi had decided to use the letter *s* because it was easy to send and to decipher. SOS was settled on for this reason, and not, as the popular press liked to suggest, because it meant "save our souls."

Although the deadline for ratification of the treaty that emerged from the conference was July 1908, most of the participating countries, including even England, ratified the treaty within a year and a half after the end of the conference. The British delegates had been stubborn during the 1906 negotiations, and anti-ratification sentiment in Great Britain was strong. British newspapers were filled with articles and anonymous letters (some obviously from the Marconi Company) denouncing the proposals and generating "a prejudice . . . against the Kaiser's proposal."[159] The Marconi Company charged that because the treaty made intercommunication between various systems obligatory, it represented "an enforced partnership to which the Marconi Companies contribute everything and the German manufacturers of wireless apparatus nothing, neither invention nor capital, nor skillful enterprise. In short, the German company proposed to obtain artificially, through international legislation, the advantage of the position obtained by the Marconi Company in open competition and by private effort at private expense."[160] Marconi and his representatives lobbied fervently against the treaty in the editorial pages and in Parliament. However, British government officials determined that not signing the 1906 treaty would be economically and strategically unwise. Bowing to international pressure, government officials negotiated with the Marconi Company and agreed to compensate the company, through a three-year subsidy, for any loss it might suffer as a result of the international agreement. After an impassioned debate, Parliament ratified the agreement by a margin of one vote.[161]

Because the American delegates had been so outspoken and influential in molding the treaty, its ratification by the United States was, to the Europeans, a foregone conclusion. The Europeans, however, were mistaken. The American delegates had to sell the resolutions to a Congress and a country that were ill disposed toward government, and especially military, control of private industry; furthermore, the U.S. Congress was not interested in expanding the new navy's influence. These were also proposals sanctioned by the Kaiser, whose image in the American press had deteriorated markedly during the first decade of the century. Although the Kaiser was a great fan of Roosevelt's, and often referred to him as the greatest American president who ever lived, the press mistrusted the Kaiser's motives and described him as the dictator or chief of police of Europe.[162] One popular magazine, *Lippincott's*, even published an article titled "Is Kaiser Wilhelm II of Normal Mind?"[163] The Kaiser's relentless militarism and imperious manner prompted many to view him, and any treaties he might endorse, with suspicion.

The 1906 treaty represented a European approach to resolving conflict in the ether. It was spawned by European suspicions and rivalries, which were exacerbated by the geographical proximity of the contestants. It embodied the assumption that the ether was a strategically vital territory to be cordoned off, a territory best patrolled by military authorities. It imposed on "the air" military perimeters beyond which civilians were not to trespass. These assumptions and solutions may have served European needs, but they did not find a very receptive audience in America. The United States was not surrounded by rival nations whose wireless transmissions were a source of annoyance or paranoia. Our ether was as open as the West had once been. Hence, the opposition to cordoning it off prematurely and making the navy its custodian was opposition to be reckoned with.

• • •

THE POWER OF MARCONI'S assumption—that the ether could be monopolized for corporate profit and British imperial desires—was evident from the quick and vehement response that assumption provoked. As the U.S. Navy and its European counterparts constructed official reactions to Marconi's business policies, the ideological battle lines about how wireless should be used were drawn and redrawn. Also emerging were starkly competing notions about whether the ether was primarily private or government property. That it was property of some sort, despite the fact that it could not be seen, touched, or measured, was becoming clear. Marconi conceived of the ether as a resource he could monopolize, whereas military men regarded the airwaves as a part of their government's territory which they had a duty, and a right, to occupy and protect. The contest, one that was to persist, was between a capitalist and a military mindset.

To the press, the choice was clear. Newspapers and magazines stood to benefit financially from the private cultivation of wireless; the last thing they wanted was for the government to gain control of such an invention and restrict its benefits to military applications. It is not surprising, then, that the press condemned the Roosevelt board's proposals for a de facto navy takeover of American wireless. Allied with the press were the independent inventors, who had as yet received little patronage from the navy, and whose hopes for financial success would be dashed by naval control. The economic self-interest of both the inventors and the press is evident; but, as ardent believers in the connections between private enterprise and progress, both groups, in truth, seriously doubted whether the navy could promote technological advances.

As foreign governments and the U.S. Navy worked out their relationships with wireless inventors between 1899 and 1906, important precedents were set. The international scope of the dilemma of managing the ether was recognized, and Western countries settled on the mechanism of the international conference to arbitrate competing claims. In the United States, the press came out squarely against any government control of America's wireless systems, and, more importantly, the U.S. Navy cultivated a strong negative reaction against the Marconi Company, a reaction that in the years ahead would shape the destiny of radio in America.

CHAPTER FIVE

INVENTORS AS ENTREPRENEURS
Success and Failure in the Wireless Business

1906–1912

IF ONE WERE TO SCAN the newspaper headlines between 1906 and 1911, and base an assessment of wireless telegraphy's technical and corporate progress on the coverage the invention and its promoters received, one would have to conclude that this was a period of little accomplishment. Such a conclusion would be quite mistaken, however, for during this period, when the press became more critical of the invention, when journalistic visions of its applications were more circumscribed, and when the press ignored fundamental developments in wireless design and management, major technical breakthroughs occurred which would lay the groundwork for radio broadcasting. Also ignored was the manner in which the various inventors linked their technical work to their business strategies between 1906 and 1911, making decisions that determined which companies gained—and which companies lost—control of radio technology.

These were not insignificant developments. Yet, changes in wireless apparatus and in the everyday behind-the-scenes operations of the fledgling wireless firms were not, according to prevailing journalistic conventions, big news stories. They did not take place in public settings, they did not involve "the people": they lacked human interest. In its constant search for the new, the unusual, the romantic, and the dramatic, the press, when it did cover wireless during this period, focused on stories of shipwrecks in which wireless saved lives, and on the growing group of "amateur operators" who adopted wireless sending and receiving as a hobby. The journalistic bias toward staged, public demonstrations, and toward framing technical change in terms of how it immediately affected the lives of middle-class consumers, meant that incremental technical and managerial changes would not receive much attention. This bias was reinforced by a newfound wariness of the claims of wireless inventors. The same press that, in 1899, had painted such flamboyant images of wireless telegraphy's promise now took the inventors to task for failing to turn prediction into reality.

In fact, by 1906, after seven years of increasingly intense commercial and technical development, the prospects for wireless telegraphy in America seemed gloomy when compared with the hopeful outlook of 1899. As with earlier inventions, such as Edison's phonograph and light bulb, public expectations as shaped by the press had outdistanced actual achievement.[1] The inventors had helped paint the visions of the future of wireless, but as yet the inventors had left these visions unfulfilled. Marconi's transatlantic wireless service, which was supposed to bring Europe and America closer together by dramatically reducing the cost of transoceanic communication, and which seemed imminent in 1902, was not yet established. Although Marconi still garnered occasional front-page headlines, as when he sent a message to the *New York Tribune* from the middle of the Atlantic Ocean in 1904, none of his public successes could compare with his debut at the yacht races or the transatlantic achievements. Instead of promoting world peace, wireless had exacerbated the prevailing xenophobia and was now the subject of international debate and rivalry. Furthermore, its potential for keeping the peace remained unproven, whereas it had already been used as a strategic weapon in the Russo-Japanese War.

The invention's performance also led to disappointment. The press had envisioned that wireless would free reporters and other people from dependence on the existing wired networks of the telegraph and telephone: with wireless, people could send messages whenever and to whomever they wanted without going through Western Union or Bell Telephone. Reporters had initially suggested, based on Marconi's own assessments, that there were enough "waves" and "tunes" available in the ether to allow plenty of room for everyone.

Ray Stannard Baker, writing for *McClure's* in 1902, predicted that a range of users, including "great telegraph companies," "important governments of the world," and "the great banking and business houses, or even families and friends," would "each have its own wireless system with its own secret tune."[2] This had not come to pass. Instead, the diplomatic and commercial contests over the transmission of wireless messages indicated that there was room in the ether not for many, but only for a few. The spark transmitters in use were still crude, sending out broad-banded waves that made very inefficient use of the spectrum and produced considerable interference. In addition, all the users agreed that some portions of the spectrum were more desirable than others. These wavelengths, 600 to 3,000 meters, were the ones most operators preferred, and the ones the Marconi Company and its opponents fought over. In the United States, with the navy and the American companies vying for a foothold in the ether, interference worsened, as did antagonism over who had priority when and where.

In the press and among potential clients of wireless, the message to inventors was clear: live up to your promises by giving customers equipment that will allow them to use the ether without having to compete for access or to contend with interference all the time. Between 1906 and 1911, as Fessenden, De Forest, and Stone sought to refine their apparatus and consolidate their businesses, they all confronted increased skepticism in the press, and ambiguous, unsettling reactions to wireless in the marketplace. Wireless was still intriguing, but now there were questions about how seriously the invention should be taken.

Such questions had been raised several years earlier in the technical press. *Electrical World* in 1902 had lectured impatiently about the need for tuning: "Truth to tell, it is about time for syntonic working to fish, cut bait or go ashore. . . . If the rival wireless systems can really avoid interference with each other and live like Christians in peace and concord, it is high time that they did it."[3] One year later, the *London Electrician,* in a sarcastic editorial about the state of the art, imagined that wireless was in use and then the cable was invented: "With what rapturous delight would that new invention be hailed the world over! The electric cable—that takes the message straight to its destination, and does not allow it to be scattered in all directions—the speedy, certain, secret electric cable! What a marvelous improvement upon the ether wave!"[4] Now the popular press, which had at first been more enthusiastic, even credulous, about wireless, also began to emphasize the invention's drawbacks. The coverage Marconi received when he finally established his transatlantic wireless service in 1907 reflected the new caution.

On October 17, 1907, the front-page headlines of the *New York Times* proclaimed "Wireless Joins Two Worlds." After six years of work, Marconi and his assistants had established a daily, 8:00 A.M. to 8:00 P.M., transatlantic wireless service, which the *Times* hailed as a monumental achievement. Transmission speed at such powerful stations was still a slow twenty words a minute, because the telegraph keys were huge and difficult to manipulate, and the operators could only receive one message at a time. But the Marconi Company charged the press five cents a word and all others ten cents a word, as compared with the cable companies' fee of twenty-five cents.[5] This dramatic savings prompted the *New York Times,* Marconi's faithful booster, to compare once again Marconi's progressive approach and the arrogance of the cable companies.

> There has been no reduction in the cost to the public of cable communication for the past score of years. This has not only been a distinct hindrance to the development of business, but it has been a hindrance to that improvement in the relations of nations to each other. The cable companies have been as incapable of improvement as the Martian canals, and were managed with about as much reference to the needs and wishes of the population on earth.[6]

The *Times* contracted with Marconi for a regular wireless press service, and the Marconi Company reciprocated by arranging for the first westward transatlantic message to be sent to the *Times* from its London correspondent. Beginning in October of 1907, the paper published a special and exclusive "Marconi Transatlantic Wireless Dispatches" section every Sunday.[7]

Except in the *Times,* however, press accounts of the achievement were more guarded, and even more critical, than previously. The *New York Tribune,* which also gave Marconi front-page coverage under the headlines "Wireless Messages Sent across Ocean" and "Marconi System in Successful Operation," balanced this story with a cautionary editorial. The story quoted John Bottomley as saying "The system as established is an absolute success, all the rumors about interference with and interception of messages are rot, for under our system the 'tuning' is such that interception is practically impossible." The paper's editors remained unconvinced, and instead of celebrating the beginning of a new era, they warned of the limitations still imposed by interference and the possible "stealing" of messages.[8]

The only two popular magazines featuring stories on the service were the *Outlook* and *World's Work.* The *Outlook* observed: "Mr. Marconi has certainly accomplished a wonderful achievement; but there are two problems which remain to be solved before long-distance wireless telegraphy will realize its greatest usefulness." Those problems were the slow speed of transmission and the fact that messages could still be overheard or stolen.[9] *World's Work* inadvertently suggested one reason the democratic visions of 1901 were not repeated. Marconi's accomplishment, the magazine opined, was "the opening of a wireless 'line' to the business of the world," making "transatlantic wireless a servant of commerce."[10] According to the technical press, the actual rate of transmission was between three and seven words per minute, because most messages had to be repeated, some at least six times. Eastbound messages were, at best, "occasional," and the service was "a case of magnificent promises and poor results."[11] This atmosphere of criticism, and demands for better performance, exerted pressure on wireless inventors to develop apparatus that would make more efficient use of the spectrum.

The lack of clients also put pressure on the inventors. Wireless was not being used as much as the inventors and the press had hoped: it was still considered a luxury, not a necessity. In 1906, the Marconi Company's gross earnings were $55,170. *Electrical World* calculated that this figure represented 25,823 messages per year, 70 messages per day, or 1 message per day per ship. The editors found this less than encouraging: "It is difficult to figure out much net revenue after the salaries of officers and operators have been paid."[12] The Americans, however, could not even count on these paltry revenues. The one clear market for wireless telegraphy was signaling over water, a market the Americans had not yet cultivated very successfully.

The economic climate was made even less hospitable to wireless inventors during this period by the panic of 1907. After what the editor of *Manufacturer's Record* had labeled "the most prosperous period in our history," when, according to the *Review of Reviews*, "everything seemed so safe and sound," America's economic boom came to an abrupt halt.[13] Referred to variously as the Rich Man's Panic, the Wall Street Panic, and even Roosevelt's Panic, the crisis that began in the New York financial community in October quickly spread to other American cities. Although the panic of 1907 was short-lived, it prompted suicides, paralyzed industry, created unemployment, and caused major banks and corporations to revamp their operations. All the leading magazines and newspapers ran lead stories on the panic, which was the primary topic of public discussion.[14] Overextended credit and scarcity of capital, reckless corporate speculation, and highly publicized government investigations into the business practices of the trolley, railroad, oil, and insurance industries all contributed to the crisis. Revelations of corporate mismanagement and corruption dampened the previously robust investment spirit, and various firms began having difficulties marketing their stocks and bonds. Public confidence wavered, and there were intermittent reports of runs on banks.[15]

The Mercantile National Bank of New York was forced to close its doors on Thursday, October 17, after depositors began withdrawing their funds. There was a run on the Knickerbocker Trust Company, also of New York; within three hours, eight million dollars were withdrawn. The company suspended further payments and was then declared insolvent.[16] "Within twenty-four hours," reported one observer, "almost every trust company in the city was under suspicion."[17] Runs on other banks occurred, first in New York and then in other major cities such as Pittsburgh and Chicago. As people hoarded their money, the currency shortage intensified, producing a "money famine."[18]

Once again assuming a leadership position, J. P. Morgan, in cooperation with other bankers, financiers, and the secretary of the treasury, coordinated the pooling of resources and the importation of European gold to shore up the financial community and forestall further damage. By February of 1908, the press could report that the panic was over, although the business depression it had triggered persisted.[19] While some members of the business community publicly blamed Roosevelt's trust-busting policies for the panic, most press accounts acknowledged that the irresponsible "gambling" and "rascality" among speculators had precipitated the fall. *Everybody's* published an article titled "Game Got Them; How the Great Wall Street Gambling Syndicate Fell into Its Own Trap." The panic directed considerable attention to the need for reforms in banking and in Wall Street practices, and illustrated how risky playing the stock market could be, even for those who allegedly knew the rules well. Magazine editorials cautioned readers not to buy on margin, and to buy only those stocks that were proven and reliable.

Wireless stocks were anything but reliable. Five months before the panic, *Success Magazine* had published an exposé on wireless stock, which emphasized the activities of De Forest's company. Advising readers that wireless was a bad investment, the magazine reported that "millions of dollars of wireless stock manufactured in the past eight years is to-day worth no more than the paper on which it is printed." "The most shameful chapter in the record of the prostitution of this great invention," the magazine contended, "deals with the network of the De Forest Companies promoted by Abraham White, a modern Colonel Sellers."[20] White's and De Forest's brand of promotion had undercut the credibility of all wireless firms; the panic seemed to cement the negative assessment in place. *World's Work,* in its December 1907 issue, reported that the very word *wireless* brought "a smile to the lips of Wall Street men." The magazine added:

> Wireless stocks, at large, are to be regarded by the public as little better than racetrack gambling. Most of these wireless telegraph stocks have been put through a long period of juggling, washing, manipulation, fraud and malfeasance that should effectively remove them, for good and all, from the field of investment. . . . Widows and orphans, poor men and parsons, all looked alike to the wireless fishermen who

spread their nets for the American public. Thousands of men and women in this country have already learned to curse the day Marconi made his first experiment.[21]

The 1907 panic, then, which prompted retrenchment and caution in the business community, and increased skepticism among investors in general, severely compromised the operations and experimentation of the wireless companies.

It was in these circumstances that Fessenden, De Forest, and Stone, as well as Marconi, sought to survive as wireless entrepreneurs. Each man faced conflicting technical and managerial requirements. On the one hand, wireless transmitters and receivers had to be refined so that waves were more defined, and the signaling more reliable. This work would require taking technical and financial risks. On the other hand, in the face of the economic recession and meager revenues, inventors were well advised to be cautious and conserve their resources. There were tensions, too, between individual creativity, which expanded technical and entrepreneurial possibilities, and the distinctive demands and opportunities of the American marketplace, which often exerted countervailing constraints. Navigating in these different tides was anything but easy.

■ ■ ■

IRONICALLY, IT WAS the man who advocated one-wavedness, John Stone Stone, who was the earliest commercial casualty of the shift to more defined waves. Although his four-circuit tuning and loose coupling had been significant contributions, Stone had not developed distinctive or competitive continuous wave transmitters and receivers. Stone's principal customer, the U.S. Navy, was adopting Telefunken's quenched spark system, which was technically incompatible with loose coupling. This shift in 1907–8 coincided with the financial panic, and the Stone Company quickly withdrew from the wireless field. Early in 1908 the technical staff was disbanded and a petition of bankruptcy filed. Lawrence Sherman, a trustee, tried unsuccessfully to interest Telefunken, Marconi, and NESCO in the Stone patents. The small scope of the Stone Company and its limited backing prevented the company from weathering these hard times and technical readjustments. Stone had had several promising marketing ideas, but none was settled on and implemented. For example, he had written in 1904, "The policy of the company which seems to me the most likely to be profitable in the future is one similar to that which has been so faithfully and successfully followed out by the American Bell Telephone Company, namely, that of licensing the manufacture of its apparatus and leasing the apparatus to operating companies." Stone did not, however, pursue this plan.[22] When business picked up in 1905 and 1906, he increased his payroll. Unfortunately, he did not hire an enterprising business manager. Nor did the company ever determine whether it was a manufacturing firm or a communications services firm, and it was unprepared to be both. In 1908, Stone returned to his consulting practice.

One of the Stone Company's officers reviewed the wireless situation and cynically described how he would run a wireless company, had he the chance to do it again. What a successful company needed was, "first, ample funds—and to command these [was] needed . . . a brazen 'pyrotechnic' exploiting of the gullible public. There was too much self respect, too much of earnest devotion to science, to enlist public interest, which *demands* that its hopes be raised to exuberance, before it will part with its money—not considering that *it* is paying for the golden visions held before it, in glowing advertisements."[23] But during this period, even glowing advertisements were not enough. Technical refinements protected and promoted by the right business strategy—these now were paramount. Stone had failed to settle on any particular strategy or to link technical developments with marketing goals. The field was thus left to Fessenden, De Forest, and Marconi.

Creative people respond differently to having to compromise their goals, to having to share their particular inspirations and visions with others less gifted and less personally invested in their dreams. Some become passive, resigned, or reclusive. Others become obstinate and combative; they fight back. So it was with Fessenden between 1906 and 1911. Believing that he had already made major compromises to suit his backers, compromises that interfered with his experimentation and that required him to fill too many roles at once, Fessenden became increasingly uncompromising and abrasive. He came to see every negotiation over every detail as a battle over preserving the autonomy and discretion he had left. His backers, who by 1905 had already invested half a million dollars in Fessenden's visions, stoked the embers of their own resentment, which Fessenden fanned with each new demand. The increased tensions within the company, which were exacerbated by external events such as the panic, left both sides feeling beleaguered and frustrated. Fessenden, Given, and Walker had never been able to agree on and pursue long-term business strategies, and the erosion of their superficial alliance during these years precluded the discovery of a remedy for the situation. They continued to pursue short-term projects that were sustained only through the first intoxicating flush of enthusiasm. When the endurance and determination necessary to sustain a strategy over years rather than months was not summoned, one short-term plan replaced another. A productive alliance can provide the sustenance a company needs, but such an alliance did not exist at NESCO, and this deficiency had major repercussions not only on the company, but also on how and by whom radio would be developed.

After Fessenden, his family, and his corps of assistants moved to Brant Rock, Massachusetts, in the summer of 1905, all work was devoted to establishing a transatlantic service. In the early winter of 1904–5, Fessenden had asked officials at the British embassy in Washington to help him obtain a license to build and operate a station in England. He no doubt hoped, if not expected, that his Canadian background and the fact that his wife was from Bermuda would make it easier for him to obtain such assistance. Meanwhile, NESCO's patent attorneys in London began

**Reginald Fessenden and his assistants at Brant Rock, Massachusetts.
Fessenden's cylindrical steel aerial is in the background.**

discussing the proposition with government officials there.[24] Because private wireless telegraphy in Great Britain had come under the supervision of the Post Office in 1904, the postmaster general could determine whether such a license would be granted and where such a station might be located. Although it is not clear what location Fessenden was hoping for in England, he was assigned a site on the northwestern coast of Scotland in a tiny town called Machrihanish on the Mull of Kintyre.[25] The location was even more remote and inaccessible than Marconi's site in Poldhu, Cornwall. Although not far from Glasgow as the crow flies, Machrihanish was at the end of a long, barely populated peninsula separated from the mainland by the Firth of Clyde. There was no railroad service and the roads were rudimentary. H. J. Glaubitz, NESCO's construction engineer, sailed for Scotland in the summer of 1905 to supervise the station's construction. Many of NESCO's men made the final leg of the journey in an open, horse-drawn cart.

Fessenden became increasingly caught up in the excitement and challenge of transatlantic work. He suggested to Walker that NESCO form a Canadian company; he believed such a company would easily obtain a license to operate between England and North America.[26] Meanwhile, Fessenden supervised the work at Brant Rock. While he waited for General Electric to deliver his high-frequency alternator, Fessenden used a rotating spark gap transmitter that was superior to stationary spark gaps and produced a high-pitched musical signal. He designed a new type of aerial that was supported by a tower made of steel tubing just wide enough for a person to climb up inside. His model was the smokestack, an easy-to-build structure that had proven to be durable and sturdy.[27] Fessenden insisted on steel because it was fireproof and a steel tower cost about four thousand dollars less than a wooden one.[28] The towers at Brant Rock and Machrihanish were completed on December 28, 1905. Neither of Fessenden's transatlantic stations was as large and powerful as Marconi's, but their construction was nearly as expensive and certainly as difficult.

All of this work was shrouded in great secrecy, as everyone in NESCO feared espionage by other companies.[29] Walker warned Fessenden: "De Forest and other obnoxious persons should be prevented from seeing what you are doing."[30] A watchman guarded Brant Rock twenty-four hours a day. Helen Fessenden complained about being "pestered by the idle and curious, who disregarded notices and signs against trespassing with the traditional aplomb of the tourist." She added: "More than once Fessenden staged a realistic tempest in a teapot to teach the public proper respect for our regulations."[31]

In December of 1905, Fessenden notified Walker about a letter he had just received from the navy. The navy's wireless operator in San Juan reported hearing messages from an unknown station. A transcript of the messages was sent to Fessenden to ascertain whether he knew anything about them. They had originated at Brant Rock, approximately sixteen hundred miles from Puerto Rico.[32] By January 1906, NESCO's transatlantic stations were exchanging messages. Service was sporadic, however, and in the summer, static interfered almost constantly. Fessenden wrote to a friend, "Sometimes the signals are very loud, so that we can hear Machrihanish with the telephones six inches away from the ear, but two or three times in every month we can hardly hear them at all, which of course is not commercial." Fessenden's approach to the static problem was similar to Marconi's: in May and June he remodeled the transatlantic stations, increasing the strength of the station ninety times.[33] But the increase in power did not help, and Fessenden was forced to suspend operation.[34] Communication resumed in the autumn of 1906.

Although Fessenden had been successfully transmitting messages across the Atlantic intermittently for months, he did not announce the achievement to the press. Fessenden wrote: "We do not intend to have anything become public until we are ready to work commercially both during night time as well as day time."[35] While he was striving for perfection, Fessenden experienced the same disaster that had previously befallen Marconi. On December 6, 1906, a storm destroyed the aerial at Machrihanish. Fessenden blamed the construction company, charging that its workers had installed defective guy wires.[36] This may have been true, but it was also true that a transatlantic station was a new venture for NESCO, and Fessenden had not had any previous experience with massive, vulnerable, long-distance aerials that had to survive often severe coastal storms. Because Marconi was more single-minded and systems oriented about wireless, he immediately improved and rebuilt any long-distance aerials that blew down. But the Machrihanish mishap caused NESCO to abandon its transatlantic work for several years. Much of the money invested in this venture was lost. Again, the company changed its business strategy, this time from establishing a transatlantic service to marketing wireless telephony.

At the same time he had been developing the transatlantic stations, Fessenden had been working on transmitting the human voice without wires. To achieve reliable wireless telephony, inventors had to redesign significantly both transmission and reception. This Fessenden had already begun to do. His work on the high-frequency alternator was driven by the insight that only continuous sustained waves could carry the undulations of the human voice. As Charles Steinmetz and other G.E. officials had observed in 1900, Fessenden's vision was beyond the technical capabilities of the electric power industry at that time. Fessenden's insistence that high frequencies were attainable was contagious, however, and G.E.'s research department gradually internalized and institutionalized Fessenden's goal. After Steinmetz designed a 10,000-cycle alternator, which Fessenden found adequate but not rapid enough, the fulfillment of Fessenden's order was assigned to a newly arrived Swedish engineer, Ernst F. W. Alexanderson. Years later, Alexanderson recalled how he obtained the assignment: "The alternator was one of the inventions that I had to make in order to hold my job! The request came in from Fessenden for a high frequency alternator. That was passed along to the regular designers. They thought it was a rather fantastic thing, and I was crazy enough to undertake it." Alexanderson designed the 100,000-cycle alternator, but he reported regretfully to Fessenden in the summer of 1906 that it could not be operated at more than 50,000 cycles.[37] One major limitation Alexanderson confronted was the enormous heat generated by the mechanical speed of the alternator, which caused parts of the machine to burn out. Also, Fessenden's own stubbornness compromised performance. Alexanderson originally designed an alternator with a stationary laminated iron armature between two rotating discs. Fessenden insisted that the armature be made of wood, because he believed that iron could not be used at high frequencies. Some of this prejudice stemmed from the fact that Steinmetz had used an iron armature in his 10,000-cycle alternator. Alexanderson disagreed with Fessenden but followed the inventor's specifications. At the same time, Alexanderson designed his own alternator with the iron armature.[38]

In the fall of 1906 the 100,000-cycle alternator was delivered to Fessenden at Brant Rock. Fessenden and his assistants had to repair the machine, which had been damaged during shipping, but by October, Fessenden reported successful voice transmission.[39] Fessenden soon discovered that the constantly damp atmosphere at Brant Rock did not help the performance of the wood armatures, and by 1907 he conceded that Alexanderson's design was superior. Because of Alexanderson's continued improvements on the alternator through World War I, the dynamo came to be named after him, but as Alexanderson himself stated, "How much of it was Fessenden's idea and how much was my idea is very difficult to disentangle. It was a productive partnership. . . . The patent on the specific way of doing this is in my name because that was my idea, but of course in the general idea—setting the aim of where we wanted to go—Fessenden was naturally the leader."[40]

The alternator, in 1906, still had major handicaps: it was big, cumbersome, and expensive; Fessenden was paying approximately five thousand dollars for each one. A reliable method of modulating the alternator did not yet exist. The carbon microphones in use at the time could not handle the energy generated by the alternator, and consequently they burned out quickly. Yet the alternator was a critical breakthrough in radio technology and elegant evidence of Fessenden's genius in synthesizing his previous work in the electric power industry with his present need for a transmitter. The alternator was developed because Fessenden badgered, nagged, insisted. He gave the men at G.E. little rest and his backers little relief from the expenses surrounding the development of the alternator until the machine was delivered and operating. These traits earned him enmity and resentment. They also produced a revolutionary new transmitter that, during World War I, became the centerpiece of America's radio network, the invention capable of sending messages directly to Europe. The alternator was considered so valuable in 1919 that naval officials, intent on keeping the invention out of Marconi's hands, helped orchestrate the formation of the Radio Corporation of America to control the invention.

By October of 1906, using the alternator, Fessenden had transmitted speech over a distance of ten miles. He established an experimental station in Plymouth, Massachusetts, twelve miles south of Brant Rock.[41] Fessenden wrote to recording companies asking for a good phonograph and several records, especially recordings of Sousa, Caruso, and violin solos. He explained to one supplier: "What I want is something that will test the talking qualities of the telephone so as to compare it with the regular wire line telephone."[42]

On Christmas Eve of 1906, Fessenden used the alternator and a microphone to transmit a special holiday broadcast from Brant Rock. Three days earlier, he had notified ships equipped with Fessenden apparatus to listen for this broadcast. There is no record of Fessenden notifying the press, and the demonstration received no newspaper or magazine coverage. The program included music from phonograph records, Fessenden playing the violin, Fessenden singing, and Fessenden making a speech. He broadcast a similar program on New Year's Eve, and many surprised shipboard operators wrote to the inventor reporting that they had received the unprecedented transmission. Although the talking and singing were not very loud, and the voice reception was intermittent, Fessenden had successfully demonstrated what would soon be called radio. The Christmas Eve program is still considered the first radio broadcast in American history, and a truly dramatic demonstration of the alternator's capabilities.[43]

Fessenden's original 100,000-cycle alternator, 1906.

It is important to note that Fessenden was not proposing that the wireless telephone be used for broadcasting; he was still trying to improve point-to-point communication. He wrote to a colleague: "The chief use . . . of the wireless telephone would be to take the place of the present long-distance pole lines . . . which are very expensive." In his 1908 address to the American Institute of Electrical Engineers, he stated that he believed "the use of wireless telephony would be seriously curtailed unless it could be operated in conjunction with wire lines."[44] Wireless telephony was still experimental, but the advantage it offered was evident: wirelessly transmitted speech was often more distinct and less distorted than speech carried over the wire lines.

Clearly with Bell Telephone in mind, Fessenden wrote to a naval official: "We have decided . . . to keep our inventions to ourselves until we can sell them to some company powerful enough to make the politicians at the heads of the Departments walk straight."[45] The time to sell seemed imminent. NESCO had been eyeing AT&T since 1904, and with the loss of the Machrihanish station and the expense of the alternator, Given and Walker were quite eager for an infusion of capital. As Walker wrote in 1905, "Our customer should be a telephone company. . . . As

Fessenden's assistants at Brant Rock testing his wireless telephone, 1907.

soon as [the wireless telephone] is perfected, we should exhibit it privately to one of the telephone companies or other strong phonical people and endeavor to sell it to them."[46] Fessenden, too, believed that once corporate officials saw firsthand the work at Brant Rock, they would want to acquire NESCO. He invited representatives from the English, French, and German embassies, *Scientific American,* the Associated Press, Western Electric, and Bell Telephone to a demonstration of wireless telephony between Brant Rock and Plymouth in December of 1906.[47] Fessenden stationed some of the representatives at Brant Rock and some at Plymouth, and encouraged them to radio-telephone each other to test the quality and efficiency of his appartus. The tests were successful, although the voices were faint and chopped up. Nonetheless, AT&T's chief engineer, Hammond V. Hayes, the man who had encouraged Stone's early work on wireless, was sufficiently impressed to advise president of AT&T Frederick P. Fish: "I feel that there is such a reasonable probability of wireless telegraphy and telephony being of commercial value to our company that I would advise taking steps to associate ourselves with Mr. Fessenden if some satisfactory arrangement can be made."[48] Before any such arrangement could be made, the panic of 1907 and organizational changes within AT&T intervened.

Although AT&T had enjoyed unprecedented growth between 1902 and 1907, expansion had been expensive, increasing the company's debt from $60 million to more than $200 million. The need for further expansion, which could not be financed through earnings, prompted the Boston-based executive committee to sell large blocks of stock to a group of backers who, it turned out, were fronting for J. P. Morgan. Morgan intended to consolidate the telegraph and telephone systems in American much as he had consolidated the steel industry. By April of 1907, the Morgan interests were in control. They appointed Theodore N. Vail president.[49] Vail had first joined Bell Telephone in 1878 and had resigned as chief operating officer in 1887 because he considered the Bostonian administration too cautious and unimaginative. Both Vail and Morgan were determined to reduce the company's debt, consolidate and streamline its operations, and concentrate on establishing the corporate hegemony of the Bell system. This meant devoting attention exclusively to refining and extending long-distance service, which in 1907 was still plagued with problems. It meant retrenchment. It did not mean investing in potentially promising but peripherally related inventions.

In 1907, Vail fired twelve thousand Bell employees. He also consolidated all of the company's research and development, which had been conducted in three separate locations, into one laboratory in New York City. Hammond Hayes, the chief engineer who had been so enthusiastic about Fessenden's work, was replaced by John J. Carty, who headed the new lab. Carty was a dedicated technical-systems builder and would pursue only those inventions which fit into and advanced the existing wired network. He and Vail defined the lab's task as standardizing Bell equipment and advancing long-distance work. As one historian has noted, "To Vail, the *system* came first, and all of his actions followed from the desire to build, integrate, and protect that system wherever and however possible."[50] Such a man would not be interested in an invention that did not promise to extend or to strengthen the network that was already in place.

After reviewing Hayes's recommendation that AT&T make an arrangement with Fessenden, Vail asked the company's chief patent attorney, Thomas D. Lockwood, to assess the potential advantages and disadvantages of following Hayes's advice. In a twenty-six-page report, Lockwood advised Vail that wireless competition was too great and commercial outlets too unpromising for the company to invest in wireless telephony. He also noted that by the time wireless telephony might be technically advanced enough to serve as an adjunct to or substitute for the telephone, Fessenden's basic patents would have expired.[51] AT&T informed NESCO that it had decided against investing in Fessenden's wireless telephone.

The panic arrived just after Fessenden had demonstrated the revolutionary potential of the alternator, just as he was transmitting the human voice without wires, just when he had AT&T interested in doing business. It was a severe blow and laid bare the tensions just beneath the surface in NESCO. Marketing wireless telephony to AT&T apparently was a strategy Fessenden and his backers had agreed on, and, given AT&T's resources and its determination to absorb or thwart all competition, the strategy was not completely unrealistic. But it rested on the favorable decision of one customer, and when that customer said no, the strategy became defunct. Fessenden, Given, and Walker began to bicker more sharply. Fessenden criticized what he believed was a distinct lack of entrepreneurial flair on the part of his backers: "The only danger which I see ahead of our company, and I consider it a great one, is that we are rapidly drifting into the position of being the owners of a perfectly operating commercial system, but shut off from any place to work it, through having allowed our rivals to obtain a monopoly of the operating licenses all over the world."[52] Given replied testily, "The most serious danger which I see ahead of our company is that our performances do not keep pace with our claims. I do not think Marconi or De Forest will stand in the way of our obtaining permits to work anywhere, if we can do the thing and they cannot."[53]

In the summer and fall of 1907, NESCO curtailed its experimental work and laid off employees. Advising Fessenden that the times were somewhat strenuous, Walker informed the inventor that because of financial conditions, the company would "allow wireless to lay low for fairer weather and then discuss any further moves."[54] Strategic planning was on hold. NESCO now had no commercial stations and only three experimental ones.[55] Brant Rock was still one of the most sophisticated and powerful stations in America; naval operators stationed in Cuba wrote to Fessenden that he came in so strong they did not have to tune for him.[56] Despite promising results, NESCO continued through 1907 and early 1908 without revenue, and little technical or financial progress was made. Except for the 1906 demonstrations of wireless telephony, NESCO stayed out of the public eye.

By the spring of 1908 the intensity of the panic had abated, and Fessenden began pushing for policy changes within NESCO. Events completely out of the company's control—the storm in Scotland, the stock market crash, AT&T's reorganization—had frustrated Given and Walker's goal of selling the entire company. Fessenden, who since the company's inception had wanted to develop a manufacturing firm that sold apparatus, began urging his backers to adopt this strategy. While not abandoning their long-term goal, Given and Walker did agree to limited sales to bring in some revenue. Fessenden's motivations were both personal and financial. As an inventor, he wanted his products out in the marketplace, on display and in use. Also, his 1902 contract with Given and Walker made their payment of $330,000 for a 55 percent interest in Fessenden's patents contingent on the company's first earnings. Since there had as yet been no earnings, Fessenden had received a monthly salary but no money for the nearly two hundred patents he had filed over the preceding ten years. He had become extremely impatient with this arrangement, and was eager to supplement his two-thousand-dollar annual salary.

Fessenden also pushed the company to hire a skilled salesman, a move he had advocated since 1904. By 1908, he knew he had exactly the right man for the job: Colonel John Firth. Walker and Given agreed to hire him. Gregarious, smooth-talking, and capable of sustaining that energetic yet easygoing persona so necessary for sales, Firth was everything Fessenden was not. He had been a sales representative for De Forest and in 1906 had formed his own company, Wireless Specialty Apparatus, to sell individual components such as condensers or receivers, primarily to the U.S. Navy. Fessenden's sales approach was to emphasize the superiority of the apparatus and expect outstanding performance to sell equipment; Firth's style was to cultivate friendships with key people and interweave chumminess and business. He played poker with potential clients, brought them cigars, slapped them on the back, and told jokes— and he sold a great deal of wireless equipment.[57] His relationships with the men in the Bureau of Equipment were excellent. He was a close personal friend of Mack Musgrave's, who was in charge of communications for United Fruit. Firth joined NESCO in June 1908 and dealt directly with Fessenden. By the end of the summer he had won orders from the navy and United Fruit totaling $152,000.[58] Selling to United Fruit was not difficult: its operators at the Caribbean stations regularly heard the high-pitched tone of Brant Rock, and the company was eager to acquire similar transmitters. His major coup was convincing the navy that with the latest NESCO equipment, especially Fessenden's rotary spark gap, it would be able to establish a high-power long-distance station capable of signaling to Europe and the Caribbean. He persuaded naval officials to experiment with the equipment at Brant Rock and talked them into paying rent while working there.[59] Thus, under the guidance of Firth, NESCO resumed doing business with the navy and was especially successful at selling transmitters.

Firth also had to assume the role of negotiator between Fessenden and Walker and Given. Fessenden considered the $152,000 "first earnings" and wanted to get paid for his patent rights. Walker refused, reportedly saying, "If you don't behave, don't do what we say you won't get anything. Mr. Given and I have the company tied up so that if you make any fuss you won't get a cent."[60] On September 11, 1908, Fessenden submitted his resignation, reminded Walker that, not having received payment, Fessenden was in possession of the patent rights, and made several suggestions on how they might compromise. On September 12, Firth mediated between the inventor and his backers and helped them draw up a new contract. Only a few minor changes were made in the existing agreement. The $330,000 due Fessenden was still to be paid out of the first profits, and Fessenden's salary was raised to $600 a month. Given and Walker agreed to advance "from time to time if it should be needed" up to $50,000 to construct stations under contract and provide the running expenses for the company. All future business decisions were to be voted on, and major differences were to be referred to an arbitrator.[61] Although Fessenden was not yet to be paid for his patents, the agreement set up a schedule of what constituted first earnings and made payment seem more imminent. The breach between Fessenden and his backers was temporarily bridged, and Firth was clearly responsible. He was bringing in the contracts that promised to make NESCO viable, possibly even profitable, and he made both parties recognize the importance of staying together.

Despite Firth's success at selling apparatus—the first strategy to succeed for NESCO—Given and Walker clung to their dreams of transatlantic service and the offer they hoped it would bring. Walker urged Fessenden again to "try to interest someone in the project" and to "get up a Trans-Atlantic Company at once."[62] He must not have found Fessenden's first attempt at this terribly reassuring. Fessenden notified Walker in late September of 1908 that an agent representing W.E.D. Stokes had offered to buy out Given and Walker and form a new company with Fessenden as its president.[63] Stokes, who had inherited eleven million dollars from his father in the 1890s, was a self-styled financier and owner of the recently built Ansonia Hotel in New York City. He controlled the Chesapeake Western Railroad Company, bred racehorses in Kentucky, wrote a book about eugenics titled *The Right to be Well Born,* and had a son who was interested in wireless.[64] Fessenden had periodically needled Given and Walker by reporting such offers to them and was obviously trying to remind them, in an unabashedly transparent way, that others with money and connections considered him and his work valuable. But he carefully added that he "would not be willing to go in with the other people without [Walker] and Mr. Given were in also." What Fessenden suggested instead was that Stokes arrange for the merger godfather of them all, J. P. Morgan, "to take charge of the entire affair and get up some plan for putting the whole business in shape."[65] No record exists of Walker's reaction to this, but the scheme was not pursued, or even mentioned, again.

Instead, while Fessenden worked on the transatlantic service Walker wanted to establish between the United States and Great Britain, he also renewed his earlier plan of establishing a Canadian transatlantic company. He corresponded with Brenton A. MacNab, editor of the *Montreal Star,* who advised him to form a separate company with local incorporators. MacNab's belief was that a British company would have a better chance of obtaining a license from the British Post Office than would an American company. MacNab also informed Fessenden that the Marconi Company was considered monopolistic in Canada and, consequently, that routing Marconi would not be difficult. Encouraged by MacNab, Fessenden planned to build a new station in Newfoundland which could communicate with Machrihanish. He took MacNab's advice and formed the Fessenden Wireless Company, based in Canada. The Americans Walker and Given, were excluded from the directorship yet they were still expected to advance funds to the new company.[66]

Fessenden now planned a trip to England, where he hoped to secure a long-term license for operating his transatlantic stations. Prior to leaving, in February of 1910, Fessenden offered the planned but not yet operating system to T. N. Vail, president of AT&T. Fessenden boasted that his stations would operate at a speed of 250 words per minute and would achieve a distance of 3,000 miles. "Two such stations, one in Europe and one in America, working duplex, are capable of handling more traffic than is at present handled by all the Atlantic cables combined," he said. Then he asserted: "The question of interference has been solved. . . . A method has been devised for eliminating interference so completely that many thousands of subscribers' stations may be located within a mile of each other and yet work independently." He further claimed that his wireless telephone now operated over a distance of 425 miles. Fessenden advised Vail to act quickly in acquiring NESCO, because after the end of the year, NESCO's rights would probably not be for sale. If AT&T failed to buy NESCO, then Fessenden would have no choice but to compete with AT&T in the future.[67] Vail must have found the naive effrontery of this letter both preposterous and amusing. Six months earlier, J. J. Carty had reported: "I have personally talked by the Fessenden wireless method from Brant Rock . . . to Plymouth. . . . The talk was very faint indeed." Vail himself wrote to a colleague, "As to the 'wireless': I can only refer you to the success of the wireless telegraph and the [negligible] inroad made by it upon the general telegraphic situation as compared with the promises and prophecies."[68]

Fessenden and his wife sailed for England on March 12, 1910. All expenses were paid by NESCO. Helen Fessenden recalled, "On reaching London a suite was engaged at Claridge's, Brook Street, that hostelry of Royalty, and here Reg remained for the entire seven months of his English visit. This was in line with his standard way of attacking any problem— to be satisfied with nothing less than the best."[69] The Fessendens clearly enjoyed their regal lifestyle, believing they were finally mixing with people of their own caliber and receiving the treatment they deserved. The trip did not have a salubrious effect on Fessenden's vanity or sense of entitlement. While in England, he corresponded primarily with his secretary at Brant Rock, Miss Bent, who kept him informed about both business and technical developments. It was she who advised him of changes in machinery and warned him that "many of the new designs [were] not good electrically."[70]

After a series of high-level and complicated negotiations, Fessenden succeeded in securing a twenty-year license for transatlantic work. He returned to the United States on November 10, 1910, and met his wife, who had returned earlier, and Walker for dinner. Relations between the business partners seemed amicable. They were not. Apparently, in Fessenden's absence, the men at Brant Rock felt free to complain about working with him. Ernst Alexanderson, who got along very well with Fessenden, nevertheless remembered him as "so domineering that people who worked with him said every week or so he fired them all when they didn't do what he expected them to do, and then he rehired them the next day."[71] George Clark, who had been testing NESCO apparatus for the navy in the summer and fall of 1910, wrote a colleague just after Fessenden's return, "I went to Brant Rock yesterday. I fear that things will be very nasty in the future. Even Mr. Kelman said Fessenden was intolerable. Hill expects to throw up the job next week. Fessenden is worse than ever."[72] An unsigned letter to Walker from Brant Rock referred to Fessenden's "erratic methods" and his "tyrannical treatment of the men." "Dissatisfaction with the actions of Professor Fessenden," the letter continued, "was general throughout the entire force and this dissatisfaction was voiced loudly by the men." John Kelman, an engineer who served as general superintendent during Fessenden's absence, was described as having brought order out of chaos.[73]

The staff at Brant Rock, Given, and Walker realized how much they had enjoyed Fessenden's absence. This drove home the extent to which the tensions between Fessenden and his backers and Fessenden and his staff had burdened and preoccupied the company. Walker and Given became convinced that NESCO could operate more smoothly and profitably if Fessenden's managerial role was circumscribed. On December 10, Walker proposed reorganizing the company and appointing a business manager who would be headquartered in Pittsburgh. Fessenden would be relieved of the title general manager and become, instead, scientific engineer of the company, so that he could devote all his time to "the more rapid development of [the company's] inventions." Walker added: "Under [this] arrangement, of course, you will be consulted fully upon all business matters, the details of which you have so often complained [were] preventing you from devoting your time to the scientific work of the company."[74] The company's Pittsburgh-based patent attorney, Francis Clay, reiterated this argument, but was more flattering: "The company has a beautiful opportunity to enter into very lucrative business and the situation is very excellent if the company could proceed in its business with your direction in scientific matters. I have long felt that you were wasting very high-priced time and attention on very low-priced work. . . . I think that there is but one Professor Fessenden and there are many business managers."[75] No record exists of Fessenden's response to this proposal, but given his dissatisfaction with his backers' business strategies in the past, and his conviction that only through his technical, marketing, and diplomatic efforts was the company surviving at all, he must have vehemently opposed such a change.

On December 28, 1910, Fessenden went to Pittsburgh at Given and Walker's request. While Fessenden attended the conference in Pittsburgh, John Kelman presented a written order to Miss Bent, signed by Walker, which notifed her that the Brant Rock office was to be shut down and all papers and documents packed up and shipped to Pittsburgh.[76] Bent immediately showed the order to Helen Fessenden, who told Kelman the order could not be carried out until Fessenden had been consulted. She tried all day to reach her husband by telephone at the Farmer's National Bank in Pittsburgh, but was repeatedly told that Fessenden was not there. By mid-afternoon Kelman returned with two men who had been hired to help him remove the office files; they intended to begin immediately, with or without Fessenden's knowledge. Helen Fessenden threw her arms around the largest set of files, but her grip was, in her words, "forcibly loosened." After about one-third of the files had been removed, all three men inadvertently stepped out of the office simultaneously; Helen Fessenden slammed and locked the door and barricaded herself in. Fessenden called in at 6:00 and, on hearing the news, advised his wife to call a lawyer immediately and do whatever she could to prevent the removal of Fessenden's papers. A lawyer in Boston instructed her to get the county sheriff to have the files attached, which she was able to do. Meanwhile, Fessenden received notice of an injunction filed by Given and Walker enjoining him from further participation in the affairs of the company. These papers were dated November 1910. Fessenden was officially discharged on January 8, 1911; he immediately sued for breach of contract.

The immediate cause for the break was a dispute over patents; Given and Walker wanted certain patents signed over to the company, and Fessenden refused. They then informed him that they considered that the company owned not only his wireless patents, but any other of his patents as well. Fessenden violently disagreed.[77] The split had been coming for years. As experimentation continued and expenses mounted, each side cultivated a proprietary attitude toward the company. As each side compromised over money, time, or corporate strategy, it became more resentful of the other side. Each came to view the other as less valuable, as replaceable. The three men had shared the common hope that wireless would bring them considerable money and prestige, but they had never truly shared a vision of how to make that happen. Without a common long-term strategy that they all wholeheartedly endorsed and pursued exclusively, they could not survive in a corporate setting increasingly dominated by determined strategists. Nor were their fortunes helped by Given and Walker's lack of expertise and flair in marketing electrical technology, or Fessenden's lack of humor and diplomacy. These men were not organization builders. As a result, they could not subordinate their differences for the good of the company, compete with Marconi, or get a foothold in corporate America. In May of 1912, a jury found in Fessenden's favor and awarded him just over four hundred thousand dollars. To conserve resources during the appeal process, NESCO went into receivership. The case was eventually settled, years later, out of court.[78]

The collapse of NESCO was certainly due to inadequate strategic planning and insufficient coordination within the firm. The failure of NESCO demonstrated how a poorly defined organizational structure could exacerbate business problems brought on by personality conflicts.

Here were strong personalities, all eager for success, yet in disagreement about how success might be achieved. The duties of the three men, particularly in the areas of marketing and promotion, overlapped too frequently; at the same time, Fessenden enjoyed too much autonomy over technical developments, on which the entire enterprise relied. It was never clear who was in charge, the men with the money or the man with the inventions.

Most importantly, the three tried to run the firm simultaneously as an equal partnership and a hierarchy, without resolving the contradictions between these two approaches. When Given and Walker sought, unilaterally, to restructure the lines of responsibility in 1910, after seven years of operation, the result was dissolution. That is how fragile the organizational structure was, and its indeterminate nature ensured that as duties overlapped, as investment of time and money increased, and as the stakes got higher, personalities would clash more frequently. These organizational problems meant that Fessenden's influence on the emergence of radio would be primarily intellectual and technological. This influence was, however, anything but slight, for Fessenden's dedication to continuous wave technology changed the course of radio history.

• • •

LEE DE FOREST'S ideas about wireless during these years were equally significant. In 1905, De Forest was living out a classic version of the American dream: through technical mastery, he was becoming rich and famous. While in St. Louis for the World's Fair, De Forest lived in a three-story brick house complete with a carriage and a coachman, cook, and butler. He did not immediately find love, however. In November of 1905 he married a woman named Lucile Sheardown, whom he referred to less than one year later as another man's mistress and a harlot. Apparently Sheardown had other allegiances; she refused to consummate the marriage, and she and De Forest parted within five months.[79]

De Forest's backer, Abraham White, issued more and more press releases, and they were more and more audacious. They apparently were also convincing. The frequency and seeming legitimacy of the claims made De Forest appear to be the most technically advanced and successful American inventor, and stock continued to sell. White predicted that someday every automobile would be equipped with wireless and began planning a transatlantic venture for the firm. In pursuit of this goal, the company late in 1905 financed a trip to Europe for De Forest so he could secure a license and begin work on a transatlantic station. He also hoped to sell some stock. When Marconi first learned of this plan, he wrote to Cuthbert Hall, "I think it would be a good thing to let the Press know (even through advertisement) if his prospectus appears, that we intend taking legal action for infringement if he sticks up any station in England."[80] Marconi also intended to thwart De Forest's efforts to get a license. In the midst of the transatlantic experimentation, which he privately admitted was a complete failure, De Forest was summoned back to America because of patent problems.

On arriving in New York in April 1906, De Forest learned that a warrant was out for his arrest, and the De Forest attorney advised the inventor to flee to Canada for a few months until White could raise the five-thousand-dollar bond. De Forest and White had been cited for contempt for continuing to market the electrolytic detector. The presiding judge also decided that De Forest and White should pay damages for using Fessenden's detector as their own during the Russo-Japanese War and the St. Louis exhibition. After extricating himself and De Forest from threatened incarceration, White demanded De Forest's resignation. White was furious over the patent suits and accused De Forest of misleading him about the rights to the receiver. He warned De Forest that failure to resign would result in White's rescission of the inventor's bond. When he left the company in the summer of 1906, De Forest had only five hundred dollars.[81] He complained of White in his diary: "He has made of me these years an office boy, a traveler about the country to meet people, to talk glowing prospects, to build and operate impossible stations, so that his stock agents might reap large commissions, while he stole the residue."[82] Demoralized and broke, De Forest asked Hay Walker, of all people, for a job. It is not difficult to imagine the responses of Walker and Fessenden to this proposition.

The late fall and early winter of 1906 were very difficult for De Forest, and he saw few possibilities for personal or financial renewal; he used the phrase "sinking into the mire" in his diary. It was in the midst of this depression and uncertainty that De Forest took solace in what had, after all, gotten him started: the technology. He resumed his experimentation on receivers, work that he had not had time to pursue while demonstrating stations and helping to sell stock.

While Fessenden concentrated on continuous wave transmission, De Forest focused on reception. His experimentation was propelled by Fessenden's successful infringement suits against De Forest's use of the electrolytic detector. Needing a new detector, De Forest returned to experiments he had conducted with his partner Edward Smythe in 1901.[83] While testing their spark gap and "responder" in his apartment, De Forest noticed that the Welsbach gas burner that lit the room dimmed and brightened in rhythm with the sparking coil emissions. Although he soon discovered that it was the sound of the spark, not the waves it emitted, which caused the burner to dim and brighten, he remained convinced that gases could detect wireless signals. Four years later, after trying several alternatives to the Welsbach burner, he settled on experimenting with an incandescent lamp. It is important to emphasize that De Forest believed the bulb would have to contain gases in order to function properly.

In October of 1906, at the monthly meeting of the American Institute of Electrical Engineers in New York City, De Forest announced the invention of his new receiver, the audion. This early vacuum tube represented a technical revision of a device patented by Marconi's scientific adviser, Professor John Ambrose Fleming. Before working for Marconi, Fleming had been scientific adviser to the Edison Electric Light Company, and in 1890 he had investigated what came to be known as the Edison effect. This set of experiments led Fleming to discover a method for "rectifying" electrical oscillations.

Using a regular incandescent light bulb, Fleming sealed a small plate inside the bulb, next to but not touching the filament. This little plate was connected by a platinum wire to the base of the bulb. When the positive terminal of a battery was connected to this plate, and the negative terminal to the filament, a current flowed across the space between the two. However, when the connections were reversed, when the positive terminal was connected to the filament, and the negative terminal to the plate, no current flowed. In the first arrangement, the negative charges were attracted to the positively charged plate. In the second, a negatively charged plate repelled the electrons.

Fifteen years later, Fleming reconsidered his results. He recognized that to produce an audible sound in wireless, the signals would have to be rectified, or made to flow in one direction only. Electromagnetic waves, consisting of high-frequency alternating current, oscillated up and down the antenna wire. The rapidity with which the oscillations flowed in both directions did not allow a telephone diaphragm to vibrate properly, which is why simply connecting a telephone receiver to a wireless aerial had been fruitless. Theoretically, the negatively charged plate should repel or cancel out the negative half of the alternating current, converting the oscillations into unidirectional, intermittent current. Fleming realized that the light bulb with the tiny metal plate sealed inside acted as a valve that suppressed the current in one direction, allowing the current of the opposite direction to act as a carrier wave. Retaining the vacuum in the bulb, he substituted a tiny metal cylinder for the plate and connected this to the antenna. The filament was connected to the ground and the telephone receiver placed in circuit. The Fleming valve was patented in 1904 in Britain and 1905 in America.[84]

After all his spadework, it is not surprising that Fleming, with considerable sarcasm, pointed out the "remarkable similarity between the appliance . . . christened by Dr. De Forest an "audion" and a wireless telegraphic receiver [Fleming] called an *oscillation valve*."[85] In the fall of 1906 the devices were remarkably similar, and De Forest admitted knowing about the valve. But shortly after writing the AIEE paper, De Forest made a brilliant addition. Between the metal plate or cylinder and the filament, he inserted a third element: a tiny grid with bars of fine wire supported by a separate connecting wire and fused through the glass of the bulb. This grid magnified the currents in motion and amplified the incoming signal enormously. *Electrical World* described the response in the telephone receiver as being "several times as loud as any other known form of wireless receiver." "A listener in the telephone," the magazine enthused, "will hear a sound whose pitch is exactly that of the spark."[86] Most importantly, the audion was capable of picking up the undulations of the human voice. The Fleming valve could also receive the human voice, but, unlike the audion, it could not amplify or oscillate. The exact date of the invention of the grid audion is not known, but it occurred late in the autumn of 1906. On December 2, De Forest met with his patent attorney and sketched out the audion on the back of a breakfast menu, and by December 21, the patent application was drawn up. De Forest, who was broke at the time, claimed that it took him nearly six weeks to raise the fifteen-dollar patent fee. The patent was filed on January 29, 1907.[87]

De Forest publicly introduced his three-element grid audion in a lecture on March 14, 1907, before the Brooklyn Institute of Arts and Sciences. The lecture was attended by a number of boys known as amateur operators, who experimented with wireless as a hobby. Several of these young men tried the audion.[88] One amateur, who claimed to have possessed one of the first audions, recalled that he used the receiver successfully for three years: "The first audions had two filaments, and when one burned out you used the other. They were a remarkably sensitive detector."[89] However, the audion, while theoretically sophisticated, had several practical drawbacks. One major problem afflicting the audion in 1906 was its low vacuum, the very feature De Forest insisted on preserving because he was convinced that the hot gases in the audion helped the amplification. Lloyd Espenschied, an amateur who later became an electrical engineer for AT&T, acknowledged that the audion was sensitive, but he complained that the filaments burned out much too quickly, and that the audion "would not hold its adjustment very long, and would glow at times and block the signal." "It did not occur to me," he said, "to wonder why the audion did not come into general use in the ensuing years. It was too mysterious and uncertain a device. No two bulbs were alike, their life short, and they would block with blue haze." The early audions were also very expensive. Espenschied paid five dollars for his, a prohibitive price for many people, considering that it equaled about half a week's wages.[90] Despite the audion's initial imperfections, De Forest's early vacuum tube was a revolutionary advance in radio and remains one of the most significant inventions of the century.

De Forest did little developmental work on transmitters and initially tried using the spark gap for voice transmission, but "with very disastrous results," as the transmission was almost unintelligible.[91] He, too, needed a method of generating continuous waves. His audion served as receiver, and for the transmitter he borrowed from the work of W. B. Duddell, an English engineer, and Valdemar Poulsen, a Danish scientist. Their arc transmitters generated pulsating or oscillating currents from a direct current.[92] De Forest added his own refinements to the arc transmitter and connected it to a carbon microphone. In these early years, the arc was not as successful as theory had suggested it would be. The frequency of the wavelength and the sound level often varied; maintaining their constancy was often quite difficult and sometimes impossible. The arc also produced a hissing noise in the telephone receiver. The radiations were "encumbered with 'mush' and harmonics," and sometimes the arc produced no voice at all.[93] Nonetheless, it was a start, and with the arc and the audion, De Forest had at least a rudimentary system of transmitting and receiving the human voice.

After resigning from American, De Forest swore that he had always deplored the use of wireless as a gimmick for selling stock. But in January of 1907, six months after his departure from White's company, De Forest formed a new company, the Radio-Telephone Company. His partner in this enterprise was James Dunlop Smith, one of White's former stock salesmen. De Forest wrote to his assistant, Frank E. Butler, "If I could get $1000 I would build a little demonstration set working ½ mile and that would make stock sell like wildfire."[94] Although early stock sales were slow, De Forest was able to borrow one thousand dollars from Harriet Stanton Blatch, the noted suffragist and De Forest's future mother-in-law. With this sum he began testing and demonstrating his radiophone.

The timing of the actual moment of insight remains uncertain, but sometime during the insecure winter of 1906–7, De Forest conceived of radio broadcasting. It was an insight fueled less by a compelling technical vision and more by the desires and longings of the social outcast. During De Forest's impoverished and lonely spells, he would cheer himself by going to the opera. Usually he could only afford a twenty-five-cent ticket, which bought him a spot to stand in at the rear of the opera house. De Forest was an ardent music lover, and he considered unjust the fact that ready access to beautiful music was reserved primarily for the financially comfortable. Of course, this feeling was more pronounced when De Forest was down and out than when he could indulge in conspicuous consumption himself. In either circumstance, however, he appreciated in a visceral way the pleasure of access to culture. De Forest was convinced that there were thousands of other deprived music fans in America who would love to have opera transmitted into their homes. He decided to use his radiophone not only for point-to-point message sending, but also for broadcasting music and speech. This conception of radio's place in America's social and economic landscape was original, revolutionary, and quite different from that of his competitors.

Unlike Marconi, who was offering institutional clients a substitute system that was similar to one they already knew, De Forest was suggesting a completely new technical and entertainment system to be marketed to ordinary people. De Forest's proposal was premature for several reasons. For one thing, the technology was not sufficiently sophisticated: the arcs hissed, the microphones burned out, and the receivers picked up a blend of music and dots and dashes. He also had only the vaguest conception of how broadcasting might generate revenue, and he had not adequately considered the issues of marketing and programming. Nonetheless, De Forest's idea of using wireless telephony to deliver entertainment to people in their homes had, as we know now, enormous social consequences. De Forest envisioned radio as a way to serve the culturally and economically excluded—and as a way to make money. Having been in his life, by turns, the ridiculed outcast and the exploiter of the gullible public, De Forest carried with him two very distinct impulses that guided the development of radio. For De Forest, radio broadcasting blended his altruistic and his self-serving impulses. It resolved his internal contradictions just as it would later straddle, and mask, contradictions in the culture at large.

De Forest began pitching his plan to reporters. He told the *New York Times*, prophetically, "I look forward to the day when opera may be brought into every home. Someday the news and even advertising will be sent out over the wireless telephone."[95] As the *Review of Reviews* explained De Forest's plan, the music would be distributed from a "central station, such as an opera house." "The music of singers and orchestra could be supplied to all subscribers who would have aerial wires on or near their homes," the *Review* continued; "The inventor believes that by using four different forms of wave as many classes of music can be sent out as desired by the different subscribers."[96] *Electrical World* hoped that such a central station might also "send out orders to the whole police force in an instant, publish election or ballgame returns, give free concerts to the whole population and accomplish a good many other things which would tend to better the social life of its citizens."[97]

The press was not uniformly enthusiastic. The *New York Times,* in a 1906 editorial titled "A Triumph but Still a Terror," commented: "There is something almost terrifying in the news . . . that attempts at telephoning without wires have already attained such success that scientists announce the approach of the time when a man will be able to speak without any conducting wire to a friend in any part of the world."[98] In an interview, the *New York Times* asked Professor Michael Pupin his opinion of the value and application of the wireless telephone. Pupin had quite a few reservations: "On land, think what would happen. There would be thousands of voices traveling in all directions. There would be a babel of voices. And what a chance for long-distance eavesdropping. . . . You see, we would never get away from it. What privacy would we have left? It's bad enough as it is, but with the wireless telephone one could be called up at the opera, in church, in our beds. Where could one be free from interruption?"[99] *Electrical World* suggested one cause of the opposition to radio: "The psychological effect of the voice reproduction is so powerful that mere telegraphic signaling is not in the same class with telephonic signaling . . . [which] can carry personality and the force of mind."[100]

De Forest first tried out his broadcasting idea with the Cahill brothers, who had invented a giant, organlike instrument called the Telharmonium, a turn-of-the-century synthesizer. The Cahills were trying to market background music. Like De Forest, they had been unable to formulate or appropriate a viable system for their invention. The Cahills had attempted, unsuccessfully, to interest the telephone company in their "synthetic music" and to run their own cables under the street to subscribers' homes. Unable to acquire a line system for their music, the Cahills agreed to try the ether. From February through May of 1907, De Forest broadcast the Cahills' music in New York City. Apparently the quality of the broadcasts was unacceptable, because after the four-month trial period, the Cahills abandoned radio.

Undaunted, De Forest pursued his broadcasting dream and was

supported, both spiritually and technically, by his fiancée, Nora Stanton Blatch. Blatch's contributions to the early development of voice transmission have been either completely ignored or dismissed; one historian has identified her simply as a lady pianist.[101] She was, in fact, the granddaughter of Elizabeth Cady Stanton and the first woman to receive a civil engineering degree from Cornell University. Her mother, Harriet, had named her after the heroine in Ibsen's *A Doll's House* and had encouraged Nora to be independent and capable of doing a man's job. Nora Blatch was just what De Forest needed: she was technically knowledgeable; her family had strong political and commercial connections; and she loved music. She and her mother lived in the same Riverside Drive apartment building as De Forest; he quickly fell in love with her and began wooing her both in person and over the wireless telephone. He even, according to one account, imagined that "destiny had brought the 'fated one' to his very door."[102] After she met De Forest, Blatch quit her job as an engineer with the New York City Water Department and began studying electrical engineering under Professor Pupin at Columbia. De Forest and Blatch worked together in the New York wireless laboratory in the evenings, and many local operators heard music or conversation interrupting the usual dots and dashes. However, no commercial outlet for the radiophone had as yet materialized, and De Forest believed that a sensational publicity stunt was needed.

In February of 1908, Blatch and De Forest were married and went to Europe for their honeymoon. While Lee De Forest tried to sell the foreign rights to the audion, Nora Blatch De Forest met with her uncle, Theodore Stanton, who was the Paris representative for the Associated Press. He helped arrange, through the French War Office, for a De Forest radiophone demonstration from the Eiffel Tower. After a few disappointing tests in which the broadcast traveled only six miles, a government wireless operator near Marseilles, 550 miles from Paris, reported receiving De Forest's phonograph music. De Forest immediately envisioned transatlantic radio broadcasts.[103]

De Forest received good publicity from his success in France, and when he returned home, the financial status of the Radio-Telephone Company had improved. De Forest had decided that his transatlantic radiophone service would transmit between the Eiffel Tower and the nearly completed Metropolitan Life Insurance Tower in New York, and "James Dunlop Smith thought the idea so wonderful that he had new stock certificates struck off, showing the Metropolitan and Eiffel Towers shooting off sparks and linked by the legend 'words without wires.'" Stock promotion was intensified, and Smith augmented his general sales strategy with carefully targeted pitches. Harriet Stanton Blatch made a speech on women's suffrage over the radiophone which Smith used to his advantage in selling stock to suffragists and their sympathizers. In the writings of Mary Baker Eddy, a passage appears which reads "Spirit needs no wires or electricity to carry messages." Smith and De Forest succeeded in selling stock to hundreds of Christian Scientists by citing this passage and assuring the investors that Eddy herself had prophesied wireless. In 1909 Smith bought a yacht equipped with De Forest's wireless telephone and christened it *Radio*. On weekends, De Forest conducted demonstrations just off the Rhode Island coast in the hopes of enticing the wealthy residents of Newport to invest.[104]

In 1908 and the first half of 1909, De Forest was once again wealthy. Like its predecessor, American De Forest, however, the Radio-Telephone Company was severely mismanaged and survived only because of stock sales. De Forest was as guilty as Smith of advocating any unscrupulous method that helped him maintain financial security. When the Radio-Telephone Company faced competition in the Midwest from a small company owned by a wireless entrepreneur named Thomas Clark, De Forest advised his assistant, Frank Butler, "I do not think we should worry over the Clark system, only spy on them and spoil their stock sales all you can." Smith and De Forest spent the company's income on publicity stunts and lavish lifestyles; little went to improve the apparatus. By the end of 1909, Smith decided to resign. He presented De Forest with a "balance sheet showing the company in debt some $40,000 with practically no cash, then walked out."[105]

Elmer E. Burlingame, Smith's chief stock-selling aide, took over and helped De Forest reorganize and form yet another new company, North American Wireless. Despite his efforts, De Forest had failed to cultivate a commercial use for the radiophone. He tried one more time to generate an interest in radio by arranging, in January 1910, for the Metropolitan Opera Company to broadcast its performance from the roof of the opera house. Eager reporters tuned in, as did ship operators and amateurs. The publicity was not good. "The warbling of Caruso and Mme. Destinn in 'Cavalleria Rusticana' and 'Pagliacci,'" according to the *New York Times,* "was not clearly audible to the reporters who were summoned to hear it at the headquarters of the inventor. . . . At the receiving station . . . the homeless song waves were kept from finding themselves by constant interruptions . . . [and] the reporters could hear only a ticking which the operator finally translated as follows, the person quoted being the interrupting operator: 'I took a beer just now, and now I take my seat.'"[106]

By 1910 De Forest was again broke. The Radio-Telephone Company was bankrupt, the New York laboratory and factory were shut, and Nora Blatch De Forest sued for divorce. Although Blatch had given De Forest invaluable help both in the lab and in the marketplace, De Forest did not publicly recognize these contributions. Nor did the press. He was the brilliant inventor; she was simply his wife. According to one biographer, De Forest was extremely difficult to work with in the lab; he was "a slave-driving taskmaster, often moody and profane."[107] Blatch, a feminist and a trained engineer, must have balked at such treatment. She was also suspicious of the company's business practices and argued with De Forest over James Dunlop Smith's trustworthiness. The couple also disagreed over Blatch's role after the birth of their first child in the summer of 1909. Blatch expected to continue working; De Forest vehemently disagreed, asserting that once a woman was a mother she was "duty bound" to devote all her time to her family. He believed that if a woman was unwilling to sacrifice her career for her family, she should abstain from marriage.[108] Clearly, the differences between the two had become irreconcilable. De Forest wanted Blatch's help, but only on his terms. He fancied himself progressive because he supported women's suffrage, but at home he expected to be the final authority in both business and domestic matters. Blatch's independence was fine, as long as De Forest could define its parameters. Her feminism extended far beyond women's right to vote, however; she expected to have control over how she worked and lived. When De Forest challenged that control, Blatch left.

De Forest and Burlingame could not market the stock of the North American Wireless Corporation, because the government was finally launching an aggressive campaign against wireless stock promotion. De Forest left New York City late in 1911 and took a job with the newly formed Federal Telegraph Company in San Francisco. However, the dreamer and the huckster had already left his mark on the way radio would be socially constructed and on how those controlling radio technology would come to view their customers.

De Forest imprinted radio with the possibilities and excesses capitalism allows. He saw his inventions as passports to the American dream, as bringing him the wealth, and especially the deference, that would elevate and separate him from his childhood in Talladega. De Forest believed desperately in upward mobility through invention, but his faith was self-serving and cynical, for he was willing to nurture and exploit others' equally desperate hopes of quick success to attain his own. He knew the American dream was only possible for a few, yet he perpetuated through his promotional material the myth that it was possible for all. In his business practices, De Forest clung to, at the same time that he saw through, the get-rich-quick myth. He, not those other un-deserving and credulous chumps, would rise to the top. Here was a conflicted, complicated man whose attitude toward his investors and, later, his audience, was a mixture of identification and contempt.

De Forest had that rare ability to believe he was above the hoi polloi while sharing their yearnings and their tastes. His conviction that people would welcome having music and speech brought into their homes through radio proved absolutely correct, and his persistence in using the invention this way from 1907 on helped pull radio out of the orbit of telegraphy and into the pulsating center of mass entertainment. Certainly De Forest and Fessenden were worlds apart, occupying different intel-lectual, ethical, and cultural spheres. But these two men's pursuit of continuous wave radio and the transmission and reception of the human voice completely redirected how radio would be used by the Americans Marconi had so beguiled at the yacht races.

■   ■   ■

BECAUSE DE FOREST abandoned wireless telegraphy for telephony, and NESCO's business strategies remained short term and ad hoc, oppor-tunities to market wireless apparatus still existed in America. The Ameri-can Marconi Company, which had concentrated initially on servicing the major ocean liners, and then on handling the American end of the trans-atlantic business, was forced in 1907 to remain conservative and unag-gressive. The marketing vacuum was filled by Abraham White, De For-est's erstwhile backer. After dismissing the inventor, White reorganized his wireless concern, dropped De Forest's name, and formed the United Wireless Company. NESCO could not collect the damages it had been awarded because American De Forest no longer existed.[109] After Fes-senden's victory in the barretter suit, however, White was compelled to find a new receiver, and he had dismissed De Forest's audion as too delicate, erratic, and expensive for mass marketing. H.H.C. Dunwoody, vice-president of the De Forest company, had in 1906 discovered that the crystal carborundum could be used as a very sensitive and rugged wire-less receiver. Armed with this inexpensive receiver, White resumed his aggressive sales campaign. In the summer of 1907, White was ousted from the presidency of United by a former Confederate army colonel and rival stock promoter, "Christopher Columbus" Wilson. According to De Forest, White's stock promotion efforts paled in comparison with Wil-son's. United Wireless quickly became the most prominent wireless con-cern in the country.

Although American De Forest had equipped some American coastwise ships between 1905 and 1906, many ships remained unequipped in 1907. Steamship company managers along the Atlantic and especially along the Great Lakes were still reluctant to invest in wireless. Wilson ensured that most American steamers would install wireless by furnishing the apparatus without cost. United Wireless also paid the salaries of the wireless operators and rarely charged the steamship companies any leasing or toll fee for the wireless service.[110] Wilson was committed to exposure, which he believed would boost his sales of worthless stock. At $150 a share, the stocks provided ample income for Wilson to give away as much wireless apparatus as he wanted. Other smaller wireless companies simply could not afford to meet United's terms and were forced out of the wireless market. By 1908, United monopolized almost all the commercial business along the Atlantic coast and controlled wireless on the Great Lakes.[111] By 1910, United had equipped 312 American ships, as compared with Marconi's 176 installations, Fessenden's 6 and De Forest's 5.[112] Thomas Clark, the head of Clark Wireless, which supplied the Great Lakes area, was forced out of business by United. He later recalled, "In New York I had seen C. C. Wilson, President of United Wireless, and I told him he was discrediting wireless for himself and everybody. He told me he didn't give a damn."[113]

Wilson overcame the initial opposition of the steamship companies by providing free wireless service, yet he quickly reestablished skepticism about the invention by installing shoddy apparatus. In some cases United apparatus contained short circuits and such bad connections that it would set the ship's woodwork on fire. In 1910, "apparatus far short of the reasonable requirements of the law had in some instances been supplied, and in some cases ships [had almost departed] with apparatus inert and useless."[114] Consequently, by 1910 most American shore stations and ships were equipped with apparatus that daily confirmed the worthlessness of wireless.

By 1910 the federal government began investigating wireless stock fraud. Magazines such as *Collier's Weekly* and *Success* advised readers not to invest in United, which was "absurdly over-capitalized." *Electrical World* observed: "It is not too much to say that no important invention has ever been the victim of more reckless and culpable exploitation, or has made less commercial returns for a large expenditure of capital." The magazine's editors sermonized: "Wireless has been sowing its wild oats for too long; it is time now for repentance and reform."[115]

One company that had not been sowing its wild oats, either technically or financially, was the Marconi Company. Between 1906 and 1911, Marconi's lab work was still that of the technological revisionist.

Unlike his American competitors, Marconi did not pursue voice transmission or continuous oscillations. He persevered in his dominant goal, a long-distance wireless system capable of competing with the cables. He would continue transmitting dots and dashes, and needed only to refine the sparking technique to reduce interference and enhance tuning. Marconi struck a technical compromise between highly damped waves and sustained oscillations: discontinuous oscillations. As Marconi explained, "It was found to be neither economical nor efficient to attempt to obtain continuous waves. Much better results are obtained when groups of waves are emitted at regular intervals in such manner that their cummulative effect produced a clear musical note in the receiver."[116] In other words, he thought discontinuous waves superior to continuous oscillations. These groups of waves whose damping would be abbreviated or "quenched" were generated by the disc discharger, patented in 1907. This transmitter was the direct result of Marconi's and Fleming's experimental work to improve the spark gap and not of any desire to produce sustained oscillations.[117]

One problem that had plagued the spark gap was the erosion of the gap at the point where the spark was emitted. Fleming reasoned that the erosion would be forestalled if the spark balls rotated "so that the point at which the discharge took place moved around the circumference of the balls." This technique distributed the deterioration, but it did not solve the basic problem.[118] In 1907, building on Fleming's previous work, Marconi assembled three parallel metal discs, the two outside discs turned ninety degrees from the middle disc. The middle disc revolved at high speed between the other, more slowly rotating, discs. To strike particular tones, Marconi lined the circumference of the central spinning disc with metal studs serving as a series of electrodes.[119] As each stud came opposite the disc, the gap became narrow enough for the spark to be struck. When the stud passed the disc, the gap lengthened and the spark was quickly extinguished. This very fast rotation produced semi-continuous oscillations at the rate of approximately one thousand per second. With this new transmitter, the pitch of the received signal became higher and much more distinct, tuning was easier, and interference was decreased. The high, whistlelike pitch of the disc discharger was a welcome improvement over the low roar of the spark gap, whose pitch was similar to that of static. The disc's tone could be adjusted to various pitches by altering the speed of rotation and the number of studs.[120] This invention was not nearly as daring as Fessenden's alternator, but it was a dramatically improved transmitter that was relatively easy and inexpensive to construct and operate.

Although Marconi's technical work during this era was not as imaginative as Fessenden's or De Forrest's, it was nonetheless audacious and increasingly expensive. While working to refine the spark gap, Marconi and Fleming had discovered that their weak link in the high-power stations was the aerial. For all their size, these aerials were still highly inefficient. In 1905, when Marconi was working on fortifying the weak signals between Glace Bay, Nova Scotia, and Poldhu, Cornwall, he inadvertently discovered a new aerial design that strengthened reception considerably. He noticed that an aerial wire lying on the ground received more strongly when its free end pointed away from the transmitter. This observation spurred experimentation that led to the inverted L or bent aerial, patented in 1905. While Marconi himself acknowledged that "the limitation of transmission to one direction is not very sharply defined," the aerial did serve to magnify and partially channel the signal. It was this amplification that finally encouraged Marconi to try to inaugurate the transatlantic service.[121]

In the winter of 1905, Marconi had moved the Glace Bay, Nova Scotia, station to a new location, five miles distant, which provided more room for the predecessor to the inverted L, the umbrella aerial, which had a diameter of twenty-two hundred feet. Only a few months after the relocation and reassembly of the Glace Bay station, Marconi discovered the directional aerial. Although the aerial had to be redone, the new Glace Bay station was advanced enough for transatlantic work. The Cornwall site, however, was too vulnerable to storms and did not have enough room for the inverted L aerial. Marconi proposed to his board of directors that an enormous station be built at Clifden, Ireland, a spot closer to North America. Cuthbert Hall, the managing director, was so concerned about the financial strain a new station would impose that he approached the British Admiralty for help. The Admiralty replied that "Whilst they would be glad to see the long distance station removed from so exposed a position as Poldhu . . . they [were] not prepared to defray the costs of removal and the other expenses involved."[122]

Thus, a reluctant board of directors agreed that the company would shoulder the burden of what they hoped would not be a speculative venture and approved construction of the station in western Ireland. The Clifden station was both remote and colossal; it covered 350 acres and employed engines of 1,100 horsepower. Its output was 300 kilowatts. Transportation to the station was provided by the company's own railway, and the boilers were fired by peat cut from the bog. The station featured a mile-long inverted L aerial and the disc discharger. Marconi wrote to his wife, "We had great trouble with the aerials at Clifden, it was so difficult to get them up in the rain and snow." Clifden was a giant and costly station, the "ultimate in spark telegraphy."[123]

After inaugurating the transatlantic service, the Marconi Company and its American subsidiary were forced by financial difficulties to economize. The Clifden station had drained the company's resources, and income in 1907 was insignificant. The company was forced to dismiss 150 factory workers, and senior technicians worked without wages. Cuthbert Hall wrote to Marconi that he was "extremely busy. Half my time is taken up in very unsuccessful attempts to get money, and a great part of what is left in seeing how we can do without it."[124] Marconi put all his own money into the company to keep it operating. In 1908, Hall resigned as managing director, in part because he and Marconi disagreed over how vigorously to litigate patent infringements; to restore confidence, Marconi himself took over the position. In addition to these setbacks, the British government in 1908 ratified the treaty of the 1906 convention, thus rendering illegal the Marconi Company's nonintercommunication policy. The Marconi-built and -maintained shore stations were supposed to communicate with apparatus of any make. The British Government had reserved the right to establish its own shore stations to compete with Marconi.[125]

John Bottomley, the general manager of the American Marconi Company, wrote to Cuthbert Hall in January of 1908. "Receipts have fallen off considerably, my cash balance is very low," he wrote, adding, "The finances of the company generally are giving me a great deal of thought and some anxiety."[126] He described the strict austerity measures he had adopted to save money, and after reminding Hall that the American company had "never asked the English company for any monetary assistance whatever," Bottomley asked for the money the parent company owed its subsidiary. Hall turned Bottomley down, asserting: "We are absolutely in the need of money for current expenses."[127] Despite the *New York Time*'s public support of the transatlantic service, its representatives privately complained to Bottomley about the company's rates. The paper was insisting on either reduced rates or no charges at all for messages received late. Bottomley, who acknowledged the value of "the great work" the paper had done for the company, was inclined to comply.

What Bottomley wanted additional money for was a high-power, long-distance station in the United States. Concerned about the skepticism surrounding wireless, which hampered business, Bottomley wrote, "I believe that it is essential for the good of the service that just as soon as possible a station for interoceanic transmission should be erected in the United States, as really it seems that the people here do not take very seriously the fact that the stations in Canada and Ireland are now in communication with each other; a station erected in the U.S. would bring the matter right home to all the businessmen in this community, and I feel confident that it would lead to a very quick enhancement of the value of the stock of the company."[128] Bottomley had difficulty raising money for such a station, and between 1906 and 1910 the American company remained in a holding pattern, duly providing ship-to-shore service and handling the New York end of the transatlantic service. The American company's chief engineer during this period was not known for his creativity, and thus little interesting or innovative work was done to challenge De Forest's or Fessenden's breakthroughs.[129] So, while the company's structure remained stable, its strategy languished.

Marconi kept technical and commercial experiments to a minimum during 1908 and 1909. He did little research, and the company did not support investigations into wireless telephony. Instead, Marconi quietly planned his ultimate marketing scheme, of which the transatlantic service was only a part. Denied through regulation the prerogative of determining with whom he would communicate, Marconi now sought to preserve and advance his company's interests through a grander plan. The Marconi Company would establish the first around-the-world wireless network; it would "girdle the globe with wireless." The establishment of the transatlantic service convinced Marconi that several more strategically located and powerful stations like the one at Clifden could link the entire British Empire.[130]

In the midst of this technically fallow period, Marconi wrote to his wife, "Some of the papers have said that I have got the Nobel Prize of £8,000. It rather makes one's mouth water to think about it just now, but I suppose it's not true."[131] But the papers were correct, and in 1909 Marconi shared the prize in physics with Ferdinand Braun of Germany. The award, which must have been extremely gratifying to a man frequently dismissed by the scientific community, further legitimized Marconi as the inventor of wireless. This was a concept Marconi was becoming eager to assert in the courts.

By 1910, Marconi was frustrated by his management tasks and was anxious to return to research. Through his wife's family, Marconi met Godfrey Isaacs, an enterprising businessman with excellent financial connections. In January 1910, Isaacs became the new managing director of the Marconi Company. Both the timing of this appointment and the appointment itself were excellent. Isaacs was determined to settle the company's commercial affairs and to enforce strictly the Marconi patents.

Within a year of his appointment, Isaacs achieved a negotiated settlement with Marconi's major European rival, Telefunken, whereby the two companies formed a new German company in which Marconi held a 45 percent interest and Telefunken a 55 percent share. The new corporation handled all the German mercantile marine business. Isaacs also worked out a patent-pooling arrangement between the two companies which eliminated patent disputes.[132]

Marconi's two-year program of fiscal austerity had prevented him from developing more sophisticated apparatus whose patents would extend and fortify his technical position. But the retrenchment policy had stabilized the company's finances, which meant that by 1910 Marconi was financially prepared to file several long overdue patent suits. Several major American and European companies were violating Marconi patents, especially the "four sevens" tuning patent, and Isaacs was determined to prosecute all infringers. Thus the company was able to compensate for its three-year experimental lull by suing infringers, some of whose innovations Marconi then acquired. The company's consolidation under Isaacs gradually eliminated competition, and now Marconi was able to buy much of what he had not invented.

In 1911, the Marconi Company won its case against British Radio-Telegraph and Telephone, which had infringed on the "four sevens." The same year, the company absorbed the Lodge-Muirhead Syndicate, came to an agreement with Sir Oliver Lodge over tuning patents, and named Lodge associate scientific adviser.[133] Buoyed by its success in Britain, the company was prepared to confront its American competitors. In 1911, Bottomley issued a portentous statment to the *New York Times:* "Our rivals must either discover a device at present undreamed of, or pay such royalties as the Marconi Company decides to require, or go out of business altogether."[134]

The American company Bottomley and Marconi most wanted to drive out of business was United Wireless. In this they received considerable assistance from the activities of the Justice Department. De Forest's and United Wireless's stock promotion activities had been described in all their lurid details in an exposé in *Success* magazine in 1907. Other publications followed suit. In addition, the U.S. Navy and the Justice Department had received increasing numbers of letters between 1907 and 1909 asking government officials for corroboration of the usually extravagant claims these companies made about their sales and net worth. The Justice Department launched an investigation of De Forest's companies and United Wireless, devoting most of its energies to the latter. In June of 1910, U.S. Post Office inspectors and police "raided the luxurious offices of the United Wireless Telegraph Company."[135] Christopher Columbus Wilson and several other officers of the company were arrested and charged with mail fraud. Asserting that United had perpetuated "one of the most gigantic schemes to defraud investors that has ever been unearthed in this country," postal officials estimated that approximately twenty-eight thousand holders of worthless United Wireless stock stood to lose nearly twenty million dollars.[136] Indictments were handed down on August 4; sixty-four-year-old C. C. Wilson, ever the flamboyant showman, chose that day to marry his eighteen-year-old secretary. The trial proceded through the fall, when Wilson was found guilty of contempt of court for failing to turn over the company's books for inspection. On May 29, 1911, Wilson and four of his associates were found guilty of mail fraud. Wilson was sentenced to three years in the Atlanta Penitentiary. United Wireless, which had managed to continue operating during the trial, was put into the hands of receivers. A committee of United stockholders organized in the hope of taking over the company.[137]

The government's action served at least two purposes: it prosecuted criminal activity and it made clear the difference between responsible, reputable wireless companies such as Marconi's and irresponsible deviants like United. The trial also played beautifully into the Marconi Company's hands. After the years of technical and economic retrenchment beginning late in 1907, the Marconi Company had successfully conserved its resources and refortified itself. United Wireless, which had grown rapidly during those very years, was by 1910 the dominant American wireless company and a major competitor. Isaacs was determined to challenge United's position as soon as possible. The trial made the company extremely vulnerable, and thus irresistible.

Striking in the spring of 1911, at "the critical time of the criminal trial," Isaacs filed suit against United for infringement of Marconi's "four sevens" tuning patent.[138] The receivers and the committee of stockholders had failed to agree on United's management; thus, they were unable at first to reach a negotiated solution with Marconi, so Marconi came to the United States in late winter of 1912 to begin his infringement suit, which clearly he would win. By March, United agreed to a merger, which actually was a takeover. The Marconi Company assumed control of all of United's assets, which included approximately four hundred ship installations and seventy shore stations. In exchange, the organized United stockholders received Marconi Company stock. John Bottomley, the general manager of American Marconi, referred to United as a "hopeless mess." At first, sorting out United's affairs was an organizational nightmare; Bottomley put it this way: "A spirit of carelessness seems to have run through the whole conduct of the business of the United Wireless Company which is almost unparalleled in business history." Once Bottomley got the paper work straightened out, he was able to write an annual report very different from the gloomy one he had submitted in 1907. Then he had complained that the company barely had any money. Now he wrote, "The condition of the treasury is satisfactory, if not plethoric. We have, together with cash on hand and investments running 4–12 months, the sum of over five million dollars, all of which is available at any time."[139]

Marconi was elated. He wrote to his wife, "They have admitted to having copied or stolen my patents and all their stations are to be called Marconi stations—this will do our company heaps of good." "Heaps of good" was certainly something of an understatement, for, as he added more definitively a few lines later, Marconi would "control all the wireless in America."[140] Marconi noted that NESCO, given its court battle with Fessenden, was virtually out of business. De Forest, who had moved to San Francisco in 1911 after the failure of his latest company and his marriage, was arrested in April 1912, also on charges of mail fraud. "So you see," Marconi confided to his wife, "Your Dick is slowly overcoming his enemies—of course it's not right to glory over it."[141] In public, he never did, but in private, he couldn't help himself. It was a momentous victory.

• • •

BY 1911, THEN, important technical and managerial developments had occurred which would guide radio's subsequent development. Two critically important components, Fessenden's alternator and De Forest's audion, the one capable of sending, the other of receiving the human voice, had demonstrated that wireless telephony was possible. Marconi had no comparable apparatus and had eschewed continuous wave transmission. This meant that within a few years, when it became clear that continuous wave technology would supplant the old spark gaps, Marconi would find himself in a technically disadvantageous position. His efforts to remedy this situation by trying to buy American alternators activated the chain of events that led to the formation of RCA. Here was a technical oversight, the absence of a technological idea, which was to have important repercussions.

These important developments occurred primarily in the private sphere, behind the scenes. The press was unsure where wireless was going between 1906 and 1911, but between the persistent complaints about interference, the exposés of stock fraud scams, and the lack of stability, and profits, among wireless firms, the invention did not seem destined to revolutionize commercial communications as was once predicted. There were Americans, however, who had embraced the invention and were putting it to uses quite different from those Marconi had in mind. These individuals, the amateur operators, took Marconi's invention to places he had never dreamed of and used it in ignorance or defiance of his carefully planned corporate strategies.

CHAPTER SIX

POPULAR CULTURE AND POPULIST TECHNOLOGY
The Amateur Operators

1906–1912

ON SUNDAY, NOVEMBER 3, 1907, two weeks after Marconi began his transatlantic wireless news service, the *New York Times Magazine* featured as its lead story an article titled "New Wonders with 'Wireless'— And by a Boy!"[1] The large headline spanned all six columns, and the article covered the entire page. The youthful star of the article was Walter J. Willenborg, a previously unknown wireless experimenter, a twenty-six-year-old resident of Hoboken, New Jersey, and a student at Stevens Institute. A large oval portrait of Willenborg in the center of the page was surrounded by photographs of his home-built wireless station, which included transmitting and receiving equipment.

Willenborg had tuned his apparatus and fitted the *Times* reporter with headphones so the reporter could listen in to the dots and dashes being exchanged between Marconi's two distant transatlantic stations at Glace Bay, Nova Scotia, and Clifden, Ireland. For the reporter, the air suddenly came alive: "Messages from everywhere to everywhere and back buzzed into our receiving instrument. Only those in cipher escaped." The reporter became intoxicated by his brief foray into the ether. Newly returned from his adventures in the "great void," he excitedly advised his readers: "For intrigue, plot and counterplot, in business or in love or science, take to the air and tread its paths, sounding your way for the footfall of your friend's or enemy's message. There is a romance, a comedy, and a tragedy yet to be written." Only those with wireless apparatus would be privy to the unfolding melodrama, however. The *Times* reporter continued: "The millions below us knew nothing of this strange intercourse through the night above"—they were unawares, left out, tethered to more earthbound discourse. They had no idea that the "folds of the night" contained "hidden, mystic jabbering." They did not know that in the ether, "ghosts tiptoe by night." They did not know they were surrounded by an invisible and mystical realm to which youthful "wizards" such as Willenborg were privately gaining access.

Willenborg, an otherwise ordinary young man, had become a celebrity through mechanical tinkering. He told the reporter that he expected to be exchanging messages with Paris and Berlin within a month. He claimed that his system was superior to all others and that his messages would be "sent and received without interference or detection" by any other system. The young man's claims were not challenged; rather, they served as secondary headlines for the article. To certify the boy's success further, the reporter noted: "Young as he is, Willenborg has been employed by the United States to perfect wireless tests aboard ship, and has been highly paid for his work." The reporter affirmed that, although Willenborg's father was well-to-do, the inventor did not rely on his

Amateur operators like Walter J. Willenborg became inventor-heroes in the popular press.

father's largesse: "He is so frequently called as an expert witness in so many important suits over electrical matters that his fees give him ample resources."

In the hands of the *Times* reporter, Willenborg became a role model for other boys. His ordinariness and diligence were emphasized: "He is no prodigy. What he has done has been done by hard work. He began at fifteen in a little closet-like room on the top floor of his house." His physical features also resembled those of a Horatio Alger hero, for he was "grey-eyed, clear cut of feature, intent," and he had a furrowed brow. He even showed admirable discretion and a touch of chivalry. At one point while he was listening in to the messages being exchanged between passing ocean liners, he overheard a "sweetheart message" sent between a man and a woman traveling in different ships. While Willenborg "enjoyed hearing it," he "said that it wouldn't be right to hand it around." Yet Willenborg was not prissy. His equipment was extremely powerful, and he had the ability to "destroy" the messages of others at will. The reporter asked for a demonstration. Willenborg scanned the ether for an appropriate target, and overheard an operator from the Atlantic Highlands beginning to transmit a message. Willenborg leaned on his transmitting key for thirty seconds or so and then switched on his receiver. " 'Lay off, New York' came the call from the Highlands man. Again Willenborg shoved out his air waves. 'Go to h- -l' came from the Highlands man. 'Certainly' [Willenborg] replied, and again began the clamor. 'O.K.' finally sent the Highlands man, meaning that he would wait and [Willenborg] could proceed with [his] message." This victory prompted Willenborg and the reporter to begin "chatting and laughing over the plight of the Atlantic Highlands man."

Willenborg made excellent copy, which was not lost on the editors of *St. Nicholas,* "An Illustrated Magazine for Young Folks." In April 1908, a story about Willenborg appeared in its pages under the headline "A Young Expert in Wireless Telegraphy."[2] The author cautioned his readers: "Even today there are young folks who make the same mistake in thinking that all great things that are worth doing have been done; all the great discoveries made; all the grand inventions finished." They became discouraged and idle and were in danger of letting opportunities and knowledge slip by. But not Willenborg: "He decided to try in his own way to learn at home all that was already known and then he would try to learn more." The writer emphasized that Willenborg was "a rather quiet young man with a pleasant face" and "simple and natural manners." But "when he speaks," the writer reported, "we find he talks like a man of science." The article described Willenborg's apparatus, the distances he was able to traverse by wireless, and his ability to listen to "the faint, as it were, whispered words spelled across the Atlantic." "Think of it!" exclaimed the author. "Only twenty and yet a man of science, an inventor and skillful operator in this new art. Could anything be more inspiring to every boy and man?"

Willenborg was the young man the press chose to represent the many other nameless boy operators in America. He was the perfect role model for young men facing the beginning of the twentieth century. His story embodied several trends: the increasingly important role popular culture and journalism were playing in identifying and reinforcing acceptable norms of behavior, the boom in instructive hobbies with their many "amateur" practitioners, and the rise of the boy inventor-hero as a popular culture archetype. His story also captured a more subtle yet profound process: the gradual redefinition of what it meant to be a man, particularly a white, middle-class man, in America.

The boy on the verge of manhood who might want to emulate Willenborg was surrounded in 1907 by vivid yet often conflicting definitions of masculinity and success. On the one hand, the physical culture movement of the 1890s, the explosion in competitive sports with their "organized physical combat," the revival of boxing, and the glorification of the "strenuous life" by the nation's president all equated true masculinity with physical strength, which one should be more than willing to test and assert.[3] A new respect, even reverence, for man's "primitive" side was revealed in the success of Jack London's *Call of the Wild* and Edgar Rice Burroughs's *Tarzan*.[4] The first Boy Scout manual, which addressed a legion of new enthusiasts, warned boys not to become "flat-chested cigarette-smokers, with shaky nerves and doubtful vitality," but to be "robust, manly, self-reliant."[5] According to this ideal, it was not enough to be physically vigorous; men had to have forceful, commanding personalities, as well.[6] All of these traits, it was believed, were best cultivated by a more active life in which men were more directly in contact with nature.

On the other hand, it was clear that in the business world, physical strength mattered little: physical combat was a metaphor for other kinds of confrontations. Increasingly, what landed a young man a good job, what gave him an edge in the race for success, was intelligence, education, and certain skills. The increase in high school enrollments, the growing popularity of adult education, and the self-improvement craze all attested to the new importance attached to education and specialized knowledge.[7] Boys educated in both academic and corporate institutions learned that having a "forceful personality" was, in reality, often either unattainable or a liability. Despite the prevailing mythology, much of a man's life was spent indoors, in urban areas, away from the enlivening and therapeutic tonic of the outdoor life. In reality, being the master of one's environment, or having mastery over other men, was, for many, simply not possible.

For a growing subgroup of American middle-class boys, these tensions were resolved in mechanical and electrical tinkering. Trapped between the legacy of genteel culture and the pull of the new primitivism of mass culture, many boys reclaimed a sense of mastery, indeed masculinity itself, through the control of technology. The boys lacking "animal magnetism" could still triumph over nature if they controlled the right kind of machine. If they failed to recognize how the desire for adventure, combat, and the assertion of strength, on the one hand, could be reconciled with the need to prepare for life in the modern world, on the other, popular books and magazines were there to remind them. Everything could be achieved through technical mastery. Playing with technology was, more than ever, glorified as a young man's game. Even the Boy Scout manual urged boys to be "handy with tools."[8] Few inventions were more accessible to the young man than the latest marvel, wireless telegraphy. Just as articles giving instructions on "Building Your Own Wireless Set" began appearing with increasing frequency, so did stories and books that celebrated boy wireless experimenters. This was no insignificant development, for the popularization of wireless experimentation had, for a time, as decided an effect on radio's development as did the inventors, the navy, or the corporate world.

The stories about Willenborg captured the many attractions wireless experimentation might hold for a young man. On a practical level, if the boy was successful, he could make extra money from his pastime. He might become the center of attention and even get his picture in the newspaper. He would have technical knowledge and skills few others possessed. He learned a code. But beyond these advantages, the boy would enter a new realm in which science and romance commingled. He became an explorer. He both triumphed over and was in harmony with nature. Through wireless, the experimenter went through the looking glass, to a never-never land in which he heard the disembodied "voices" of ships' captains, newspaper men, famous inventors, or lovers passing in the night. This was an invisible, mysterious realm, somewhere above and beyond everyday life, where the rules for behavior couldn't be enforced—in fact, were not yet even established. The boy who entered it could, without detection, eavesdrop on the conversations of others. He could participate in contests of strength, power, and territory, and win them without any risk or physical danger. He heard things others did not, and he did things maybe he should not have done. He could please his parents by acquiring this instructive hobby, and he could defy them by using it, without fear of being discovered, to misbehave. In this realm, in the "folds of the night," by mastering a new technology while letting his imagination and his antisocial inclinations loose, he could be, simultaneously, a boy and a man, a child and an adult. He could also straddle old and new definitions of masculinity.

Willenborg was the latest incarnation of the boy-hero, a central figure in popular literature for decades. In the early 1900s the boy-hero remained a stock character; only the basis of heroism changed. Willenborg's early predecessors, in the 1870s and 1880s, were the young cowboy heroes of dime novels who, like "Deadwood Dick," triumphed over the wild frontier with its animals, Indians, and foreign terrain. Dick, "a youth of an age somewhere between sixteen and twenty," could count on his "muscular development and animal spirits" to achieve his victories.[9] He asserted his masculinity directly through physical endurance and conquest. Thus, he was quite different from the other boy-heroes so popular during the same era, the protagonists of Horatio Alger's stories. These heroes were hard working, morally upright, self-reliant, and physically sturdy. Instead of fighting train robbers or "redskins," they sought to overcome villains ready to foreclose on the mortgage, supercilious rich boys, or the lure of the pool hall. Being a successful man in Alger's stories meant overcoming poverty and a chaotic, corrupt environment. More importantly, it meant becoming middle class, making money, and joining a business in which a boy could work his way up. Masculinity here meant solvency and respectability; it was measured by material success. It was made possible, in Alger's stories, through a fluke accident or lucky break, in which the hero saved a young girl from peril and then learned her father was an extremely grateful millionaire. By 1900, with the obvious consolidation and hegemony of the corporate sector, the prospect of the mythic lucky break seemed dated and naive. Young men had seen real rags-to-riches stories in the newspapers, but the vehicle had not always been luck: it had been, quite frequently, technical mastery.

In dime novels and other juvenile literature, heroes such as Ragged Dick and Deadwood Dick gave way to Nick Carter and Frank Merriwell, two heroes popular in the late nineteenth and early twentieth centuries. Carter, a young detective, possessed "strong muscles and terrible fists." But, more importantly, he was highly intelligent. He was a master of disguise and spoke "almost every known language, as well as many that are comparatively unknown." His cases took him to hidden valleys in Nepal, to undersea kingdoms or lost civilizations in the Amazon. Frank Merriwell, who first appeared in 1896, had been to prep school and was attending Yale, although he occasionally had to leave Yale for financial reasons. He was an excellent athlete, always leading his school teams to victory. Of course he showed "courage, push, determination and stick-to-it-iveness." He and his friends had numerous adventures as they tracked down various rascals.[10]

These characters had much in common with the earlier heroes. They were brave, resourceful, valiant, morally decent, cheerful, and hard working. Yet several important characteristics distinguished these boy-heroes from their predecessors and set the stage for the popularization of wireless. These turn-of-the-century boys were distinctly middle class. Specific mention was made of their educational background. Physical combat, when it did occur, took place more frequently on the playing field. Brute force or violence was less essential to their success than sharp analytical skills and mechanical flair. What set these boys apart from others was not luck, it was skill.

Concerns about technical mastery and the stock figure of the boy-hero came together in a new character: the boy inventor-hero. Tom Swift, of course, was the apotheosis of the boy inventor-hero. His roots lay not only in earlier juvenile literature and dime novels, but also in the science fiction of Jules Verne and H. G. Wells. Like the earlier heroes, Tom mastered his environment, but not with his fists or guns. He used machines. Like other turn-of-the-century heroes, Tom possessed foresight and vision and the power of thought and will.[11] He made a social contribution through his inventions. The stories about him and his inventive capacities were extravagant, but Tom was still a hero boys could emulate on a more modest scale, the way Willenborg had.

The boy inventor-hero, like the inventor-heroes constructed by the press, exemplified how mass entertainment symbolically made sense of technical change. Would-be Willenborgs were surrounded by popular culture celebrations of technical mastery. As Wild West shows and circuses declined in popularity, vaudeville, a genre oriented more toward the dilemmas posed by urbanization, mechanization, and immigration, reached its heyday. Vaudeville contained its share of animal acts and western motifs, but it also confronted technology head on. For example, one famous and favorite vaudeville routine performed during the first decade of the century consisted of a team of men taking apart and reassembling a Ford on stage in eight minutes.[12] At Coney Island, machines brought pleasure, wonder, and excitement.[13] Early films featuring automobiles, trains, and airplanes celebrated those who could control the machinery and made fun of those who could not. Through a range of action-filled tales, formulistic plots, and broad and sometimes vulgar humor, American culture symbolically addressed the tensions and contradictions brought about by the seeming flood of new machines and gadgets. The man who was befuddled by all this machinery was a clown, emasculated; the man who made technology his slave, a genius newly empowered.

The emergence of the boy inventor-hero is important to the early history of radio because the genre of popular juvenile writing surrounding this new hero provided information about wireless and encouraged boys to experiment with the invention. It also placed wireless work within the larger context of contemporary heroism. Most importantly, the popular writings glorifying boy experimenters presented ideas about wireless and the ether which were totally at odds with those held by Marconi and military officials. The stories about Willenborg cast wireless as a young man's toy and the ether as his playground. In the magical, almost other-wordly realm described by the *Times* reporter, the concepts of corporate monopoly or military preemption seemed alien, mean spirited, and completely unenforceable.

The growing audience of wireless enthusiasts mattered little to an entrepreneur such as Marconi. He had fixed his gaze firmly on the communications grid established by the cable companies, and he sought to mimic, elaborate on, and compete with it. The aspirations of many middle- and lower-class Americans, fanned by the country's democratic myths and participatory ideology, and symbolically represented in popular culture, were relatively foreign to Marconi. Certainly they did not fit into his corporate calculations. For him, the popular arts had nothing whatsoever to do with wireless telegraphy. It was this oversight that began to drive a wedge between Marconi's original vision of his invention's applications and the ultimate use to which it was put in the United States. Within fifteen years, radio would become the vehicle through which popular culture was imprinted on electrical communications and was brought more directly and brazenly into the home than ever before. The year 1906 marked the beginning of this revolutionary trend, for, just when stories such as the ones about Willenborg began to proliferate, certain technical opportunities emerged which made emulating the boy wonder much easier.

· · ·

FROM 1906 TO 1912, when American wireless companies were on the verge of declaring or had in fact declared bankruptcy, and when the corporate sphere publicly expressed indifference toward the invention, America experienced its first radio boom. Thousands of people, believing in a profitable future for the invention, bought hundreds of thousands of dollars worth of stock in the various fledgling wireless companies. Others took even more decisive action: like Willenborg, they began to construct and use their own wireless stations. These Americans came to be known as the amateur operators, and by 1910, their use of wireless was being described in newspapers and magazines around the country. The *Outlook* outlined the emerging communications network: "In the past two years another wireless system has been gradually developing, a system that has far outstripped all others in size and popularity. . . . Hundreds of schoolboys in every part of the country have taken to this most popular scientific fad, and, by copying the instruments used at the regular stations and constructing apparatus out of all kinds of electrical junk, have built wireless equipments that in some cases approach the naval stations in efficiency."[14]

Shortly after Marconi introduced his invention, the press predicted that eventually Americans would communicate with each other using their own apparatus and would not have to rely on the telegraph or telephone. The amateurs began making these predictions a reality. Unlike their doubting elders, who thought wireless too impractical or too unremunerative, these boys believed earnestly in the new marvel and were eager to explore its possibilities. Through the popular culture, these youngsters witnessed, unhindered as yet by acquired disbelief, the unrefined and unself-conscious aspirations of the culture, especially the hope that technology could serve as the vehicle for individual and societal progress. Businessmen and military men were not part of this world, and they no doubt considered such visions of wireless unrealistic. But the amateurs were captivated by the idea of harnessing electrical technology to communicate with others, and they were not deterred by a lack of secrecy or by interference from other operators. In fact, these features, considered a major disadvantage by institutional customers, increased the individual amateur's pool of potential contacts and the variety of information he could both send and receive.

How were the amateurs able to master this particular technology? The first and most tangible development was the availability, in 1906, of an inexpensive and simple radio receiver. This was the crystal set, a device that could, for reasons that had not yet been explained, detect radio waves. The events at a receiving station were the same as those at the transmitting station, but in reverse sequence. At the transmitting end, inventors had to devise the most efficient method of generating very high frequency alternating current from a direct current source. At the receiving end, the problem was "rectifying" these oscillations: translating high-frequency alternating current back to a unidirectional pulsating current that could flow through a telephone receiver. Hertzian waves are of such a high frequency that the telephone diaphragm alone could not handle their speed or rapid reversal of direction. By 1906, both Fleming and De Forest had established the importance of rectifying incoming electromagnetic waves. Fleming's valve and De Forest's audion were sophisticated and expensive receivers that allowed the current to run in one direction only. The introduction of these receivers represented a major advance in wireless which rendered the invention less accessible, intellectually and financially, to the scientific dabbler. Wireless was becoming less simple. However, this apparent progression toward increasingly complex and expensive components did not continue unswervingly.

General H.H.C. Dunwoody of the army, also affiliated with American De Forest and then United Wireless, discovered in 1906 that carborundum (the compound of carbon and silicon), when used as a wireless detector, suppressed half of the incoming wave frequencies. A few months later, G. W. Pickard, co-founder (with John Firth) of the Wireless Specialty Apparatus Company, patented his silicon receiver based on the same principle. Inventors did not understand how the crystal worked, but they did know that it was a sensitive, durable, inexpensive receiver that was simple to operate and required no renewal of parts.[15] At the time, how and why the crystals worked as receivers was not as important as their simplicity and very low cost. Some spots on the crystal, especially the sharper edges, were more sensitive than others. The crystal was placed between two copper contact points that were adjustable so the pressure could be regulated and the most sensitive portion of the mineral selected. To keep the contact as small as possible, often a thin wire (known popularly as the catwhisker) made contact with the mineral. Because the catwhisker was "springy," it was less easily upset by vibrations or a ship's roll. Like its more sophisticated counterparts, the crystal could detect voice transmissions.[16]

The ramifications of the introduction of the crystal detectors cannot be overemphasized. The crystals contributed more than any other component to the democratization of wireless, the concomitant wireless boom, and the radio boom of the 1920s. The new receiver provided access to the airwaves to the new group of would-be Willenborgs, the amateurs. They were primarily young, white, middle-class boys and men who built their own stations in their bedrooms, attics, or garages. Although they existed throughout the country, they were most prevalent in urban areas. They earned no money as operators and had no particular corporate or professional affiliation. For them, wireless became a hobby—one that would shortly have national significance.

The amateurs' ingenuity in converting a motley assortment of electrical and metal castoffs into working radio sets was quite impressive. With performance analogous to that of an expensive detector now made available to them in the form of the inexpensive crystals, the amateurs were prepared to improvise the rest of the wireless set. They had to in the early days, before 1908, for very few companies sold equipment appropriate for home use. Also, one of the crucial components, the tuning coil, was not supposed to be available for sale because it was part of the patented Marconi system. As the boom continued, children's books, wireless manuals, magazines, and even the Boy Scout manual offered diagrams and advice on radio construction. One author instructed: "You see how many things I've used that you can find about the house."[17]

Boys also exchanged technical information with one another at school and over the air. They were especially interested in information on improved reception and accurate tuning. The amateur measured his success by how many different and faraway stations he could pick up. The basic tuning coil, based on variable inductance, consisted of a cylinder wound with wire. Mounted around the wire were variable contacts called sliders, which could be moved to make a connection at any point along the coil, thus matching the inductance of the receiving station to that of the transmitting station. By moving the sliders back and forth, the amateur could listen to one station and exclude the others. Finding suitable cylindrical objects was not always easy; some boys used broken baseball bats or old curtain rods for lack of anything else. Later, when Quaker Oats began packaging its oatmeal in cylindrical cardboard containers, these tubes became the standard core for the tuning coil.[18]

In the hands of the amateurs, all sorts of technical recycling and adaptive reuse took place. Discarded photography plates were wrapped with foil and became condensers. The brass spheres from an old bedstead were transformed into a spark gap, and were connected to an ordinary automobile ignition coil–cum-transmitter. Model T ignition coils were favorites. One amateur described how he made his own rotary spark gap from an electric fan. Another recalled that he "improvised a loudspeaker by rolling a newspaper in the form of a tapered cone and filled the room with the Arlington time signals. Everyone in the house gravely set their watches at noon each day by this means." Another inventor's apparatus was "constructed ingeniously out of old cans, umbrella ribs, discarded bottles, and various other articles." Amateurs used these umbrella ribs as well as copper or silicon bronze wire to erect inexpensive and relatively good aerials. Some amateurs, dissatisfied with the limited power that batteries provided, stole their power from the electric companies by tapping into outside electrical lines.[19] The one component that was too complicated for most amateurs to duplicate, and too expensive to buy, was the headphone set. Consequently, telephones began vanishing from public booths across America as the amateurs lifted the phones for their own stations.[20] Thus, the amateurs did not just accept and use this new technology; they adopted it as their own. They experimented with it, modified it, and sought to extend its range and performance.

The size of this burgeoning wireless network is hard to gauge. Estimates vary, but Clinton De Soto, in his history of amateur radio, asserts that "it was the amateur who dominated the air."[21] In 1911, *Electrical World* reported: "The number of wireless plants erected purely for amusement and without even the intention of serious experimenting is very large. One can scarcely go through a village without seeing evidence of this kind of activity, and around any of our large cities meddlesome antennae can be counted by the score." An operator for United Wireless wrote that one Boston manufacturer alone was selling thirty complete sets every month. The *New York Times* estimated in 1912 that America had several hundred thousand active amateur operators.[22]

Increasingly, magazines, newspapers, and popular fiction celebrated the wireless dabbling of these young men. Francis A. Collins, in his children's book *The Wireless Man,* wrote: "On every fair night after dinner-time and when, let us hope, the lessons for the next day have been prepared, the entire country becomes a vast whispering gallery."[23] In an

Two early crystal detectors. The contact point could be adjusted or changed to find the most receptive spot on the crystal.

article he wrote for *St. Nicholas* titled "An Evening at a Wireless Station," Collins changed his metaphor without sacrificing vividness or drama: "Imagine a gigantic spider's web with innumerable threads radiating from New York more than a thousand miles over land and sea in all directions. In his station . . . our operator may be compared to the spider, sleepless, vigilant, ever watching for the faintest tremor from the farthest corner of his invisible fabric."[24] These operators, "thousands of miles apart," wrote Collins, "talk and joke with one another as though they were in the same room."[25] In *Tom Swift and His Wireless Message,* Tom saved himself and his companions who were shipwrecked on a volcanic island by devising a wireless set and sending for help: "Would help come? If so, from where? And if so, would it be in time? These are the questions that the castaways asked themselves. As for Tom, he sat at the key clicking away, while, overhead, from the wires fastened to the dead tree, flashed out the messages." Finally, "from somewhere in the great void," a reply came back and all were rescued.[26] Short stories with titles such as "In Marconiland," "Wooed by Wireless," and "Sparks" appeared in popular magazines.[27] Their young male protagonists saved property and lives and won the love of a previously unattainable and beautiful young woman, all through their skill with wireless. They also got some financial reward and a better job. It was through his mastery of this technology that the protagonist's true heroic qualities of courage, selflessness, and chivalry were ultimately revealed, while he simultaneously proved himself to be an invaluable organization man.

The most avid promoter of the hobby was Hugo Gernsback, most frequently referred to as the father of science fiction. Gernsback, an immigrant from Luxembourg who came to the United States in 1904 to market a dry battery he had invented, was also a wireless enthusiast.[28] By 1906 he had opened "that great emporium of the amateur world, the Electro-Importing Company" on Fulton Street in New York; this shop was probably the first in the United States to sell wireless apparatus appropriate for home use directly to the public.[29] To promote both the hobby and his sales, Gernsback began publishing the magazine *Modern Electrics* in 1908. The magazine contained technical information and wireless boosterism. Gernsback also wrote letters to newspapers such as the *New York Times* praising "the ambition and really great inventive genius of American boys." He advised parents to encourage the hobby: "This new art does much toward keeping the boy at home, where other diversions usually, sooner or later, lead him to questionable resorts; and for this reason well-informed parents are only too willing to allow their sons to become interested in wireless."[30] How could the middle-class parent, concerned about pool halls or nickelodeons, resist such an argument?

Such celebrations of the boy wireless operator were powerfully reinforced whenever professional operators became heroes. On January 23, 1909, a dramatic and highly publicized accident emphasized the importance of wireless at sea. The White Star liner *Republic*, which was taking vacationing well-to-do Americans to the Mediterranean, was cruising in a thick fog twenty-six miles southeast of the Nantucket lightship when it was rammed by the Italian Lloyds ship the *Florida*, which was bringing hundreds of Italian immigrants to New York. The *Republic*'s engine room was pierced and immediately filled with water.[31] Two people whose sleeping berths were at the point of collision were killed instantly. Others were injured. The two ships were carrying, between them, more than twelve hundred passengers. The *Republic* began to sink and the crews from both ships started transferring the *Republic* passengers to the *Florida*. It was not clear whether the Italian ship, which was half the size of the *Republic* and had a severely damaged bow, would be able to reach port with such a heavy load. There was panic and chaos on both ships. But "then came the wonder of modern knowledge," reported *Harper's Weekly*: "Out of the heart of the fog, far and wide, to all points of the compass, Captain Sealby flashed by wireless telegraphy word of the peril and requests for assistance—a general ambulance call of the deep sea." Several ships responded; they "turned in their tracks and headed for the far-off voice that summoned them!" "What a wonder-tale it is," continued the *Harper's* editorial, "and how deeply moving— the cry for help thrown out into the air from a mast-tip, and caught, a hundred miles and more away. . . . It is a new story; there was never one quite like it before."[32] The *Baltic* reached the *Republic* and *Florida* first and took on all the survivors from both ships. The *Republic* sank, and the other two ships returned to New York. The "marvels of wireless telegraphy" were front-page news for four straight days, dominating the first three pages of the *New York Times*, which called the accident "the greatest shipwreck in years."[33] Wireless, announced *Harper's*, "has robbed accident by sea of half its terrors. No longer need the passengers of a wrecked ship scan the horizon hopelessly while the sea pours into the hold and, inch by inch, Death gains his footing. For an invisible network of ethereal communications unites ship to ship."[34]

But the real hero of the hour was a "youngster of twenty-six who became famous in a day," the *Republic*'s wireless operator, Jack Binns. The *Florida*, which was carrying poor Italian immigrants who were

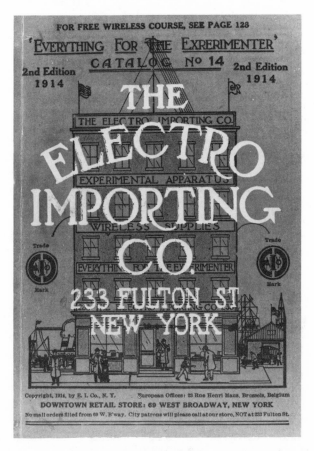

**Advertisement for Hugo Gernsback's Electro-Importing Company,
which specialized in amateur wireless apparatus.**

fleeing the recent disastrous earthquake in Sicily, was not equipped with wireless apparatus, so the duty of calling for help fell solely on Binns's shoulders. "With the wall of his metal cabin splintered and shattered by the knife-bow of the Italian liner," reported *Putnam's*, "Binns stuck to his instrument all through the dreary day, sending, sending, sending the hurry call of the sea—CQD! CQD!" When the *Republic*'s engine room flooded and the ship's power went off, Binns had to shift to reserve storage batteries for transmitting, and this cut the radius of his messages by one-half. Yet, he still "sat at the key for many hours without respite, till help arrived."[35] He communicated primarily with H. G. Tattersall, the operator on the *Baltic*, who sat at his key for fifty-two hours until he collapsed. In the end, Binns was "drenched and hungry," "exposed to the weather, his hands so stiff from the cold that it was with difficulty that he could hold the key of his instrument."[36]

By the time he reached New York, Binns was lionized. Reporters hounded him for interviews; crowds followed and cheered him from the moment he set foot on dry land. Congressmen made speeches urging that he be immortalized.[37] Managers of music halls and museums "vainly tempt[ed] him to go on the stage and pose."[38] One vaudeville chain offered him one thousand dollars a week for ten weeks to appear on its billing.[39] Two days after Binns returned to New York, his friends dragged him onto the stage of the Hippodrome, where he was forced to make a speech and was then swarmed over by all the chorus girls, who tried to hug and kiss him.

Binns, like Marconi ten years earlier, was the perfect media hero. He fit perfectly the literary conventions and cultural expectations of his times. He was a real-life Tom Swift. It was through his technical mastery that he had proved himself a man. He had been selfless and vigilant, impervious to the threat of failure or the risks of danger. Physical combat had not been important; physical endurance had been critical. He was "young, boyish, quite immature in appearance, but possessed . . . iron nerve and dauntless resolution."[40] And he was modest, even self-effacing. He refused to admit "that he did any more than any of his fellow-operators would have done in like circumstances."[41] Under a large portrait of him in *Harper's Weekly* appeared the caption "Why, I didn't do anything."[42] He even refused the many offers to benefit financially from his newfound celebrity, which prompted one *New York Times* reader to write, "Good for you, Jack Binns! May your example be followed by future newspaper 'heroes.' "[43] After the flood of attention, *Putnam's* reported that Binns confided to a friend, "I can't stand any more of this. I never want to see my own picture again."[44] By refusing "to become an object of gaping and vulgar curiosity," editorialized the *New York Times*, Binns showed "the delicacy and dignity which not a few real heroes have lacked."[45] "The true test of a hero," lectured *Putnam's*, "is the manner in which he takes his ovation."[46] Binns's aw-shucks approach served to fan the embers he hoped to douse.

Several publications used the incident to romanticize the work of all wireless operators. They noted that many amateurs went on to become professional operators, and that the variety and excitement to be found in wireless work made the job especially attractive to youth. These operators were reportedly educated men who had "technical college training" and came from a "superior class." On board ship, the operator was "an important personage" who ranked as an officer and took his orders directly from the captain. But what operators liked best was "the sensation of sitting in a quiet cabin, with untold ohms of power beneath their fingers."[47] According to a *Harper's* reporter, all the operators were heroes: "Going around among them and hearing them talk of their work convinces one of the incalculable service they render daily to modern navigation. And it is a service that is rendered with an admirable *esprit de corps.*"[48] It was repeatedly said that the operators formed a brotherhood.

Joining such a fraternity of tough, independent men who were always potentially at the edge of danger possessed no small portion of romantic allure. One could become part of the brotherhood without actually going to sea; taking to the air was sufficient stimulation for many. For, even though a young man might be secure in the comfort of his own home, his life did become more exciting through radio. The amateurs came to feel that their lives were intertwined with truly significant events, as they overheard messages about shipwrecks or political developments and transmitted these messages to others. As one amateur recalled, "We were undoubtedly romantic about ourselves, possessors of strange new secrets that enabled us to send and receive messages without wires."[49] Amateurs who heard Jack Binns's distress signals became celebrities by association. One remembered: "The few boys in school in the area who claimed to have received the distress call were local heroes for a time, and they made a number of converts to the radio amateur hobby among the more technically minded youngsters."[50] Hearing any news first, the night before other Americans would read it in the newspapers, imbued the amateur with an aura of privilege, of being "in the know." As Francis A. Collins wrote in *The Wireless Man,* "Over and over again it has happened that an exciting piece of news has been read by this great audience of wireless boys, long before the country has heard the news from the papers. . . . A wide-awake amateur often finds himself independent of such slow-going methods of spreading the news as newspapers or even bulletin boards."[51] Many amateurs learned of the outbreak of World War I from the Marconi Cape Cod station hours before the newspapers announced the story.[52] The amateurs were tapping point-to-point messages meant only for certain ears, not broadcasts intended for everyone. They could feel part of an inner circle of informed people because they heard the news as it happened.

As important as being privy to such spectacular but rare transmissions was the novelty of contacting other people, strangers, through space. One operator claimed that amateurs using only "a wire strung up

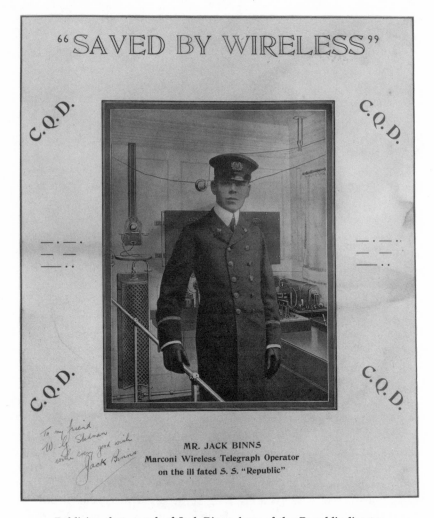

Publicity photograph of Jack Binns, hero of the *Republic* disaster.

like a clothesline between trees" were able to pick up "many long-distance messages." He added, "I know one boy who catches messages sent out from Panama."[53] Although many operators hoped to hear dots and dashes coming from "thousands of miles away," making contact over a distance as short as ten or fifteen miles was reportedly a "thrilling experience."[54] In a culture that was becoming more urbanized, and whose social networks were becoming increasingly fragmented, many strangers became friends through wireless. The amateurs loved the contact with invisible others in a realm where one was faceless and yet known at the same time. The fraternity that emerged possessed the fellowship felt among pioneers. These young men were exploring and comparing their findings on a relatively uncharted and mysterious territory. As one amateur explained, "The eagerness and frankness in distributing the results of our findings undoubtedly molded the form of fellowship which is such a striking quality of the amateurs."[55] Yet the fraternity was also exclusive. Working-class boys with neither the time nor the money to tinker with wireless could not participate as easily. Neither could girls or young women, for whom technical tinkering was considered a distinctly inappropriate pastime and technical mastery a distinctly unacceptable goal. This fraternity, whether self-consciously or accidently, brought together roughly similar men in a region uninhabited by those so different from them: women, immigrants, blacks, and boys ignorant about electrical technology. That the amateurs engaged in contests of strength and speed with one another did not undercut the mutuality of the enterprise; in fact, the contests simply strengthened the bond.

A revolutionary social phenomenon was emerging. A large radio audience was taking shape whose attitude and involvement were unlike those of other, traditionally passive, audiences. Collins summarized the development in *The Wireless Man:* "An audience of a hundred thousand boys all over the United States may be addressed almost every evening by wireless telegraph. Beyond doubt this is the largest audience in the world. No football or baseball crowd, no convention or conference, compares with it in size, nor gives closer attention to the business at hand."[56] This was an active, committed, and participatory audience. Out of the camaraderie of the amateurs emerged more formal fraternities, the wireless clubs, which were organized all over America. One of the first of such clubs, the Junior Wireless Club, Ltd., was formed in January 1909, just after the *Republic* disaster. It began in New Jersey with only five members; elected as president was eleven-year-old W.E.D. Stokes, Jr. In the same year Hugo Gernsback started the Wireless Association of America, which by 1910, Gernsback claimed, had ten thousand members. The Children's Museum, under the direction of the Brooklyn Institute of Arts and Sciences, erected its own wireless plant so youngsters could investigate wireless, learn how to build their own sets, and meet other young operators. Many high schools established wireless clubs to promote the hobby.[57] By 1912, the *New York Times* estimated that 122 wireless clubs existed in America.[58] Most of the club meetings took place "in the air," on a prearranged wavelength. The chairman called the meeting to order by sending out his call letters, and the members signified their attendance by answering with their own. During these meetings, the amateurs usually shared technical problems and solutions, discussed technical advances, and drilled one another on transmission skills. A Chicago wireless club broadcast a program every evening, "as a matter of practice for amateur operators in receiving. The bulletin usually consisted of an article of some electrical or telegraphic interest . . . [and] sometimes the program was varied by sending passages in foreign languages, to quicken the receiving ears of the amateur operators."[59]

Gradually, an informal wireless network was established as the different clubs relayed messages for one another to points too far to reach with most amateur sets. One historian has noted: "Message handling— for pleasure, for friends, in time of emergency—was rapidly becoming the predominant theme in amateur radio." In March of 1913, a severe windstorm in the Midwest blew down the telegraph and telephone lines, but local amateurs were able to maintain the region's communication by relaying messages in and out of the area. Such impromptu public service gestures led some amateurs to advocate better organized intercommunication among operators. One radio enthusiast, the inventor Hiram Percy Maxim, believed that the amateurs needed a national organization to establish a formal relay system or network to serve all amateurs. Through his Hartford Radio Club, he contacted amateurs in March 1914, inviting them to join a league and to convert their stations into official relay stations. The name of his organization was the American Radio Relay League (ARRL). The response to his invitations was so enthusiastic that within four months the league boasted two hundred official relay stations across the United States.[60] Thus, in 1914, there existed in America a grass-roots, coast-to-coast communications network. When ARRL was formed, *Popular Mechanics* proclaimed "the beginning of a new epoch in the interchange of information and the transmission of messages." Wireless telegraphy had "made it possible for the private citizen to communicate across great distances without the aid of either the government or a corporation."[61]

For the first time in America, men were being invisibly bound together by and in the airwaves, not by necessity, but for fun, to learn from and to establish contact with others. Those involved in the new hobby saw larger-than-life reflections of themselves in popular books, magazines, and newspapers. We cannot tell whether popular culture helped increase participation, but we do know, from the reminiscences of amateurs operating at the time, that popular culture articulated the hopes and dreams invested in wireless. Technology gave the amateurs the means to communicate without wires. Popular culture sustained their visions of being on the cutting edge of technological progress and of being the kind of men truly prepared for modern society.

. . .

THE INCREASED PRESENCE of amateurs in the airwaves led to a struggle for control of the ether, a struggle that pitted the amateurs against the U.S. Navy. The nature of this struggle was only hinted at in the stories about Walter Willenborg. Willenborg's conquest of the Atlantic Highlands operator, who was forced to stop trasmitting until Willenborg was good and ready to let him proceed, was protrayed by the *New York Times* reporter as amusing and even enviable ethereal jousting. It was boyish prankishness, and boys, after all, will be boys. As long as the boy wireless heroes were cast in the mold of fictional characters like Frank Merriwell and Tom Swift, a few good-natured pranks could do no harm. However, the sort of deliberate interference that Willenborg practiced began to get out of control, and to the military, in particular, it ceased to be in any way innocent or amusing. Not all amateurs, it turned out, were selfless heroes like Jack Binns.

It is important to remember that by 1910, the amateurs surpassed both the U.S. Navy and United Wireless in numbers and, often, in quality of apparatus. Between 1906 and 1910, the navy had installed wireless on many of its ships, and while the invention was still not well integrated into naval operations, it was being used more frequently. There were hundreds of high-power amateur stations, however; navy and commercial stations comprised only 15 or 20 percent of the total.[62] Many homemade stations could transmit only up to fifteen miles and pick up messages from forty miles away, but some commercially produced apparatus, which cost thirty dollars or less, was guaranteed to receive and send messages up to three hundred or four hundred miles.[63] In addition to this power, the amateurs, who were able to ignore patents, also had tuning. As one historian has noted, "The fact that amateurs had tuners prior to 1910—even though they were nothing more than long coils of wire with a sliding contact—lent them a tremendous advantage over the inadequately equipped stations of United Wireless and the U.S. Navy, which used obsolete untuned, or at best direct-coupled equipment for several years thereafter."[64]

The spectrum became more crowded as too many enthusiasts, many of them beginners, clogged the air with all sorts of transmissions. They "gossip about everything under the sun," reported one operator. "They ask each other for the baseball or football scores, make appointments to meet the next day, compare their lessons. And they quarrel and talk back and forth by wireless in regular boy-fashion."[65] Amateurs spoke with one another and with willing commercial operators. The United operators aboard the coastal ships worked long, uneventful hours, and they were more than happy to have the amateurs keep them company. As one amateur recalled (painting an image contrary to the glamorous one of Binns and his compatriots), "The operators on those ships put in boresome hours. . . . People were not great users of wireless ship-to-shore messages as they are today. . . . There was absolutely no limit to the conversation service we would get from the ships."[66] Francis Hart, a wireless operator in New York City from 1907 to 1911, described the congestion in his logbook: "The different kids around here raise an awful noise, all try to talk at once, call when anybody is in and never use any sense, half can't read 4 words a minute and sit calling everybody within 20 miles and can't hear 800 feet from another station." He commented on one amateur's conversations: "FH is a very good reader, but he tries to say too much at one time, then the poor reader makes him repeat it and they keep that blooming business up for hours."[67] While this sort of interference increased, so did "malicious" interference, which began to give the amateurs as a group a bad reputation.

Some amateurs deliberately sent false or obscene messages, especially to the navy. The temptation to indulge in such practical joking was enhanced by the fact that detection was virtually impossible. Amateurs would pretend to be military officials or commercial operators, and they dispatched ships on all sorts of fabricated missions. Navy operators would receive emergency messages about a ship that was sinking off the coast. After hours of searching in vain, the navy would hear the truth: the "foundering" ship had just arrived safely in port. In January 1910, approaching the Nantucket lightship, the steamliner *Bremen* asked for a report on weather conditions and location. The *New York Times* gave this account of what ensued: "She was answered, and was much shocked and surprised to find that the reckoning as computed by her navigator was apparently in error. Information that was untrue was also given regarding weather conditions. On reaching New York the matter was looked into, and it was discovered that the lightship had been sunk a week prior to the sending of the messages, and the work was that of amateurs." Navy operators at the Newport Naval Yard complained that amateurs sent them profane messages. *Electrical World* reported: "Attempts to report the recent naval experiments in Chesapeake Bay by wireless met all kinds of trouble from the omnipresent amateur, with his often crude, but sometimes perniciously effective, sending apparatus."

The *Outlook* warned: "The efficiency of a number of the coastal stations of the Navy has been cut in half because of the presence of dozens of small amateur stations. Boston is the head of some five hundred amateur operators, a number of whom can put the Charlestown Navy Yard completely out of commission as regards the receiving of messages."[68]

Some amateurs charged that commercial operators helped create the malicious interference and were taking advantage of the amateurs' increasingly bad press. *Electrical World* repeated the claim: "Much of the present day trouble arising from interference is attributable less to the activities of amateur operators than to the studied attempts of certain of the operating companies to interrupt the continuity of messages being sent by [their] rivals."[69] The chief electrician of the New York Navy Yard maintained that the signals from some commercial stations were so loud that if the receiver was "anyway near on their tune the telephone [could] be laid on the table and the sound heard at any distance in the room away from them."[70] Hart's logbook for 1909 contains many complaints about interference by Marconi operators. A Marconi operator aboard the *St. Louis*, Hart recorded, had been "breaking everybody for the last three nights, while lying at dock sending 'Vs' and general nonsense which is not uncommonly done by the Marconi people in the harbor." Hart resented being treated in that "absolutely pigheaded English way." He was especially irritated by the Marconi interference, because the Marconi Company, he wrote, already had a monopoly.[71]

The situation was untenable: too many people wanted access to the airwaves at the same time, and there were no guidelines for establishing priority. The commercial wireless companies did not take this contest into the public arena; interfering with one another was part of business, and they could often tune out or overpower the amateurs. When technical might failed, gentlemen's agreements prevailed. For example, the one hundred members of a wireless club in Chicago worked out an air-sharing agreement with the local commercial operators which was designed to reduce interference for both groups.[72] For the private companies, especially Marconi, public arguments over interference and congestion were undesirable because they suggested that wireless was still unreliable and prompted mention of regulation. The amateurs resented getting blamed for commercial operators' mischief, but their argument was basically not with the private wireless companies. Rather, as the battle lines over ethereal right of way became more clearly drawn, the amateurs began to defend themselves against a public campaign waged by the U.S. Navy.

Military officials began lobbying in Washington for stricter regulation of or elimination of the amateur. Their justification for such suppression was difficult to dispute: safety at sea and national security. In its cover story on the U.S. revenue cutter *Gresham,* "life-saver of the rocky New England coast," *Scientific American* echoed the military position that the ship "depends on her wireless apparatus to keep informed of the location of wrecks, and her urgent dispatches have at times been delayed for hours by the working of amateur stations." Naval operators asserted that because they transmitted important official messages, they should have priority and should not be interfered with by children. A New York City naval operator considered having to yield the airwaves to youngsters when he was trying to conduct government business preposterous. He was prevented from transmitting one evening while "a couple of kids in New Jersey compared the results of their arithmetic lessons."[73] The navy now had another chance to try to take over the operation of America's wireless stations. Amateur meddling in "official government transmissions," which included emergency messages, provided an excellent argument for military control. The navy continued issuing statements about the grave danger posed by the amateurs, and cited many instances of unpatriotic interference. The Navy Department's 1909 *Annual Report* noted the increased numbers of "seemingly semi-intelligent and wholly irresponsible operators" who "at any time through carelessness or stupidity may render hopeless the case of a shipwreck." The only solution, the navy argued, was for "laws governing the conduct of all wireless stations" to be passed.[74]

The anonymity made possible by wireless had a leveling effect on the status and power of naval officials: in the airwaves, rank was irrelevant; only technical strength mattered. The wireless amateurs expressed traditional American delight in flouting authority, and their target was irresistible. If you wanted to challenge authority and show how tough, how manly, how unafraid, you were, who better to take on than the new navy?

The amateurs fought the navy's efforts to restrict their activity. They argued with navy operators in the air and tried to embarrass the navy in the press. The *Outlook* reported that during what naval operators claimed was an emergency situation, amateurs refused to clear the air, "some of the amateurs even arguing with the Navy men over the ownership of the ether." In another instance, a Boston amateur, "told by a naval operator to 'butt out' made the following classic remark: 'Say, you navy people think you own the ether. Who ever heard of the navy anyway? Beat it, you, beat it.' "[75]

Many amateurs were justifiably rankled by the navy's charges. The press agreed that most amateurs were polite and considerate and did not transmit false or obscene messages. Many amateurs had, in fact, helped to save lives or relay critical messages, only to be blamed for interfering. The May 1912 issue of *Electrician and Mechanic* recounted one such incident, in which the naval ship *Terry* transmitted distress signals to the New York area: "The first news that the Brooklyn Navy Yard had of the *Terry*'s distress was picked up by an amateur in Bayonne, New Jersey, and relayed to the government operator." Rather than appear incompetent because he had failed to pick up the *Terry*'s initial call himself, the naval operator blamed the local amateurs for interfering.[76] A wireless amateur wrote to *Scientific American* to cite another instance in which amateur aid would have been crucial: "While experimenting with my receiving set at a point on Buzzards Bay, Massachusetts, I once heard a ship call every naval and commercial station from . . . Fire Island, New York to Boston, for an hour, without getting a reply. *Suppose this had been an urgent distress call?* That ship would have gone down with the call unanswered unless I or some other 'meddler' had given the alarm."[77]

The amateurs issued charges of their own. They claimed that the navy relied on antiquated apparatus and that most naval operators were incompetent. The amateurs believed that they were becoming scapegoats for the navy's internal problems. To illustrate that the navy was slow to update its equipment, amateur operators made comparisons between the performance of amateur and commercial apparatus and that of the navy equipment. Demanded one amateur, "Why can the majority of amateurs in Washington, D.C. pick up messages from longer distances than the naval station, which is equipped with an aerial five to ten times larger than theirs? . . . Why do the majority of amateurs here get long distance signals louder than the government station?"[78] The answer, the amateurs charged, was that the navy's equipment consisted of "non-receivers"—outdated transmitters of insufficient power—and that naval apparatus had only the crudest form of tuning.

The amateurs were also quick to point out that navy operators were amateurs themselves, who were often very slow and were unfamiliar with both the wireless code and the apparatus. While many of the amateurs had been transmitting as well as tinkering with their own apparatus for years, and then had trained for five years to become first-class commercial operators, naval operators only went through a two-month training course before manning a wireless station. (Prior to 1912, the training period for naval operators was often less than two months.)[79] John Purssell, a Washington, D.C., amateur, sent his account of a navy operator's ineptitude to *Scientific American*. One evening Purssell was listening to communication between Baltimore and the Washington Navy Yard. Purssell had received Baltimore's message without any difficulty.

> After some time the navy operator said "Please go ahead a little slower," and Baltimore thereupon repeated the message slowly in perfect Continental. The navy yard then got part of it, but asked him to repeat a portion of the middle. This time the navy operator got a word or two more, but had to have the message four times before he got it all. . . . On another occasion the navy operator at Norfolk, after trying several times to get a message through to Washington at a rate of about five words per minute, gave up in disgust, and said: "For the love of Mike, get an operator."[80]

Purssell alleged that the naval stations operated at an average speed of twelve words a minute, a rate Marconi had established at the turn of the century. Another *Scientific American* reader wrote to the editor to corroborate the claim that naval operators used the amateurs as scapegoats: "Whenever an operator in one of the government stations has any difficulty with his apparatus, or misses out on a message through his own inability, a very simple and efficient excuse is, 'Couldn't get a thing through the amateurs.' "[81]

Finally, the amateurs, who often eavesdropped on "official" messages, challenged the navy's assertion that it should have automatic priority in the air. The amateurs charged that many of the government's messages were no more urgent or weighty than anyone else's. Francis Clay, NESCO's attorney, supported this charge with an example of his own. His clients were trying to transmit crucial information concerning an important lawsuit involving a considerable sum of money. When the operator sought to send the message, "he was held up and informed that the League Island Navy Yard was using the air in sending a message and that it had priority. It was afterward learned that the wife of an officer stationed at the League Island Navy Yard, who had gone downtown to do some shopping, had a message sent asking for her Mackintosh and rubbers. She feared that she would get wet, as a rainstorm was coming."[82]

The amateurs received some support in their complaints, which were not without merit. The wireless companies, which for years had been complaining about the ineptitude of naval operators, echoed the amateurs' criticisms. Since 1904, NESCO had been urging the navy to train its operators properly so they would stop damaging the apparatus and would send messages coherently to other operators. And, while the press did not condone the amateurs' mischief, it did not consider the navy to be blameless. *Electrical World* described, for example, how "some high school boys . . . threw the Newport naval wireless station into a succession of fits with an old Morse key, a broken incandescent lamp and a few batteries." These youngsters, continued the journal, "gave incidentally a valuable demonstration of the need of improvement, to which we trust the Navy Department will give due heed. If such rudimentary equipment accomplished so considerable a disturbance, what would happen if a hostile fleet went deliberately to work with powerful and skillfully devised apparatus? . . . The Navy system as now used could be hopelessly tangled up, without going to much trouble."[83]

Corroboration of these criticisms came from within the navy itself. In 1911 and early 1912, Lieutenant Stanford C. Hooper was assigned to observe the use of wireless during naval target practice and submit a report to the Navy Department. Hooper had many criticisms of the navy's failure to integrate wireless into its strategic operations, and the operators were not spared: "About one-third of the operators are not operators and delay the general business about one-half," reported Hooper. "The speedier operators often 'burn it in' a little too hard just to show off, now and then, causing needless repetition." He urged that the operators master the Continental code and increase their transmission speed from their current ten to eighteen words per minute, noting that the navy's standard was about half that of commercial operations. Hooper blamed this low standard on insufficient training, lenient examinations, slowness in weeding out inept operators, and the great disparity between naval and commercial operators' salaries. Hooper, turning the navy's complaints back on the navy, warned the department: "The wireless is running away from us in certain regards. Three times during the past week have I heard the commercial stations legitimately complaining about the fleet's interference."[84] Hooper could not believe that the navy had not yet adopted cipher, especially given the lack of secrecy in wireless transmissions and the ability of any youngster to send messages claiming he was an admiral. He also acknowledged that the national security argument, which claimed that naval messages should have priority because they were official messages, was a fig leaf covering naval misuse of the invention. Hooper later recalled that officers exerted little control over wireless and, thus, that each operator was free "to send whatever he pleased. . . . There were more personal than official messages and more operator conversation than messages."[85]

The amateurs could not accept the navy suddenly stepping in and claiming the airwaves for itself in the name of national security when the navy had done little to ensure that wireless would help preserve that security. According to the amateurs, several intermediate steps, such as adopting cipher, improving the quality and power of the apparatus, training the operators, and developing tuning, should be taken by the navy before it asked for regulatory measures to ban individuals from the air. But the navy did not base its claim on its contributions to or investments in wireless. Rather, the navy argued for priority on the basis of its assigned role as a defender of American society. The amateurs, on the other hand, asserted that they had as much, if not more, right to explore the ether because they had worked and experimented to earn that right. The ether was a national resource, they argued, a newly discovered environment, and the amateurs considered that their enthusiasm and technical spadework entitled them to a sizable portion of the territory. Much as nineteenth-century pioneers had obtained squatters' rights by cultivating the property on which they settled, the amateurs had developed a proprietary attitude toward the airwaves they had been working in for the past five years. There were a few outlaws in the group, but their alleged violations should not mean, argued the amateurs, that all individual operators were to be excluded by the government.

· · ·

THE EMERGENCE OF this grass-roots network of boys and young men marks the introduction of yet another way of using and thinking about wireless and the ether which contributed to the social construction of broadcasting. To the amateurs, the ether was neither the rightful province of the military nor a resource a private firm could appropriate and monopolize. The ether was, instead, an exciting new frontier in which men and boys could congregate, compete, test their mettle, and be privy to a range of new information. Social order and social control were defied. In this realm the individual voice did not have to defer to the authority of business or the state. This realm, argued the amateurs, did not belong to hierarchical bureaucracies: it belonged to "the people." Thinking about the ether in this way, and acting on such ideas on a daily basis, was a critical step in the transformation of wireless into radio broadcasting.

As their battle with the navy intensified between 1910 and 1912, the amateurs and their spokesmen relied increasingly on democratic rhetoric that described the air as being free and the property of the people, for whom the amateurs tried to suggest they were the proper surrogates. But the amateurs ignored the fact that the spectrum was a common property resource with boundaries both around and within it. As overpopulation and overuse of this resource intensified, its value to all users diminished, but at the time, no standards existed for allocating property rights in the "folds of the night." The enthusiasm with which legions of amateurs took to the air alerted some in the government to the fact that such standards would have to be established soon. The nation would have to come to terms with the question "Who owns the airwaves?"

CHAPTER SEVEN

THE *TITANIC* DISASTER AND THE FIRST RADIO REGULATION

1910–1912

BY 1910, WIRELESS had been a part of America's cultural and economic landscape for a decade. Its use had not, as yet, been regulated in any way. Unlike the European nations, which had agreed in 1903 and 1906 to endorse international treaties regarding wireless communications, the United States had not considered such regulation pressing. For six years, from 1904 to 1910, the wireless companies and the amateurs successfully lobbied against wireless regulation in America. These groups' objections to the treaty of the 1906 International Wireless Conference—that it was premature, technically naive, and restrictive; that it was overly generous to the Germans and exploitative of American inventors; and that it transformed wireless into an instrument of warfare—persuaded congressmen to vote against any wireless legislation that resembled the 1906 treaty. But the power of these lobbying efforts should not be over-emphasized. That the regulation of wireless was a low legislative priority is an understatement: Congress was wrestling with antitrust legislation, child labor laws, the Pure Food and Drug Act, and a host of other major legislative controversies during the first decade of the century. All of these issues either had been instigated by books or magazine articles or had been accompanied by ongoing intense journalistic scrutiny. Many important laws were designed by well-organized and entrenched industries seeking state intervention and support. By contrast, the wireless companies in America were small, disorganized, and fiercely competitive, and they had no common interests that regulation might preserve. Also, wireless had received attention in other contexts, but not as a regulatory issue of any import. Its importance as an item on the regulatory agenda paled beside child labor or the meat-packing industry.

Only when wireless was connected to many more lives in a much more critical way would it be a major item on that agenda.

The fact that radio was still so technically and economically undeveloped was also influential in forestalling congressional action. Scientists and academics continued to argue over wireless theory and practice; congressmen could hardly have felt comfortable trying to regulate a young science that was still so poorly understood. Henry Cabot Lodge noted when explaining his "grave doubts as to the wisdom" of ratifying the 1906 treaty, "Personally I confess I do not understand the questions involved and I certainly should not be willing to vote until I am fully informed."[1] Most importantly, no precedents existed for bureaucratic management of such a method of communications. The telegraph, wireless's nearest technical relation, had remained in private hands in America. Again, the American record was quite different from Europe's. In Britain, France, and Germany, for example, telegraph systems were owned by the government; these countries had regulatory precedents to help them confront the regulation of wireless. The United States did not.

At first, what most frequently prompted talk of regulating wireless were complaints about interference. The wireless companies continued to promise technical solutions to interference which would render regulation superfluous. Inventors had finally stopped claiming that the number of available wavelengths was infinite; they now acknowledged that the spectrum was in fact a limited, finite resource. They asserted, however, that through technical advances, each user would soon be taking up less space in the spectrum. They claimed that new and refined transmitters, which emitted more defined, less damped wave trains, took up "narrower" bands, allowing room for more users in the airwaves. In 1909, the *New York Times*, envisioning a technical solution, asserted: "That the difficulty of interference, like that of confining each message to a straight line between the sending and receiving instruments, will finally be overcome nobody much doubts."[2] Three years later, the paper claimed that regulation was already anachronistic because, "as everybody knows, Mr. Marconi has devised a method of preventing 'interference' between different stations by suitable modifications of the wave lengths."[3] A congressman reading this statement might well have asked himself why he should enact legislation if a technical allocation of property rights in the ether was imminent.

Legislators no doubt preferred to wait for the promised technical arbitration because they were faced with an issue that was extremely complicated and emotionally charged—so much so that it is still being debated today: What criteria should Americans use to assign and protect property rights in the spectrum? The first intellectual leap required in addressing such a question was thinking about something that was invisible, all pervasive, seamless, and still quite mysterious as property. This was not easy for most Americans to do. Americans understood all too well that tangible things—machines, raw materials, a piece of land—were property people vied for, bought and sold, and used to get still more property. But all of these could be seen, touched, and measured. The ether could not. In addition, there was considerable confusion about what, exactly, the ether was; many referred to it simply as the air. The air was an element Americans had traditionally associated with freedom, even transcendence. As Congressman Ernest W. Roberts of Massachusetts put it, "We have been brought up with the idea that the air was absolutely free to everyone."[4] How could something people thought was free and impossible to partition actually become property?

Having made the leap to thinking about the ether as property, one confronted the next question: What kind of property? Exclusive, private property rights could not be established in this domain the way they had been in real estate, for the ether could not be broken up into discrete plots. It was intangible and could not be bought or sold in quite the same way. Rather, the ether, like the oceans or wilderness areas, was a resource held in common in which all Americans potentially had an interest and in which walls or fences could not be built. This may seem quite obvious to us today, but coming to the realization that the ether was both a resource and one Americans had collective rights to was just as difficult as thinking of the ether as property at all. Some tried to resolve the dilemma by arguing that because everyone had a stake in the ether, no one person or group could be assigned overarching rights. Congressman Roberts maintained: "It has always been understood that a man owning real estate owned to the center of the earth and the heavens above and controlled everything above and below the surface of the piece of land he happened to own."[5] Thus, if every property owner in America also owned a small tract of the spectrum, his property rights would be violated if the ether was used without his consent or contrary to his wishes or interests.

Common property resources pose very special and vexing problems. Although the uses to which a common property resource is put can affect entire populations, and although many people believe that because it is a resource held in common, everyone has the right to exploit it, if a common property resource is opened to all, its value is destroyed. As individuals or institutions try to increase their enjoyment or use of the property, it becomes overpopulated, polluted in a variety of ways, and of less value to everyone. Freedom in the commons, so ideologically appealing, in fact "brings ruin to all."[6] Given this reality, which was being powerfully demonstrated in America's congested airwaves, the dilemma became clear. Who decides who will gain access to the commons, and what will the criteria for access be? These were not easy questions.

In 1910 the groups vying for access to the ether were the military, the wireless companies, and the amateurs. The military, citing national security reasons, had a socially and politically valid claim, but bureaucratic control was unpopular in the press and in Congress. The amateurs represented independent, individual access, which was sentimentally appealing but increasingly disruptive. Companies such as Marconi and United Wireless had commercial claims on the ether, some of them politically persuasive and others less so. All of these interest groups had to be considered in any political solution to the interference and overpopulation problem. Congress was being asked with increased frequency between 1908 and 1912 to limit admission to this common property resource and to decide whose claim to the ether was valid, who had a right to transmit in a given area. Establishing such a hierarchy was an unprecedented and unwelcome task, and Congress postponed acting until events forced its hand.

The two radio-related issues confronting Congress were increasing interference and shipboard safety, and the two were intertwined. Interference was not a problem one hundred miles out at sea, but in American ports the cacophony was frustrating and dangerous. Not all ships were wirelessly equipped, which meant that some passengers were unprotected in case of an accident. Yet to equip every ship was to increase ethereal population, and passengers would not necessarily be better served, because ships equipped with wireless would still be helpless if they could not get their distress calls through the clamor. Congressmen began introducing bills to codify the use of radio. From 1910 to 1912, wireless successes and failures at sea provided the catalysts for legislative action.

The collision of the *Republic* and the *Florida* in January 1909, in which wireless played a central role in saving people's lives, precipitated the first government regulation of wireless in America, the Wireless Ship Act of 1910. If Jack Binns had not had his wireless, more than twelve hundred people might have died. Wireless had been installed early on the large, luxurious ocean liners such as the White Star liner *Republic* which catered to the more privileged classes. At sea, wealth assured access to the resource. But ships less grand, those that transported immigrants or steerage passengers, like the *Florida,* usually had no wireless aboard. If

Congress could not agree on how to assign rights in the spectrum, it could appreciate the merits of making wireless mandatory equipment aboard ship. Congress assumed its protector role, noting that the ingenuity of American inventors and America's "open door . . . to hundreds of thousands of immigrants" annually were compelling reasons "to take the lead in legislation."[7] On February 8, 1909, President Roosevelt sent a message to Congress urging, in light of "recent events," the quick passage of legislation making shipboard wireless mandatory.[8] By Feburary 18, the House Committee on Merchant Marines and Fisheries had favorably reported out such a bill, but Congress did not act on it prior to adjournment on March 3. Not until more than a year later did the Sixty-first Congress enact wireless legislation.

On June 24, 1910, the Wireless Ship Act was passed. It provided that any oceangoing steamer sailing in or out of United States ports, carrying fifty or more persons, and plying between ports two hundred miles or more apart be equipped with "efficient apparatus for radio-communication, in good working order, in charge of a person skilled in the use of such apparatus." The apparatus was to be capable of receiving and transmitting messages over one hundred miles, day or night. Intercommunication between competing systems was mandatory. The law was to go into effect in July 1911, giving shipowners one year to equip themselves.[9]

Arguing against the Wireless Ship Act was difficult. The measure sought to democratize the advantages of wireless at sea, and it provided enterprising wireless concerns with additional business, because many ships still needed to be equipped. But rather than limit access to and movement within the ether, this legislation mandated access for still more people. The 1910 act officially recognized the importance of wireless as a life-saving device, yet that usefulness was being eroded by the ever-increasing number of wireless transmitters, an increase the 1910 act fostered. Thus, while passengers on most ships now had potential access to the ether, that privilege was made less valuable to all by the overpopulation. The law exacerbated interference.

Congress had addressed the issue of using wireless to improve shipboard safety. As reports describing malicious interference increased, military officials continued to appeal to Congress to enact legislation to remedy that problem as well. Secretary of the Navy G. V. L. Meyer charged that "vicious" California amateurs had "tapped" official messages originating from Mare Island and leaked them to "sensational newspapers" for publication. Were there no rights to privacy in the ether which should be backed up by law? Charles Norton, acting secretary of the treasury, submitted excerpts from the wireless logs of revenue cutters to document the altercations and standoffs in the air waves. The log of the *USS McCulloch* for November 4, 1909, read:

> 3.20 P.M. called TI, sent him an official message; when I listened in for acknowledgement or OK for our message, CH (United Wireless) operator CX, maliciously broke in on us and said "we will show RCH (McCulloch) that our spark is stronger than his and drown him out." 3.35 P.M. told CH to please keep out, as our message was a rush government message. He said "you needn't think you are so damned much; wait until 4 P.M." His station being the stronger TI received our message at 4.10 P.M.[10]

Because the ether was still a frontier, might made right, and the military often did not have the technical might. Other testimony recounted amateurs sending false CQDs "for fun" and to get attention. The parameters that circumscribed free speech in a public place such as the proverbial movie house had not yet been imposed in the ether, yet an anonymous cry of "Fire" in the airwaves was equally, if not more, dangerous.

A Telefunken operator aboard the SS *Bremen* submitted a letter to the House of Representatives complaining about interference that, judging by its content, originated from a source other than the amateurs. "First," the operator reported, "I heard some very profane language: 'God-damned Slaby-Arco, rotten louse, humpbacked monkeys,' and other slang."[11] While this sort of transmission may have been amusing to Marconi Company operators, and may even have been useful corporate propaganda, it was hardly a noble exploitation of a limited natural resource. As each company and interest group tried to maximize its use of and position in the ether, the negative effects of the overuse and jockeying for position hurt all users. The government was becoming increasingly concerned about pollution of the ether. By 1910, six bills addressing these problems were circulating in Congress.

The three bills that received the most press coverage in 1910 were the Greene, Depew, and Roberts bills. The Greene (House) and Depew (Senate) bills were similar: they were intended to legislate away the interference afflicting the government stations. They sought to license wireless operators, to impose fines for malicious interference, and to establish the priority of distress signals and official messages. As an additional safeguard against private interference with government stations, the president of the United States would be empowered to "establish from time to time regulations by designation of wave lengths or otherwise to govern said private or commercial stations."[12] The wireless companies and operators were to be licensed by the Department of Commerce and Labor, and if any provisions of the law were violated, the department could revoke licenses. The secretary of commerce and labor was selected to oversee wireless because he already had general control over the regulation of life-saving appliances on shipboard.[13] The transmission of fraudulent messages was to be punishable by a fine of not more than $2,500 or imprisonment for not more than five years.

Although the clause establishing presidential power to regulate the private stations by "wave lengths or otherwise" was attacked as too vague and impracticable by the wireless companies, Congress did not know what other standard to use; wavelengths seemed to the layman the most equitable and least arbitrary way to allocate the resource. Technical guidelines seemed more rational and fair than economic and political ones. If, however, the president chose to impose the wavelength allocations selected during the 1906 conference, the assignments would not, in commercial eyes, be equal. The inventors argued that those wave allocations would exacerbate interference rather than reduce it. NESCO representatives pointed out that the 300-meter assignment created a party line for all maritime business and that in a busy port, some ships would wait for hours to get on this communal wavelength: "No way could possibly have been devised better calculated to give the maximum of interference and the minimum of practical service than this proposed rule requiring all ships to use the same party line."[14] Congress was careful not to spell out wavelength allocations. The words *or otherwise* were included because the House Committee on Merchant Marines and Fisheries sensed it did not yet have all the pertinent information. The committee anticipated that in the future, standards other than wavelengths might be used to sort out competing claims. "What those standards may be," the committee maintained, "cannot be forecast in fact, much less in the terms of a statute, for the advance of the art may add new words to the language."[15]

Hearings began in 1910 on both bills, in the House committee and in the Senate Committee on Commerce. Each side of the wireless regulation argument produced evidence to support different criteria for establishing priority in the airwaves. The amateurs claimed that official messages should not have priority because the messages were rarely authentically official. Their position was endorsed by the *New York Times,* which asked: "Must the splendid wireless operations of the transatlantic liners and of the radio telegraphic companies on land . . . be suspended whenever two subalterns choose to greet each other through the ether?"[16] In 1910, America's amateurs were not sufficiently organized to orchestrate a coordinated lobbying campaign against the proposed legislation, but members of the Junior Wireless Club of New Jersey traveled to Washington in April to argue against the Depew bill. Their testimony was described in headlines such as "Senators Hear Boys' Plea." The *New York Times* reported that W.E.D. Stokes, Jr., president of the club, who was still in his early teens, testified "on behalf of the inventive genius of the American boy."[17] Hugo Gernsback encouraged members of his Wireless Association of America to mail in protests to Senator Depew and Congressman Greene.[18] The amateurs advised Senator Depew that if the navy modernized its equipment and sent all its messages in cipher, much of the interference could be reduced without resorting to legislation. "Any skilled government operator knows the touch and tone of every other government operator," the amateurs claimed, "just as you know the voice of your wife from the voice of your son. . . . If our government used only certain wave lengths, they should be able to tune out all other interferences."[19] The amateurs were not objecting to the establishment of a party line in the ether; they simply wanted the navy, rather than the private sector, to be obliged to use it.

The press had begun to denounce the irresponsibility of some of the amateurs; nevertheless, editorials generally did not support the proposed solutions. The last thing the press wanted was regulation that would transfer some of the prerogatives of private enterprise to the state. The features to receive the most criticism were the automatic priority granted to official messages and the power bestowed on the secretary of commerce and labor. The militarization of the ether was especially unpopular. *Electrical World* opposed the "exaltation of the military over the other classes of American people" and questioned the assumption "that the government has some sort of prescriptive right to an art which a long line of scientists and inventors has endowed the world."[20] The *New York Times* maintained: "This Nation and the pursuits of its people are not maintained for the sake of the army and navy, and their officialism." "The pathways of the ether should not be involved in red tape," the newspaper added.[21]

Editorials charged that the Greene and Depew bills would make the secretary of commerce and labor a wireless czar, a position deemed incompatible with American democratic principles. The *New York Times* claimed that the bills were "doubly mischievous" because they conferred on the secretary, "an official who can know nothing about the technical demands of private wireless business, practically unlimited power of determining its conduct."[22] *Scientific American* agreed, maintaining that the power such bills gave to the secretary was "excessive" and might lead to "gross abuses." "To suppose that one person in a short term of office could gain a comprehensive knowledge of so large a subject—one which he must handle as a dictator—is absurd," the journal asserted.[23] The *Times* warned, "As for permitting government bureaus to issue, modify, amend, and revoke the rules that shall govern wireless, the idea ought not to be tolerated, save in time of actual war. Official control of commercial wireless business would be always in the bureaucratic interest, and not in the interest of progress and enlightenment."[24] The technical journals complained that the wireless bills were drafted by people who knew nothing about wireless.[25] *Scientific American* lectured, "This question of radio-telegraphy is too big for settlement through legislative 'jokers.' . . . The reports of wireless committee conferences, ludicrous in the extreme from a scientific standpoint, prove this fact."[26] United Wireless representatives wondered if the secretary of commerce was going to exercise any discretion in granting licenses. Otherwise, he would have to issue them to all applicants, and this was not a policy that would reduce interference.[27] Repeatedly, the press endorsed and legitimated commercial claims to the airwaves, equating those claims with democracy and progress, and government claims with inefficiency and inequity.

In supporting the Greene bill and the proposed priority of official messages, the Department of Commerce and Labor noted that the "priority of government messages by ordinary telegraph lines [had] been guaranteed in the United States since the Act of July 24, 1886."[28] The report of the House Committee on Merchant Marines and Fisheries maintained that the bill would not "deprive many bright American boy amateurs with a scientific turn of mind of a harmless and improving pastime, from which the country may hope to reap the benefit in future inventions." The committee wanted the government, through licensing, to police the amateurs, not eliminate them: "The police regulations of nearly all large cities prescribe a permit before a boy is allowed to carry a revolver, which at most could shoot a few hundred yards and possible hit one man. Amateur wireless . . . may readily interfere with messages from a ship in distress with hundreds of lives on board."[29] The committee also noted that the legislation would be instructive: "The Committee means fair play for industrious, inventive American boys. . . . In learning wireless these boys may well at the same time study their duties to others and the obligation of an American citizen to obey the law."[30] While the committees felt obliged to address the amateurs' concerns, they were not won over by amateur or commercial testimony. On March 28, 1910, Merchant Marines and Fisheries reported out the Greene bill by unanimous vote. The Committee on Commerce favorably reported out the Depew bill on April 28, 1910, and the bill passed the Senate on June 16, 1910. The House, however, did not act on the Depew bill, and no consensus legislation emerged.[31]

The 1910 bill that was more comprehensive and proposed a long-term approach to radio regulation was the Roberts bill, House Joint Resolution 95. This bill proposed that the president create a board of seven members with one expert each from the war, navy, and treasury departments, three experts representing the commercial wireless interests, and one scientist "well versed in the art of electric-wave telegraphy and telephony." The board would, throughout 1910, "prepare a comprehensive system of regulations to govern the operation of all wireless plants afloat and ashore . . . with due regard alike to government and commercial interests." The board's report would be submitted by December 1, 1910.[32] The bill anointed no one as wireless czar, but sought to distribute influence among seven "experts." Amateurs and those who manufactured apparatus for the amateurs opposed the Roberts bill because the proposed board had no member representing the independent operator. The amateurs tried to present themselves as representing the general public to imbue their position with more legitimacy and import. They claimed that to consider only the needs of government and business was undemocratic. As one amateur wrote in opposing the Roberts bill: "To vest legitimately in a wireless board, then, the proposed jurisdiction, it would seem but proper that all those individuals who are rightful owners of the atmosphere over their respective properties transversible by wireless messages should be consulted in the matter."[33] Although the Roberts bill set an important precedent by suggesting an alternative mechanism for wireless regulation, it did not pass in 1910. Yet it helped establish the legislative choices: Should wireless be administered by one man or by many? Did administration of wireless require autocratic powers, or bureaucratic management informed by experts?

With each new congressional session, bills to regulate wireless were introduced. None passed. In 1912, thirteen such bills were submitted.[34] The chairman of the Committee on Merchant Marines and Fisheries, Congressman J. W. Alexander, and the chairman of the Committee on Commerce, Senator Knute Nelson, introduced similar bills designed to reduce interference. The Nelson bill, introduced on December 11, 1911, was like its predecessors in that it provided regulation of wireless by the secretary of commerce and labor. It was referred to a subcommittee of the Senate Committee on Commerce. Early in 1912, the subcommittee reported that it had become "convinced that the bill bestowed too great powers upon the departments of Government and gave too great privileges to military and naval stations, while it did not accurately define the limitations and conditions under which commercial enterprises could be conducted." In February of 1912 the subcommittee began revising the legislation.[35] The one significant regulatory development that occurred at this time was the ratification of the treaty of the International Wireless Convention on April 3. A third convention was scheduled for June 1912, and the United States was informed that its delegates would not be welcome unless it ratified the treaty.

The conflict in the airwaves did not appear to involve the general population, and the military services, the wireless companies, and the amateurs were not, in 1910, politically powerful interest groups. Although congressional willingness to regulate wireless was clearly increasing, the interference problem would have to touch many more people before both houses of Congress would get together to enact more comprehensive wireless legislation. Overpopulation in the spectrum would have to affect those who barely knew the spectrum existed or that activity in it could profoundly affect their lives. A 1910 House report warned that waiting for such a time could prove irresponsible: "If the use of wireless is not to be regulated, it may in the future result in disaster."[36] In 1912, that disaster occurred.

· · ·

ON APRIL 10, 1912, the world's largest and most sumptuous ocean liner set sail from England for New York City.[37] The *New York Times* carried photographs of the ship's elegant interior and listed the luminaries who had booked passage on the *Titanic* for its maiden voyage. The ship represented technological audacity and arrogance taken to their limits. The owners proclaimed the ship unsinkable. It was what men dreamed of when they worked on machines: it was the biggest and the fastest, and it was impervious to the whims of nature. Guglielmo Marconi had booked passage on the ship, but a change in plans forced him to cancel. The captain of the ship, appreciating that the owners wanted to set a new transatlantic speed record, sought to make the crossing as quickly as possible.

On April 15, the *New York Times* reported that it had learned from the Marconi Cape Race, Newfoundland, station that the *Titanic* had hit an iceberg, but the article expressed no alarm. The newspaper reassured its readers by listing all the other ships in the *Titanic*'s vicinity and all the other liners that had in the recent past hit icebergs and nevertheless arrived safely in port. Supporting this sanguine tone was a wireless message reading "All Titanic Passengers Safe; Towing to Halifax," which was picked up by stations on both sides of the Atlantic, as well as by Lloyd's and the London *Times*.[38]

Few were prepared for the next day's horrifying headlines. The *Titanic* had sunk in less than three hours, at approximately 2:30 A.M., taking more than fifteen hundred passengers with it. Between eight hundred and nine hundred survived, mostly women and children. Although the ship had been drastically underequipped with lifeboats, and the captain had taken the ship too quickly through an ice field, wireless emerged as the invention that had both permitted many to survive and caused many more to die. As the story unfolded in the press during the next few weeks, the status of wireless and wireless regulation were permanently altered.

As soon as the *Titanic* struck the iceberg, Jack Phillips, one of the ship's wireless operators, began sending distress signals and the ship's position. The Marconi Station at Cape Race received the news Sunday night at 10:25 New York time—almost immediately after the collision occurred. Two other liners, the *Parisian* and the *Virginian,* also received the news immediately, but they were twelve hours away from the *Titanic.* Tragically, ships in the *Titanic*'s vicinity never heard Phillips's call. The only nearby ship that received the repeated CQD and SOS messages was the *Carpathia,* which caught the message only "by a lucky fluke." Like most other ocean liners, the *Carpathia* had only one wireless operator, who worked for twelve or sixteen hours straight. When he retired for the evening, the wireless apparatus was unattended. On this particular night, the *Carpathia*'s operator, Harold Cottam, had finished his work for the evening but had returned to the wireless room to verify a "time rush," which was a comparison of two ships' times to check the agreement of their clocks. When he put on his headphones, he heard the *Titanic*'s call for help. Had Cottam not returned to his wireless set, no help would have arrived until late the following morning. Such were the consequences of not having a loudspeaker, a relief operator, or a distress alarm for the sleeping operator. The *Carpathia* was fifty-eight miles from the *Titanic,* and when it arrived at the scene three and a half hours after hearing the distress call, it could only rescue those who had managed to get into the lifeboats.[39]

The *California* was less than twenty miles from the *Titanic* when the accident occurred. But the *California*'s only wireless operator was asleep when the *Titanic* broadcast its distress calls. Also, because the *California* was traveling through the same ice field as the *Titanic,* its captain, as a matter of safety, had shut down the engines and decided to

The *Titanic* disaster made wireless telegraphy front-page news.

wait for daylight before proceeding. Captain Lord explained: "With the engines stopped the wireless was, of course, not working, so we heard nothing of the *Titanic*'s plight until the next morning. . . . Had we only known of the *Titanic*'s plight all the . . . passengers could have been saved."[40] Another ship, the freight steamer *Lena,* was within thirty miles of the *Titanic.* But it was not equipped with a wireless outfit. The tragedy exposed how very inadequate shipboard use of wireless had been. To have only one wireless operator providing communication for only half a day was gambling with very high stakes. The lack of auxiliary power to operate wireless apparatus in the event the ship's main boiler plant failed was equally dangerous and easily remedied.

Although this somewhat cavalier attitude toward wireless use aboard ships caused concern, no aspect of the tragedy outraged people more than the ceaseless interference, cruel rumors, and misleading messages that filled the air from unknown sources during the disaster. Friends and relatives were desperate for information. Marconi, in New York, wrote to his wife, "I've witnessed the most harrowing scenes of frantic people coming here to me and to the offices of the Company to implore

and beg us to find out if there might not be some hope for their relations."[41] Shortly after the *Titanic* struck the iceberg, wireless stations
along the northeast coast of North America clogged the airwaves with
inquiries and messages. The *New York Times* described the Sable Island,
Nova Scotia, station as "the storm centre of a great battle for news of the
missing passengers and crew. . . . The wireless operators at Sable Island
are overwhelmed with messages which have come from all quarters
from relatives of passengers craving for news." The Marconi Company
complained about the interference Marconi operators were subjected to
by "outside unrecognized stations." Out of this early "congestion of inquiries" emerged the message reporting that the *Titanic* was moving
safely toward Halifax. When the American and British press learned that
this news was completely false and that the *Titanic* had, in fact, sunk, its
editors were appalled. The amateurs were accused of manufacturing the
deception and were universally condemned.[42] *Electrical World* wrote,
"Someone, perhaps in carelessness, perhaps in fear or in greed, sent false
messages of rescue. Such a person . . . ought to serve a long term in a
federal prison. No measures of repression are too severe for the emergency before us."[43] *Literary Digest* referred to the false message as "essentially the act of a coward." "That persons of sufficient education and skill
to operate wireless apparatus will stoop to such things," the *Digest*
lamented, "is almost unbelievable." The *Times* of London described such
messages as "inventions of a cruel and heartless kind." President Taft
denounced the malicious interference as "perversion."[44]

On April 21, Captain Haddock of the *Titanic*'s sister ship, *Olympic*,
offered an explanation for the erroneous report of the *Titanic*'s safety. As
soon as Glace Bay transmitted news of the *Titanic*'s plight, operators
from all over asked the question "Are all *Titanic* passengers safe?" At the
same time, the steamship *Asian*'s operator sent the message "Towing oil
tank to Halifax." Captain Haddock "suggested that the two Marconigrams quoted above had been tapped in transit by amateurs or otherwise
unskilled operators, who omitted the interrogatory 'are' in the first message, and caught the words 'towing' and 'to Halifax' in the second,
making the whole cloth message." The Halifax station tended to confirm
this explanation. Its operators stated that "the air was full of wireless
flashes from ship and shore stations, and . . . it was very difficult to piece
together connected statements." However, by now intent and motivation were irrelevant. The false messages had been transmitted; interference had reached a dangerous level. In the eyes of the press, the ether
could no longer be used as a playground for youngsters.[45]

After the newspapers had established that the *Titanic* had sunk, and the *Carpathia* was en route to New York City with the survivors, communication between the *Carpathia* and the shore stopped. The *Carpathia*'s wireless range of eighty-five miles was not nearly as great as the *Titanic*'s or the *Olympic*'s, and its operator had relayed news of the rescue to New York via the *Olympic*. Without this relay, the Siasconset station could not pick up the *Carpathia*.[46] Because so many people were desperately awaiting the publication of the survivors list, President Taft sent two navy scout cruisers to intercept the *Carpathia* on its way back to New York so that the names of the survivors could be wirelessed in advance of the ship's arrival. The wireless range of the two cruisers was claimed to be 1,500 miles, and the headlines describing their mission read "Wireless Search of the Seas for Further News." Inability to receive news from the *Carpathia* illustrated another deficiency. Despite recent legislation mandating 100-mile performance, the *Carpathia*, like many ships, had an extremely limited range—had it been the ship to hit the iceberg, her distress signals would never have reached the shore. Communication with the *Carpathia* remained elusive despite the scout cruisers, and when the liner arrived in New York, the *Titanic*'s surviving wireless operator, Harold Bride, who helped man the *Carpathia* station, explained why: "The navy operators aboard the scout cruisers were a great nuisance. I advise them all to learn the Continental Morse and learn to speed up in it if they ever expect to be worth their salt. The *Chester*'s man thought he knew it but he was as slow as Christmas coming."[47]

Like Jack Binns of *Republic* fame, Bride became a national hero. After being rescued by the *Carpathia*, where he was hospitalized, he went on crutches to the ship's wireless room to begin sending messages. When the *Carpathia* arrived in New York, Marconi himself went to the wireless room to see Bride. The *New York Times* offered this romanticized account of the meeting: "Slowly the youth turned his head around, still working the key. The hair was long and black and the eyes in the semidarkness were large—staringly large. The face was small and rather spiritual, one which might be expected in a painting. It was clear that from the first tragic moment the boy had known no relief. Mr. Marconi asked the operator how his feet were. Both were in bandages and he was working seated on the edge of his bed. A plate of food at his side told how he had eaten." Bride's partner on the *Titanic*, Jack Phillips, had died while sending the distress calls. Bride told the *New York Times* how Phillips had heroically continued to send distress signals even after the captain told him to abandon ship. Phillips became a legend, and statues were erected on both sides of the Atlantic to commemorate his courage.

Marconi was deified in the press: some editorials gave him sole credit for saving the lives of the *Titanic*'s survivors. The *New York Times* wrote: "If Guglielmo Marconi were not one of the most modest of men, as well as of great men, we would have heard something, possibly much, from him as to the emotions he must have felt when he went down to the Cunard wharf, Thursday night, and saw coming off the *Carpathia,* hundred after hundred, the survivors of the *Titanic,* every one of whom owed life itself to his knowledge as a scientist and his genius as an inventor."[48] While presenting an address to the New York Electrical Society on April 17, he was continuously interrupted by "tumultuous applause."[49] He confided to his wife, "Everyone seems so grateful to wireless—I can't go about New York without being mobbed and cheered—worse than Italy."[50] The *New York Times* viewed the event this way: "To realize what the wireless did in this case one must think, not of those who were drowned, but of those who were saved."[51]

Yet people could not get those who were drowned out of their minds. Newspapers and magazines were filled with wrenching eyewitness accounts of husbands and wives parting, of women refusing to leave their husbands' side, preferring to die instead, and of the horrible screams of death the shocked and freezing survivors would never forget. Other stories told of people in packed lifeboats who were forced to refuse to let another survivor in because one more person would sink the boat. It was a hideous choice to have to make, and those in the lifeboats sometimes watched the one they had denied die in the sea. These, the press lectured, were the costs of technical arrogance, of the quest for speed and luxury instead of safety, of the desire to be biggest and fastest, of the belief that machines could make men impervious to nature. They were the costs of unregulated industrial capitalism writ large and indelibly.

In every leading newspaper and magazine, the reaction to the tragedy was the same: the "permanent cure . . . should be, and no doubt will be, fixed government regulations."[52] *Electrical World* editorialized, "The recent disaster to the *Titanic* points with terrible and fateful directness to the absolute necessity of a controlling power to regulate wireless telegraphy."[53] The press advocated that the number of lifeboats, the use of wireless, and even the speed ships could travel through ice fields, all be fixed by law. Journalistic rhetoric emphasized the two ills such regulation would address: corporate lack of conscience and the vulnerability such disregard imposed on innocent people. As *World's Work* put it, "A disaster that shocked the whole civilized world was necessary to awaken us from a false sense of security."[54] What Americans had to be awakened to was not that corporate control of transportation or communication was in and of itself bad, but that such control had to be monitored better. The press laid out the terms under which state intervention in corporate activity—in this case, wireless—could take place, and under what circumstances it was justified.

The regulation of wireless now was framed in the same terms that had framed earlier social or antitrust legislation. The *Titanic* disaster happened, after all, during the Progressive Era, when the call for regulating many aspects of American society was incessant and insistent. The Progressive Era marked the ascendancy of the conviction that the state had to assume a more interventionist role in the marketplace as a way of making individual Americans less vulnerable to institutional forces beyond their control. When scandals broke out in the meat-packing industry, or a crisis occurred in the oil cartels, or graft and corruption was discovered in city government, the response was the same: correct the ills through regulation. Newspapers and magazines cast the federal government as the agent of "the people" whose duty was to give the people more control over the trusts, to circumscribe corporate arrogance and hegemony, and to make people less vulnerable to business's self-serving agendas. This idealistic prose often disguised the fact that many of the newly regulated industries benefited from, and in fact helped design, Progressive Era legislation. Journalistic rhetoric surrounding new laws made certain legislation seem onerous to business, whereas these laws often brought much desired predictability and stability to corporate activities.

The press distilled and articulated the ideological debates surrounding regulation, and made clear which long-held American values and traditions were being threatened by corporate combination. Through political cartoons, inflated and flowery language, and sheer expressions of outrage, the press maintained that values, ethics, and aspirations did matter, and that there were certain things in which Americans believed, certain images Americans had of themselves and their country, which could not be sacrificed on the altar of industrial capitalism. Although corporate agendas may, in many cases, have overridden the plea to preserve traditional values and ideals, that such rhetorical protests played a critical role in the regulatory process—that these pleas had to be taken into account, even if they were co-opted later—was a significant aspect of America's regulatory process.[55] Certainly with a question such as "Who owns the airwaves?" sentiments, dreams, and ideology had as much to offer the debate as did legal precedent, which was extremely skimpy, or corporate intent, which was as yet ill defined.

As soon as the *Carpathia* arrived in New York, the Senate Committee on Commerce began investigating the *Titantic* disaster, holding its preliminary hearings at the Waldorf-Astoria on April 20 and 21, and moving the hearings to Washington on April 22. Within four months, American radio would come under government supervision, and transmitting in the ether would be not a right, but a privilege assigned by the state.

■ ■ ■

AFTER THE *TITANIC* tragedy, the perceived value of the ether as a resource increased immeasurably, and the resource had to become more serviceable. The necessary reforms were now obvious to the press and to Congress. Mandatory shipboard wireless was insufficient; the wireless had to be manned at all times. Auxiliary power in case of engine failure was essential. A strict and formal procedure for the transmission and reception of distress calls had to be officially established. Most importantly, the amateurs had to be purged from the most desirable portion of the broadcast spectrum. They had to be transformed from an active to a passive audience, allowed to listen but not to "talk." "Private stations of all kinds should be rigorously limited in wavelength and power," maintained *Electrical World,* "particularly the amateur stations which have no need for anything more than trivial energy." The magazine added that "the wireless meddler" would have to be repressed."[56]

The amateurs tried to exonerate themselves from blame for the false messages and interference. One amateur, in a letter to *Scientific American*, claimed that the amateurs had become scapegoats: "A reason had to be given the public for the delay [of messages]. The blame was laid to the parties least able to defend themselves, as is usually the case. At once great headlines flared forth the atrocities of the 'Wireless Meddler.' "[57] But what the amateurs had to face was that in the aftermath of the *Titanic* disaster, an interest group more important than the amateurs or the military emerged to stake its claim to the ether: the general public. Regulation was necessary "to insure to the people of the United States an uninterrupted wireless service twenty-four hours a day for every day in the year."[58] The ability of institutions, both government and corporate, to serve the public, particularly in life-or-death situations, surpassed any other claims to the ether, especially those voiced by seemingly scattered and unorganized individuals.

Harold Cottam, the *Carpathia* operator, had testified at the Senate hearings that there were no regulations specifying what hours an operator was to be on duty and that there was "nothing in the Marconi system at present to detect signals if the operator [was] not present." The first act by Congress was to revise the 1910 law, now requiring ships carrying fifty or more persons to carry at least two skilled operators, with someone on duty at all times, and to have an auxiliary power supply available for the wireless. Shipping on the Great Lakes was included in the legislation. President Taft signed this bill on July 23, 1912.[59] The more sweeping bill that was to regulate wireless, and then radio broadcasting, until 1927, was the Radio Act of 1912, passed on August 13. It took effect four months later, on December 13.

The Radio Act required that all operators be licensed, that stations adhere to certain wave allocations, that distress calls take priority over all other calls, and that the secretary of commerce and labor be empowered to issue licenses and make other regulations necessary to sort out the wireless chaos. Congress mandated that stations use undamped waves and issued specific technical guidelines for transmitters. Amateurs were relegated to a portion of the spectrum then considered useless: short waves of 200 meters and less. They could listen in on any frequency but could transmit only in this short-wave portion of the spectrum. The amateurs had been exiled to an ethereal reservation.

To protect distress calls from interference, the law required that a station suspend all other work whenever it picked up a distress signal, and not return to its other work until the station could no longer be of service. If the station could not help in the rescue effort, it was to remain silent. Americans were to use the wavelength designated for distress calls at the International Conference: 300 meters. All shore stations were to listen in on the 300-meter band at intervals of fifteen minutes for at least two minutes. Shipboard stations had to have a transmitting capability of 100 nautical miles. The United States formally adopted SOS as the official distress call. Intercommunication between systems was compulsory. Fines were established for irresponsible transmission: up to $500 for "malicious interference," $2,500 for sending false distress calls. The secretary of commerce and labor had the power to suspend licenses for up to one year for violation of the law. Repeated disobedience was cause for license revocation.

Most importantly, the new legislation secured for the navy increased hegemony in the spectrum. Wavelength allocations conformed to those assigned during the 1906 International Conference. Private stations were to use wavelengths below 600 meters or above 1,600 meters. That portion of the spectrum between 600 and 1,600 meters was reserved for government use. Ships within fifteen nautical miles of a government station were to reduce their transmitting power to one kilowatt. Because the act sought to ensure that ships' passengers would have access to wireless services, even if they were not near a commercial station, naval stations were now required to transmit and receive commercial messages if there was no commercial station within a 100-mile radius. Many naval stations had to be upgraded and modernized to meet this provision. Armed with the new legal mandate and improved technology, the military would continue to increase its influence on activities in the airwaves.

Congress also legally protected the privacy of wireless transmissions; fines were set for "broadcasting" private messages. During the congressional committee hearings of 1910, United Wireless had complained about the difficulty Americans encountered when trying to secure wireless licenses in foreign countries. The 1912 act provided that licenses would be issued only to citizens of the United States. In time of war or disaster, the president was empowered to close private wireless stations, or to authorize the government to take them over.[60]

The only evidence describing the Marconi Company's role in shaping the 1912 act is John Bottomley's statement from 1912 in which he reports, "The greatest care has been taken that no bill detrimental to our work or to the system generally should be permitted to pass."[61] But both the company and Marconi supported regulation. The American Marconi Company enjoyed a virtual monopoly in the United States, having bought out or driven out of business its major competitors. The company was more consolidated and entrenched than ever, and interference from the amateurs was becoming highly costly. The amateurs were interfering both with ship-to-shore work and with Marconi's wireless news services. In news dissemination, they provided unwelcome competition. In an interview published immediately after the *Titanic* disaster, Marconi advocated regulation and "control of amateur experimenters."[62] With the amateurs consigned to short waves and the navy to the 600 to 1,600 meter range, the regulation ensured that in America, the Marconi Company would have portions of the spectrum entirely to itself. Thus, while it is difficult to determine whether Marconi had a hand in designing or promoting the legislation, it is clear that neither he nor his company suffered from it.

By October, the secretary of commerce and labor had issued the regulations governing amateur stations and operators. The United States was divided into nine wireless districts, each district having its own office to manage the wireless affairs of that district. To get their wireless license, amateurs had to pass an examination that involved assembling a wireless outfit, determining if the set was faulty (and, if so, repairing it), sending and receiving messages at twenty words per minute in Continental code, and passing a written exam. The government did not yet have facilities for administering the exams, so, ironically, the navy conducted them at nine stations throughout the country. The *New York Times* reported that "the general knowledge of wireless matters and the skill displayed [was] a surprise even to the Navy experts. More than 90 percent of all applicants . . . passed the exam." Amateur stations within five nautical miles of military stations were "rigidly restricted as to length of their wave and the power of their sending apparatus."[63]

Amateur response to the 1912 law varied. Some amateurs dismantled their apparatus.[64] Others continued operating as before, but they were more courteous and deferential toward the government and commercial stations. Some amateurs did not stay below 200 meters and got away with the trespassing because the appropriations for administration and enforcement of the 1912 law were insufficient.[65] One amateur from the Pittsburgh area recalled: "Nobody in radio knew anything about licensing. We knew that the commercial stations, by which I mean ship and government stations, had call signs, but I think there were very few people who had even heard of the license regulations, let alone read them . . . [and] no one thought the regulations applied to him, as an individual. It certainly didn't apply to the listener."[66] Amateur activity in the ether, thus, was circumscribed, but it was not eliminated. Hundreds of amateur stations around the country were licensed, as were thousands of operators. The amateurs began exploring their new slot in the spectrum, and adjusting to but not acquiescing to institutional hegemony.

The Radio Act of 1912 represents a watershed in wireless history, the point after which individual exploration of vast tracts of the ether would diminish and corporate management and exploitation, in close collaboration with the state, would increase. The American spectrum was partitioned: another frontier was partially closed. The 1912 law as a legislative artifact reveals American society's early struggle to come to terms with an invisible, enigmatic, communally held resource whose potential was still only partially appreciated.

The law acknowledged that property rights could be established in the ether and that the main claimants to those rights were institutional users. The amateurs, by exploiting democratic rhetoric, had tried to argue that they represented "the people" and that the public had very legitimate interests in how the spectrum was used. The state acknowledged the latter point but maintained that the *Titanic* disaster had demonstrated all too well that the amateurs did not serve the needs of "the people" but in fact obstructed them. Thus, one critical precedent this law established in broadcast history was the assumption that only consolidated institutions—in this case, the navy and the Marconi Company—could anticipate, implement, and protect "the people's" interest in spectrum use. At the same time, it was clear that government control of America's airwaves, foreclosed for the time being by the 1912 law, was unlikely in the future, as well. Harold Bride, the *Titanic* wireless operator and Marconi Company employee, by complaining about the incompetence of naval operators, suggested what a predicament America might get into if the navy controlled wireless communications. It was a complaint that reaffirmed stereotypes about the pitfalls of bureaucratic management.

Another precedent established was that the state would assume an important role in assigning property rights in the spectrum. In other words, access to particular wavelengths would not be bought and sold on an open market. Rather, the state would determine priority on the basis of claimed needs, previous investment, and importance of the messages. Those claims would be acknowledged by wavelength allocations. What established merit in 1912 was capital investment or military defense, coupled with language that justified custodial claims based on an invaluable service to humanity. This, too, was a significant precedent. For, under the guise of social responsibility, of protecting the lives of innocents, and of managing a resource more efficiently, the military and a communications monopoly secured dominant positions in America's airwaves.

The dilemma of who had a right to transmit and who did not was tackled in newspapers and magazines before it was worked out in Congress. Government control of any communications system was anathema to the press: it threatened their news-gathering capabilities and their organizational prerogatives. It opened the door to censorship, to making the press subservient to the state's agendas. Thus, it is not surprising that the press, relying on the rhetoric of "progress" and "enlightenment," kept asserting that the American people would not tolerate government control of wireless. Editorials against premature regulation assumed an accusatory tone and focused on the evils of military priority in the ether. Commercial claims, on the other hand, particularly Marconi's, were cast as altruistic and forward looking. Marconi's transatlantic wireless service, heralded as a boon to mankind, served the press and continued to cheapen and quicken news gathering. Marconi was used to personify the press's preferred image of itself: socially responsible, concerned with the safety of others, efficient and profitable without being greedy. With the *Titanic* disaster, the press pointed to Marconi as the man responsible for lives being saved, and as the symbol of why commercial management and hegemony had to be protected and maintained. Commercial control simply had to be regulated so that irresponsible capitalists too greedy to think of others would be compelled to do so in the future.

In other regulation of the period, business leaders often had to listen carefully to bitter and angry critiques of corporate activities, and then figure out how to co-opt this criticism and eventually exploit it to their own ends.[67] This process involved developing public relations departments and learning how to improve press coverage. In the case of wireless, however, the press had a vested interest in how the invention was managed and, thus, in how it was portrayed. By 1912, Marconi had little public relations work to do with the press: newspapers saw his interests and theirs as one and pulled out all the democratic, "common man" rhetoric at their disposal to make readers—the public—see that these interests were also theirs.

The amateurs, who had made such good copy between 1907 and 1911, were less amusing when, instead of toying with the navy, they threatened commercial news-gathering networks. The press unanimously denounced the amateurs after the *Titanic* disaster for interfering with "legitimate" message handling. What caused the amateurs to lose their freedom to roam the ether at will was not so much that the government would no longer tolerate that freedom, but that a very influential business, the press, found their activities a disruptive encroachment on its turf. This violation was cast, in journalistic rhetoric, as a selfish flouting of the safety and freedom of all Americans, as a challenge to basic ideals and values about right and wrong, good and bad. Thus, with wireless as with other regulatory issues of the Progressive Era, journalistic language that asserted the sacredness of certain American values dovetailed very well with and supported selected commercial priorities and investments.

By the summer of 1912, the shape of American broadcasting was beginning to change, setting the stage for subsequent developments. Marconi now monopolized American wireless service and planned to build several major high-power stations on the East Coast. Fessenden and De Forest, both involved in court cases, had ceased to be major actors in the wireless story, but their inventions, particularly the alternator and the audion, were now in corporate hands. The navy was more centrally involved in wireless, but to preserve the hegemony it had gained through regulation, it would have to become more technically and organizationally efficient. And then there were the amateurs, confined to their ethereal ghetto. One law of spectrum use maintains that "relatively deprived users are virtually forced to innovate spectrum-economizing, spectrum-developing technology."[68] Did this hold true for the amateurs?

Would they have any influence at all now that the deck was stacked in favor of institutional users and clients? The 1912 law did not deny the amateurs' assertion that "the people" had a major stake in spectrum use. Many amateurs, seeing themselves, rather than institutions, as representatives of the people, still believed fervently in this position. Did one disaster now mean individual Americans had no rights of access to this resource allegedly held in common? Even after the 1912 act, there were unresolved tensions between individual and institutional claims on the ether. The amateurs had lost their freedom to roam in and out of the airwaves. But whether their vision of how wireless might be used was also lost in 1912 remained to be seen.

CHAPTER EIGHT

THE RISE OF MILITARY AND CORPORATE CONTROL

1912–1919

"WIRELESS SENDS VOICE over Atlantic," proclaimed the front-page headlines of the October 22, 1915, issue of the *New York Times*. Despite "adverse conditions," which included those imposed by the European war, "the human voice was projected across the Atlantic for the first time in history." The fanfare accompanying the achievement was by now familiar, even predictable; the hero, however, was not. For the moment of glory did not belong to Guglielmo Marconi, or to any other inventor-hero. It belonged to AT&T and its ally, the U.S. Navy. The story was particularly emblematic of several changes transforming radio between 1910 and 1920: the transfer of continuous wave radio technology from individual to institutional control, the increased role of the navy in wireless affairs, and the tendency of the press to legitimate corporate visions of how radio should be managed, thought about, and used.

John J. Carty, chief engineer for the telephone company, had orchestrated the new tests beautifully; he saw to it that all of the company's claims were witnessed and verified by independent onlookers. Carty and his men had, with the government's permission, installed a transmitter at the navy's recently completed high-power wireless station in Arlington, Virginia. Two AT&T engineers brought receiving apparatus to France, where they had to persuade the otherwise preoccupied French government to let them use the Eiffel Tower for their tests. While imposing certain restrictions, the French government granted AT&T access to the tower. Meanwhile, Lloyd Espenschied, an AT&T engineer based in Pearl Harbor, readied his receiving equipment. B. B. Webb, one of the company's engineers at Arlington, did the talking. His voice reportedly was heard and clearly recognizable both in Paris and in Pearl Harbor;

French military officials verified the Paris results. The goal so long dreamed of by inventors and journalists alike was now an "accomplished fact."[1]

For AT&T, 1915 had been a spectacularly successful year, marked by several public relations coups. The transatlantic feat, which had earned the largest headlines, had been preceded by other equally revolutionary and well-publicized achievements. In January, the company completed a transcontinental telephone line from New York to San Francisco in time for the February opening of the Panama-Pacific Exposition. To publicize this feat, the company was shameless in its reliance on telephone legends. Alexander Graham Bell at the New York end called his famous assistant in San Francisco and, repeating the historic line, asked, "Mr. Watson, are you there?" The *New York Times* reported that the two men "heard each other much more distinctly than they [had] in their first talk thirty-eight years [earlier]."[2] Within eight months, the company again sent the human voice across the continent, but this time without wires. AT&T had by now installed its wireless telephone transmitter at Arlington, and had receiving equipment in place at Mare Island, California. On September 29, Secretary of the Navy Josephus Daniels, on behalf of AT&T and the navy, announced the success of the transmission. Naval officials on the West Coast reported that the messages from 2,500 miles away were clearly audible. It was "the first time such a distance [had] been covered by wireless telephony."[3] In addition, the Arlington transmitter was connected by telephone lines to AT&T's main office in New York City. From there, Theodore Vail, the company's president, spoke into his phone; his message went to Arlington and then "through the air" to Mare Island, where Carty recognized his boss's voice. The next day, the front pages announced that the achievement was even greater than initially thought, because "a lone operator in a frame hut at the foot of a towering mast on the shore of Pearl Harbor, Hawaii, knew that the human voice had been heard almost twice the distance, for he had listened to words spoken in Washington, 4,600 miles away." The *New York Times* pointed out that this was greater than the distance from New York to the North Pole.[4] The distance covered by these tests was amazing, almost unimaginable. With most eyes on Europe in 1915, it was the transatlantic message and the possibilities for regular wireless telephony service between New York and London, Paris, or Berlin that gripped the public imagination.

How had AT&T achieved this coup of being the first to send a man's voice over the Atlantic? How had AT&T edged Marconi off center stage? First and foremost, AT&T had gained control over critical patents. While

AT&T in 1907 had declined to negotiate with Fessenden, citing the still rudimentary operation of his wireless telephone, the company did not ignore his invention. Vail, who publicly disparaged wireless telephony as crude, impractical, and years away from commercial application, privately monitored the invention's progress. J. J. Carty, as head of research, kept a close watch on radio developments around the world.[5] The publicity wireless telegraphy and telephony received, especially newspaper stories suggesting that eventually everyone would have his or her own private wireless sets, concerned regional Bell companies and AT&T stockholders alike.

This concern reached a new high when the *Titantic* disaster pushed wireless telegraphy onto the front pages as no other previous event had; wireless was now considered a necessity on ships. Subsequent disasters, such as fire aboard the *Volturno* in October 1913, which might have killed all the passengers if not for wireless, or the midwestern blizzard that pulled down telegraph and telephone lines but could not stop wireless transmissions, made wireless seem an indispensable part of modern life. When Marconi, his image burnished by the *Titanic* disaster, began publicly predicting long-distance and even transoceanic voice transmission, Vail was faced with the specter of a major system with which AT&T would have to interconnect. This was completely unacceptable to Vail; he would not countenance connecting the Bell system with those of "outsiders." He was willing to spend what was needed to preserve the purity and extend the reach of his system. Any technological system, whether it used wires or not, which sent the human voice over distances should, in Vail's view, be firmly under AT&T control.

This determination not to allow wireless telephony to become a competitor of telephone dovetailed with another of AT&T's technical goals and within a few years made AT&T a major corporate force in the development of radio. Under Carty, AT&T researchers worked to extend long-distance service. At the turn of the century, the company had acquired the patent rights to Michael Pupin's loading coil, which amplified telephone signals and reduced the cost of long-distance wiring. With Pupin's coil, the company had been able to extend its long-distance service as far west as Denver, but no farther. Vail was committed to establishing AT&T as a national monopoly, which meant providing coast-to-coast telephone service; the company thus had to find a "repeater" capable of relaying signals across the country. Carty saw a possible solution in the work then being done on wireless telephony, because voices transmitted without wires were still faint and various inventors were working to amplify these transmissions. In a telling and somewhat prophetic memo, Carty wrote, "Whoever can supply and control the necessary telephone repeater will exert a dominating influence on the art of wireless telephony." Carty then urged the company to adopt "vigorous measures" to develop such a repeater before people not working for AT&T beat the company to it. He added: "A successful telephone repeater, therefore, would not only react most favorably upon our service where wires are used, but might put us in a position of control with respect to the art of wireless telephony should it turn out to be a factor of importance."[6] AT&T was under increased pressure to find such a repeater, pressure exerted by its own press releases. In 1909, Carty had announced to the press that AT&T would have a transcontinental telephone line open in time for the Panama-Pacific Exposition, and he was determined to make good on his prediction. Research intensified to develop such a repeater, but by 1912 little progress had been made.

At that point John Stone, who began and ended his career working for AT&T, suggested that Carty, his assistant Frank Jewett, and Harold Arnold, a recently hired physicist, consider experimenting with De Forest's audion. While working at Federal Telegraph in California, De Forest experimented with combining three audions to increase amplification. He fed the output of the first tube into the input of the second and the output of the second into the input of the third; the audions "in cascade" produced much greater amplification than one audion alone.[7] De Forest demonstrated the audions before the AT&T men in October of 1912 and left several behind for study at their request.

Since his earliest experiments in 1900, De Forest was convinced that the gas in the audion was essential to the detector's operation. Harold Arnold, however, after extensive tests, determined that the audion operated more effectively if the gas was exhausted from the tube. It was this step that transformed De Forest's audion into the vacuum tube. Thus, the corporation found that it had its much desired repeater, and it was in possession of patents on its improvement, but an outsider still held the basic rights.

Some controversy surrounds AT&T's acquisition of the rights to use the audion as a telephone repeater. According to De Forest, after several weeks of waiting in the fall of 1912 to hear whether AT&T was interested or not, he was told that the company had not found the invention as promising as it had hoped; it was not interested, after all, in acquiring any rights to the audion. In the spring of 1913, De Forest, who had by this time left Federal Telegraph and was once again trying to manage his own company in New York, was approached by an attorney. The lawyer said he represented a client interested in the audion but that the client insisted on remaining anonymous. He assured De Forest "on the word of a gentleman" that his client was not AT&T. He offered De Forest fifty thousand dollars for the exclusive rights to use the audion for telephone and telegraphic purposes. De Forest believed these rights were worth half a million dollars and found the offer disappointing. On the other hand, he desperately needed money; he reluctantly accepted the offer. About six weeks later, De Forest learned that these rights had in fact been transferred to AT&T; he also heard that the company had been prepared to go as high as De Forest's estimate of five hundred thousand dollars.[8]

AT&T had been able to take advantage of De Forest because he was in rather desperate straits. In April of 1912 he and his Radio-Telephone Company associates James Dunlop Smith and Elmer Burlingame had been arrested for using the mails "to defraud in the sale of wireless telephone stock."[9] He claimed he did not have enough money to hire a lawyer, so Yale classmates raided a defense fund of $2,500 and hired as his attorney Harold Deming, who had worked in the U.S. attorney's office in New York. The trial began in New York in November of 1913.

The government pursued its case against Radio-Telephone with the same fervor it had brought to bear against United Wireless. Assistant U.S. District Attorney Robert Stephenson never missed a chance to express outraged indignation over the inflated advertising claims and sleazy sales methods used to replenish the company's coffers. He called more than one hundred witnesses "from nearly every state in the Union" to describe how they had "lost all they had invested" or to testify that De Forest's radiophone was a worthless invention.[10] Much of the testimony was quite damning. Stephenson showed that through stock sales, the Radio-Telephone Company had raised more than $1.5 million. Only $345,000 of this found its way to the company's treasuries; the remaining $1,161,000 cushioned the pockets of the company's officers, agents, and promoters.[11] Frank Butler, De Forest's assistant, portrayed the inventor as being above such shenanigans. But Butler's portrait of the selfless, abstracted scientist was undermined when Stephenson produced a letter De Forest had written to Butler in 1907. In it, De Forest urged Butler to continue pursuing a potential buyer of Radio-Telephone stock and to "get the $1,500 if you have to live with him." The letter was signed "Yours for the rocks, Lee De Forest." "Rocks," it turned out, meant cash.[12]

In his summation, Stephenson portrayed all the defendants as equally guilty. In one of his more dramatic flourishes, Stephenson held one of De Forest's audions aloft in his hand and, gesturing with it, charged that with "this worthless piece of glass" De Forest had claimed that soon it would be possible to send the human voice across the Atlantic. He urged that De Forest be sent, with the others, to the Atlanta Penitentiary.[13]

The case went to the jury just after noon on December 31, 1913. While others in the city celebrated New Year's Eve, De Forest paced the courthouse halls, knotted with anxiety. Deming, his attorney, recalled that when a court attendant came to tell the two men the jury had returned, "De Forest collapsed and I caught him."[14] At 1:00 A.M., New Year's Day, 1914, the jury filed in. James Dunlop Smith and Elmer Burlingame were found guilty on two counts of mail fraud. De Forest was found not guilty on two counts of mail fraud. De Forest was found not guilty on the first three counts and on the fourth, conspiracy to defraud, the jury disagreed. He was a free man.

De Forest resumed his radio work, concentrating on voice transmission and reception, and renamed his company the Radio Telephone and Telegraph Company. His continued work in wireless telephony made him a man AT&T would keep an eye on. He rented a small factory and laboratory at Highbridge on the Harlem River, where his main business was manufacturing apparatus for amateur operators.[15] He began experimenting with the audion as an oscillator, as a generator of radio waves. It was not the first time he had conducted such tests, but it may have been the first time he grasped the significance of what he was doing.

Back in 1912, when De Forest was still working for Federal Telegraph and was mindful of AT&T's need for a signal repeater, he spent a considerable amount of time trying to increase the audion's ability to amplify sound. This is when he connected three audions "in cascade," which produced increased amplification. During these tests, De Forest and his associates discovered a new phenomenon. When signals were "fed back" from the output to the input of the same tube, the tube produced a "howl" in De Forest's headphones. He jotted this observation in a laboratory notebook on August 6, 1912, and then worked to eliminate this irritating noise that could well compromise amplification. Across the country in New York, a young wireless amateur and student at Columbia University, Edwin Armstrong, made the same discovery. Armstrong, who was not constrained by the needs of AT&T, pursued the phenomenon and realized that the audion's use was not limited to signal detection or amplification: it could generate radio waves, as well.[16] He described this arrangement as a regenerative circuit or feedback circuit, which could be used either to increase the audion's sensitivity as a detector or to generate oscillations. Because his father would not loan him the money to apply for a patent, Armstrong at first simply had his drawings of the circuit notarized on January 13, 1913.[17] In October he filed for a patent, and later in the year he delivered a paper before the Institute of Radio Engineers describing his discovery.

It is not clear at what point De Forest learned of Armstrong's work, but it must have been shortly after the trial ended. De Forest had discovered the phenomenon first, but there is very little evidence that he understood its implications. Only after Armstrong's work did De Forest appreciate the significance of that mysterious howl. He applied for a patent on the audion as oscillator on March 20, 1914, and a patent on the feedback circuit on September 23, 1915. No radio patents have generated more controversy than these. The subsequent litigation between De Forest and Armstrong lasted until 1934, went to the Supreme Court twice, and reportedly cost one and a half million dollars in lawyers' fees. A prodigious amount of bile was expended as well, with acrimony high on both sides. After Armstrong won one of the early suits, he strung a banner with his patent number on it across his Yonkers home, which De Forest had to pass every day en route to and from his factory.[18] Armstrong lost the case twice before the Supreme Court, yet many radio engineers and most radio historians consider him the bona fide inventor. He may have made the discovery months later than De Forest, but at least he knew exactly what he had. The discovery and the litigation were so momentous because the implications for radio transmission were revolutionary. Here was a compact and relatively inexpensive generator of radio waves—the tube—which, in a few years, would make Poulsen's arc and Alexanderson's alternator look like dinosaurs.

By the summer of 1914, De Forest was in a different bargaining position than he had been in before his trial. Not only was the legal ordeal behind him, but also he had a small, viable company that was manufacturing apparatus, and he was wise to AT&T needs and tactics. When AT&T approached him about acquiring the wireless telephony rights to the audion, the company reportedly offered De Forest twenty-five thousand dollars and eventually paid ninety thousand dollars for a nonexclusive license.[19] This sum allowed De Forest to expand his manufacturing operations and resume the nouveau riche lifestyle he had come to know and love ten years earlier. He also got a taste of what AT&T had to offer and resolved to make an even larger dent in its treasury. De Forest was outraged that during the transcontinental telephone tests, neither Carty nor Vail mentioned him or his pathbreaking work in tubes. De Forest's audion had been basic to AT&T's success, yet when pressed by reporters, Vail had commented tersely, "As far as Mr. De Forest's lamp goes, if it played any part in the wireless conversation with

Mare Island it is news to me."[20] De Forest was also incensed to find no mention of his pioneering work in AT&T's huge exhibit at the Panama-Pacific Expo. He vowed to retaliate. He examined the patents the company filed on vacuum tubes and whenever he found circuits resembling his own, he filed a patent covering similar claims. In this way he sought to harass the firm in an area it took extremely seriously.[21]

AT&T was interested in dispensing with De Forest once and for all, and the company wanted to be in the strongest possible position for its anticipated battles with American Marconi and General Electric over the control of wireless telephony. De Forest still held important patents, especially those covering the feedback circuit, and while these had not yet been tested in court, AT&T's patent attorneys believed that De Forest had an excellent chance of defending them successfully. De Forest was looking for a windfall, and negotiations ensued. De Forest wanted a quarter of a million dollars. AT&T wanted exclusive rights under all of De Forest's patents and under all vacuum tube inventions he might make over the next seven years. The company also wanted him out of the wireless telephony business. De Forest would agree to this last condition only if AT&T allowed him to continue manufacturing equipment for amateurs and to use his transmitters "for the distribution of music and news." Since using the wireless telephone for broadcasting held no interest for AT&T, the company agreed. Although he complained in his diary that the deal would compel him to abandon "all the ambitions" of his "struggling years," he did not hesitate, in the spring of 1917, to sign on the dotted line.[22] The transfer of technological control from independent inventor to corporate research lab was complete.

By acquiring De Forest's patents and transforming the audion into the vacuum tube, AT&T had gained control over one of the most important inventions of the era. But Theodore Vail, concerned as he was with corporate preeminence, knew that technical control was not enough. Public legitimacy mattered, too. So, just as Vail sought to link technical developments with AT&T's business strategy, he also worked to shape his company's public image in the pages of the press. The highly publicized staging of the transcontinental and transoceanic tests in 1915 provided him with the perfect opportunity to legitimate, and even romanticize, what was, after all, one more step in the building of a national corporate monopoly.

Vail and his engineers lost no time in interpreting the significance of the 1915 tests for reporters. Wireless telephony would never replace the existing wired telephone system, insisted Vail and Carty. Rather, the wireless would serve "as an extension of the telephone system to ships at sea" or to remote areas not yet served by telephone.[23] The wireless telephone would bring the benefits of the Bell system to those who were physically isolated from it. AT&T personnel in any way connected with these experiments reiterated this position at every opportunity, and it appeared as the central assumption in articles about the wireless telephone tests. Noting that "the wireless telephone adds to the wire telephone but does not take its place," Carty added the public relations coup de grace: the wireless telephone represented, really, a "humanitarian rather than [a] commercial" venture.[24] The possibility of using the wireless telephone for broadcasting was never mentioned, certainly not by AT&T personnel, who were not even thinking along those lines.

Vail succeeded in having an extended press release printed as an interview in the October 17 *New York Times Magazine*. Titled "What Transcontinental Wireless Phone Means," Vail's lengthy statement was preceded by a reporter's three-paragraph introduction.[25] "The white-haired, venerable magnate" was not particularly interested in the experiment's "possible effect on the values of telephone company stock," noted the credulous journalist. He was an altruist, more concerned with the "probable economic and social significance of the scientific miracle to the whole people." Then Vail proceeded on his own. He began on a patriotic note by extolling "the one great nation in the world in which the people rule" and saying, "What impresses me is that we, in America, are doing this work, this big, constructive scientific work, at a time when all Europe is at war. That's pretty fine." He downplayed the possibility that the wireless telephone might make a lot of money for a few thousand people. What was more important was that "tens of thousands, hundreds of thousands, millions, the whole race, will draw from it a profit more desirable than dollars." That profit included reduced isolation and "free communication between people," which Vail maintained would educate Americans and "do away in this country with prejudice." Those most in need of this "interchange of ideas and thought" were members of America's "unassimilated mass." The comments contained no small dose of paternalism and suggested that AT&T was the perfect custodian not only of America's communications systems, but also of its deepest cultural aspirations. Vail skillfully used references to democracy, equality, altruism, and brotherhood to legitimate and elevate AT&T's corporate activities. To Vail, corporate goals and social goals were one.

Vail was confident he knew what America wanted: it wanted cultural homogeneity. It wanted to achieve "nonpartisanship" and to agree on a "common course of action." Technology, specifically communications technology, would be the instrument through which Americans would ultimately achieve consensus, and Vail and AT&T were going to provide the machines to make this happen. They were also not loath to provide political direction; Vail spent the last two-thirds of his statement extolling the virtues of big business and outlining what the government's role in the marketplace should be. He then made a comment that reflected the rise of public relations agents and revealed the mindset of those who were coming to control radio technology. "Instead of listening to everyone seeking notoriety by attacking property or corporations," Vail lectured, "every constructive mind in this country should endeavor to aid in getting accurate information concerning actual facts before the people. Many of them have been wrongly informed." He added: "Ninety percent of the attacks on wealth are made by those who do not comprehend wealth. . . . Why should business continually be attacked? . . . Every one of those condemning [success] is himself trying to achieve it." While the connections were not made explicitly, they were there: if those (like Vail) who were correctly informed had access to a technology that could reach the "unassimilated mass" and deliver "actual facts," then there would be much less tension between the people and the corporations, and between the rich and the poor. There would then be a "common course of action." Not only had a major corporation demonstrated its control over wireless technology; it also sought to reshape how Americans thought about that technology. Vail's statement unwittingly offered the core ideas that eventually would come to guide the structure and content of American broadcasting.

That the press promoted, without question, AT&T's interpretation of its recent success is evidenced by magazine coverage of the wireless telephone feat. *Literary Digest* simply reprinted major portions of the *New York Times* stories that quoted Vail and Carty extensively.[26] *World's Work* praised the achievement as a cooperative rather than an individual endeavor that relied on "men of many callings": "It is an admirable example of the fine helpfulness and long-visioned patience of enlightened business working with pure scientists to achieve results that gratify both equally and that serve the public as well."[27] The magazine envisioned a time when a man in Cleveland, using his own telephone, could have his speech transmitted first by wire, and then by wireless, to a man in London. The *Independent,* after marveling over the distance the human voice was sent, offered its view of the event's significance: "Wireless telephony means that the billion and a half people living on this planet have been virtually gathered into one room where they can listen to one man's voice. The human race has snuggled together like a family about a fireside on a cold evening and can chat comfortably with one another."[28]

Presumably, this "one man's voice" would act as a unifying and homogenizing agent; a great but distant patriarch would be helping to make these billion and a half people less different from and more in tune with one another. They would become, through the auspices of monopoly capitalism, as harmonious as the mythical happy family.

Several aspects of this press coverage are noteworthy and represent a major change from the way Marconi's wireless achievements had been covered fifteen years earlier. First of all, it was not the inventor-hero who was tapping nature's mysterious potential; now it was a corporation and its team of scientists and engineers who were mastering the great void. They were obviously, even emphatically, working to promote institutional goals. Vail and Carty received the most coverage, but neither was celebrated as an inventor-hero the way Marconi had been. The achievement seemed more anonymous, a testimony to teamwork and organizational resources rather than to one man's lonely quest. The press applauded such experiments as the proper province of responsible capitalists whose success would promote social progress for all. Articles about the tests and the tests' significance demonstrated the extent to which corporations had learned the skills of news management and public relations. Coverage also revealed changing attitudes toward big business. The hated trusts of the late nineteenth and early twentieth centuries were now becoming, in the pages of the press, farsighted and committed public benefactors. Gone were the visions of individuals speaking to whomever they wanted whenever they wanted without having to go through a corporate or government network. Also gone, for obvious reasons, were the predictions about world peace. Fanciful visions of a decentralized and democratized communication system that would take people into mysterious realms were replaced by corporate-defined applications emphasizing public acquiescence to corporate control of the technology. The visions of 1900 had depicted the invention as the instrument of individual freedom; now, with the possibility of millions listening to "one man's voice," radio was cast as the agent of conformity and passivity.

This change in the portrayal of radio's progress and potential was part of a larger conservative swing in the popular press away from muckraking and toward accommodation with and even admiration for American business. As a 1914 editorial in *Collier's* saw it, there was in America "a swing toward reaction, a fatigue with tumult, . . . a growing distaste for the harsher and noisier leaders of reform, a tolerance, almost a sympathy for their victims."[29] Certainly on the editorial boards of leading magazines and newspapers, such a swing had occurred. After

1912, advertisers and creditors pressured newspapers and magazines to tone down or eliminate muckraking. Those who failed to comply found themselves suddenly short of revenue. Advertising had become increasingly important to the press; between 1914 and 1919 alone, advertising expenditures in newspapers and magazines doubled. For some publications, revenue from advertising began to outdistance revenue from sales.[30] Offending important advertisers, or critiquing the role of big business in America, was simply no longer a judicious journalistic style. In fact, magazines and newspapers themselves were becoming more consolidated, centralized, and bureaucratized every year, resembling in production method, distribution, and outlook the other major corporations of America. Organization men with backgrounds in business were taking over editorships previously held by crusading journalists or upper-middle-class reformers. Mark Sullivan, for example, a writer who became editor of *Collier's* in 1914, had to answer to an advisory manager who believed the magazine's purpose was to give Americans "the kind of instruction . . . which comported with the well-being of business."[31]

In the service of this goal, magazine heroes were, after 1914, men who presided over a large, bureaucratic organization and made it operate efficiently while at the same time getting "results." They were not rugged individuals who achieved their success alone; they were team players, genial, self-effacing, who made their organization's success top priority. They were men like Vail and Carty, not untamable individuals like De Forest or independent inventors like the young Marconi. Often, the portraits of these new heroes had been suggested, or even created, by the growing cadre of press agents now being retained by major companies. It was progressive reformers who had most strongly advocated the use of publicity to educate the people, primarily about society's ills and their possible cure. The great irony, of course, is that the progressives' most attentive students turned out to be business firms, who sought to co-opt this faith in publicity to their own ends. Because adverse publicity and public agitation had proved a time-consuming nuisance in the past, corporations now wanted to "engineer public consent" so that middle-class Americans would see corporate goals as being consonant with, and even furthering, their own.

The ethic of organizational efficiency, which gripped press management and filtered into the normative messages of stories and articles, showed how in the public, journalistic arena, values and attitudes were coming to terms with the bureaucratization of American life. For in the more private, behind-the-scenes world of corporate America, as well as throughout the federal government, the elaboration of internal hierarchies, the rationalization of work that produced countless new rules and procedures, and the obsession with order and efficiency in general were transforming how business was done and how people spent their workdays. All of these changes, which more clearly articulated the lines of authority and responsibility in large organizations, allowed institutions to extend their reach and power. And now this quest for institutional hegemony was endorsed, even promoted, by the popular press.

Wireless telegraphy, its use recently regulated by the federal government, would be transformed by the bureaucratization of American life, which was itself profoundly accelerated by World War I. The invention's relationship to the lives of everyday, ordinary Americans would be explained and promoted by a press more in tune with corporate America and the federal government than ever before. When Theodore Vail said Americans wanted to pursue a "common course of action," he had not misread his audience: he had, in fact, anticipated the obsession with cultural consensus and the intolerance of divergent viewpoints which World War I would unleash in America. He also anticipated how the war years would transform radio from a device for transmitting navigational or commercial messages between specific senders and receivers to a tool for exerting political, economic, and cultural power.

Radio changed in many important ways during these years, and many of the most dramatic changes were due to the establishment of corporate control over radio technology. In 1910, individual inventors and their companies still held the patent rights to the continuous wave components they had invented. Between 1912 and 1917, these rights were transferred to large corporations.

AT&T was not the only company that had gained control of continuous wave technology. General Electric, after Fessenden's departure from NESCO, found itself heir to the alternator, the transmitter that Alexanderson had improved and made more powerful. Fessenden was no longer interested in the radio-frequency alternator; once he broke with his backers, he abandoned wireless telegraphy and telephony. Thus, in 1911, G.E. was left with a 2-kilowatt alternator for which it had no customers. Alexanderson wished to pursue radio work and recognized that there were potential clients for G.E. apparatus. He interested his coworker Irving Langmuir in testing and improving the audion.[32] Langmuir, like his counterpart at AT&T, Harold Arnold, realized that the audion amplified more efficiently once it was converted to a high-vacuum tube. Langmuir patented the device, and Alexanderson incorporated it into what he regarded as a "complete system" for continuous wave transmission and reception. But G.E. and AT&T were very different companies: G.E. was primarily a manufacturing firm, supplying products to others, while AT&T operated its own communications network. While AT&T saw the wireless telephone as a potential competitor, G.E. did not. Thus, G.E. did not have the same internal incentives as AT&T did to stake its claim publicly to the technology.

As a result, little radio work was done at G.E. between 1911 and 1914. Alexanderson continued to press for further experimentation, and, after 1914, external events also worked to redirect the company's attention to wireless. Shortly after World War I broke out, England cut Germany's transatlantic cables and imposed strict controls over its own cable traffic. Thus, there might now be a market in Europe and the United States for a wireless transmitter capable of regular transatlantic work which could provide an alternative to the cables. More importantly, the American Marconi Company in December of 1914 expressed interest in the alternator. This interest justified Alexanderson's efforts to make the alternator more powerful and farreaching. Marconi inspected the alternator in the spring of 1915 and was impressed with its performance.[33] The alternator was powerful, it emitted a high, distinctive pitch, and it was capable of covering great distances. In July the two firms signed a tentative agreement whereby G.E. would supply alternators to the Marconi companies. G.E. now had a customer for its alternator, and the Marconi companies were on the verge of acquiring their first continuous wave transmitter. This acquisition was by now critical to both the British and the American Marconi companies, for the American Marconi Company's dominant position in the marketplace belied its vulnerable technological position.

In 1912, American Marconi was the preeminent wireless company in the United States: it controlled virtually all of the country's civilian ship and shore stations and handled the bulk of the country's commercial and press messages. Press accounts of the *Titanic* disaster, in reaffirming the importance of intership and ship-to-shore wireless communication, never mentioned Fessenden or De Forest. The hero of the hour was Marconi, and he was once again legitimated as the true and only inventor of wireless. If anything seemed inevitable in the American wireless industry in 1912, it was that American Marconi would continue to grow and consolidate its position, technically, organizationally, and politically. Yet this did not happen. Rather, from nearly the moment the Marconi Company vanquished its competitors in the American courts, it began its decline in the United States.

After 1912, the Marconi Company of America sought to consolidate its position and more firmly entrench itself in American soil. In October of

1913, Edward J. Nally, former Western Union messenger boy who had worked his way up through the telegraph business to become the vice-president and general manager of the Postal Telegraph and Cable Company, resigned that position to become the president of American Marconi.[34] The company's primary mission was to build several high-power, long-distance wireless telegraph stations in the United States capable of transoceanic message handling. Nally and Bottomley announced plans for a station in Massachusetts that would communicate with a sister station in Norway. Telegraph lines would connect the Massachusetts station to Boston and New York.[35] The company also began work on a receiving station in Belmar, New Jersey, and a transmitting station in New Brunswick which would communicate with Carnarvon, Wales.[36] It also announced plans for a transpacific link between California and Hawaii. By this time, however, the technical evidence was in: only with continuous wave technology could truly reliable long-distance work be achieved. American Marconi simply lacked the technology necessary to implement its strategic vision.

Marconi and his assistants had belatedly begun exploring wireless telephony, which relied on continuous wave transmission, in 1913 and 1914. But the European war interrupted this work, and Marconi was forced to suspend his experiments with wireless telephony. On August 2, the British government took control of all wireless transmission, and the various British Marconi stations came under the control of the Admiralty or, in a few cases, the Post Office. The Marconi Company continued to operate the stations, but on behalf of the government. The Clifden–Glace Bay transatlantic circuit continued operating commercially, but priority was given to military messages. Research and development had to meet military rather than commercial needs. Government demands seemed both endless and urgent: thousands more operators; stations to intercept enemy transmissions; more stations installed with direction finders, which could determine the origins of enemy transmissions; portable wireless sets; wireless for airplanes and dirigibles; high-power, long-distance transmitters; and a host of other requests, large and small. When Italy declared war on May 24, 1915, Marconi returned home and served as a lieutenant of the army General Staff in charge of organizing the army's wireless service. Experiments with continuous wave apparatus were now completely out of the question.[37]

While the Marconi Company clung to spark technology, one other company, much smaller than G.E. or AT&T, also gained control of important continuous wave technology. Formed in 1911 to exploit Poulsen's arc transmitter, the Federal Telegraph Company of California offered a commercial wireless service along the West Coast and between San Francisco and Honolulu. De Forest, needing a continuous wave transmitter, had used a version of the arc during his 1907 and 1909 voice broadcasts in New York with limited success. The man responsible for refining the arc transmitter and making it Federal's most important asset was Cyril F. Elwell, an Australian-born engineer educated in California.[38] By 1912, when Elwell first demonstrated the arc for navy officials, it was using less power to transmit over greater distances than the rotary spark gaps designed by Fessenden or Telefunken. At this point both Federal and the navy were using the arc to transmit dots and dashes. The navy's successful experiments prompted it to abandon spark technology for major long-distance work and to embrace the newest continuous wave transmitter. Thus, the arc made Federal, a successful regional firm far away from the centers of power, a company of national importance.

By 1914, a critical shift in the balance of power had occurred away from the Marconi companies and toward G.E., AT&T, and, to a lesser extent, Federal Telegraph. While outward appearances such as long-distance stations and volume of ship-to-shore business suggested Marconi's firm control over radio in America, the distribution and control of patents told another story indeed. Marconi had failed to see, until it was too late, why developing continuous wave apparatus might be important. The failure was primarily intellectual, but it had severe economic consequences. As the switch from spark to continuous wave technology became inevitable early in the century, Marconi was without the requisite apparatus. This made him, for the first time, dependent on American inventions and an American corporation, and after 1912, it was corporations, not individuals, who controlled continuous wave technology.

One of the tragedies of this transition from individual to institutional control over wireless was the way it affected the brilliant pioneer of continuous wave transmission, Reginald Fessenden. Just at the historical moment he had been impatiently waiting for—when customers (including his rival, Marconi, and his nemesis, the navy) realized, more than ten years after he did, that continuous wave transmission was vastly superior to intermittent spark transmission—he would get little credit. Just when the orders for continuous wave apparatus were coming in, when Marconi himself wanted exclusive rights to the alternator, Fessenden would receive no contracts. The spoils of all he had worked for were going to others. His work with wireless was now so permeated with feelings of bitterness and betrayal that experimentation became unbearable for Fessenden. He enjoyed a brief moment of vindication in May of 1912 when a jury awarded him more than four hundred thousand dollars for the patent rights he had assigned to NESCO. NESCO appealed, and the next decision, in August of 1913, went against Fessenden. While he and his lawyers continued the battle with NESCO, now in receivership, Fessenden explored other areas, including power transmission and storage, engine design for automobiles, and turbo electric engines for ships. In 1912 he became a consulting engineer for the Submarine Signal Company of Boston and developed his oscillator, which greatly extended underwater signal transmission and reception.[39] He also developed apparatus that would determine, through a submarine echo, the location and size of icebergs. This work stood him in good stead during the war, when the demand for submarine signaling and detection equipment skyrocketed.

Although he was able to find other work, and to redirect his creative energies, Fessenden never regained the optimism or the technical daring that had characterized his early work with NESCO. Now he was more pompous, even imperious, a manner that failed to veil his deep depression and his unremitting sense of persecution and paranoia. The biography written by his wife reveals much. Chapters covering the years between 1911 and 1927 are little more than accounts of inventions unjustly credited to others but actually, according to Helen Fessenden, all invented much earlier by her husband, who received no credit. These chapters also describe tension between Fessenden and his new employers, which was never Fessenden's fault but was due to the duplicity and incompetence of others. Everyone was out to get him; no one could be trusted. Most pathetic is the story about the Medal of Honor awarded to Fessenden in 1921 by the Institute of Radio Engineers.[40] The medal was meant, in radio circles, to be a great honor, and Marconi was the 1920 recipient. Fessenden came to suspect that Marconi's medal was solid gold while his was only a plated imitation, so he sent it off to Washington to have it assayed. When his suspicions were confirmed, he returned the medal to the IRE. The institute had to launch an investigation into the medals, and only after it convinced Fessenden that all the medals had been the same, that no slight had been intended, and that simple cost considerations mandated the medals be goldplate rather than solid would Fessenden take the medal back. Even gestures meant to honor, or simply flatter, now had for Fessenden dark, hidden motives that he felt compelled to expose.

Of all those who worked with wireless in its formative stages, Fessenden was the most technically farsighted. When what he was most proud of in himself was not truly valued by others, but was, he believed, squandered for some fleeting and insignificant gain, he became irreparably hardened and cynical. While Fessenden worked in other areas, he continued to press for restitution for his patents. His claim was simple. In his earliest agreements with Given and Walker, he had assigned his patents to NESCO in exchange for a salary plus a fee that was to be paid out of first earnings. When he was fired, he had yet to receive this fee. It was irrelevant to him that Given and Walker had paid all his research, construction, and travel expenses, which by 1910 totaled over two million dollars. After the court cases he kept pressing NESCO, which was in receivership and unable to pay. Fessenden's quest was to last fifteen years, and the stress it produced took its toll on the inventor and his wife. Helen Fessenden recalled those days when her husband would ransack the house, collecting all the documentation he believed he needed to confront the opposition: "Usually the strain was tense until he drove off with his case of papers, then I would find myself weeping into the breakfast dishes from sheer relief."[41]

Once Fessenden left his company and filed suit, NESCO dramatically curtailed its operations. It remained an insignificant company doing little business, and its only assets were Fessenden's patents. On November 28, 1917, NESCO's receivership came to an end and all its assets were transferred to a new company, the International Radio Telegraph Company. International manufactured apparatus for the military during the war, but with the end of hostilities, its business declined and its main assets once again were the Fessenden patents. In 1920, Westinghouse, trying to challenge RCA's control over radio, negotiated a deal with the International stockholders and gained control of the patents. After a short but intense competitive scramble between Westinghouse and RCA, the two companies reached an agreement in the spring of 1921 which made Westinghouse a partner in the corporate alliance known as the Radio Group and brought Fessenden's patents to RCA. At this time, Fessenden decided to press RCA for his money. Early negotiations fell through, and he filed a sixty-million-dollar suit against RCA. Two-and-a-half years later, in March of 1928, Fessenden and RCA settled out of court for what his wife called an "inadequate amount": half a million dollars.[42] Fessenden retired to Bermuda, where he died in July 1932.

Fessenden's story epitomized what had happened to radio by the early teens. It was no longer the province of the independent inventor; now it was in the hands of several large-scale corporations. Yet it is important to emphasize that neither AT&T nor G.E. had at the time devised a long-term strategy for entering the radio communications business themselves. G.E. controlled the alternator, at that time a continuous wave transmitter with enormous potential. But G.E. wanted to sell the machine, not use it to establish its own communications firm. AT&T's acquisition of De Forest's audion patents was as much a defensive move as anything. The company wanted the audion rights, first, to extend the reach of its wired network and, second, to prevent others from using wireless telephony to compete with Bell's existing and expanding network. AT&T was not working toward establishing a wireless communications business either.

There was another institution, however, that was, after 1912, moving very much in this direction: the U.S. Navy. Nothing had a greater impact on American radio during the teens than the navy's increasingly proprietary attitude toward America's wireless system. By 1912, several key naval officials began pushing for better integration of radio into naval strategy and operations. These men had a consistent and influential leader in Josephus Daniels, Woodrow Wilson's secretary of the navy from 1913 to 1921. Daniels was an outspoken advocate of complete naval control of American wireless, and he missed few opportunities to convert that goal into an established fact. Naval control of American wireless went through two stages during these years. The first stage, from 1912 to April 1917, was marked by organizational and attitudinal changes within the navy. As the United States struggled to remain neutral during these years, the navy was charged with ensuring that the American ether was not used by any of the belligerents to further their own military goals. This assignment increased the navy's institutional interest in and power over wireless transmitting and receiving. The second stage began once America entered the war, when naval control over American radio became complete. Throughout both stages Daniels worked to make this control permanent.

When the navy had first tested wireless, its officers had proved unable to, and in many quarters uninterested in, fully integrating the invention into naval operations. The department's decentralized structure and the attitude of many navy men made the navy especially ill suited to manage this new technology. Yet several changes, both within and outside the navy, began to pave the way for a keener appreciation of radio's value. For decades the various secretaries had recommended that the number of bureaus be reduced and their duties consolidated. While this much-needed reform was not enacted until World War II, in 1910 the Bureau of Equipment was abolished and its duties distributed among the remaining bureaus.[43] The Bureau of Steam Engineering, in existence since 1862, had long been the department's center for engineering—first steam, and then electrical—and it was this bureau that assumed control of acquiring and installing radio in 1910. The Bureau of Steam Engineering was responsible for designing, constructing, maintaining, and repairing the machinery on board naval vessels. It was not just a procurement bureau, it was a bureau actively involved in the building and successful mechanical operation of the ships. This bureau, with its strong engineering tradition and greater influence within the fleet, provided a more propitious organizational niche for radio's deployment.

Most importantly, the Radio Act of 1912 required the navy to increase its radio activities. Naval radio stations now had to exchange messages with civilian ships if no commercial station was operating within a 100-mile radius.[44] As Secretary of the Navy G.V.L. Meyer observed in his *Annual Report* for 1912, "The radio work and expenses of the department will be largely increased. It will be necessary to modernize and improve the apparatus of coast stations so that the commercial work may be successfully handled. . . . The added work will undoubtedly prove an incentive to increased efficiency."[45] To achieve this efficiency, the navy needed an office to coordinate and manage its new quasi-commercial service, and thus established the Naval Radio Service as part of the Bureau of Navigation.

Improvements in transmitters between 1907 and 1912 reduced technical uncertainty and made the navy less wary of rapid obsolescence. John Firth, NESCO's sales representative, had convinced the navy in 1909 that with Fessenden's new rotary spark transmitter, the department could begin planning a network of high-power, long-distance stations between the continental United States and the Caribbean. Firth promised that Fessenden would deliver what De Forest had failed to when he worked in the "hellhole of wireless" during 1905. The first such station was to be located in Arlington, Virginia, close to the center of federal power; others were planned for the Panama Canal, the West Coast, and at various locations in the Pacific. Construction began in 1911, and early in 1913 Arlington went on the air. By this time, Cyril Elwell of Federal Telegraph had refined the arc transmitter and convinced naval officials to test it at Arlington and compare its performance with that of the rotary spark. As Elwell had maintained, the arc, a continuous wave transmitter, achieved distances the rotary spark could not come close to matching.[46] Furthermore, the arc was quieter, required less power, and was more reliable. The success of this demonstration marks the navy's acceptance of continuous wave technology and its widespread adoption of the Federal arc. The West Coast firm was awarded contracts to equip the rest of the navy's planned high-power stations with transmitters.

Thus, by mid-1912, a new constellation of technical, legal, and organizational changes confronted the Navy Department. But they did not ensure that radio would be efficiently integrated into naval operations. Radio, with the potential to establish new and strong channels of communication in the navy, had been forced to operate within a nineteenth-century organizational structure. Only the efforts of an officer with entrepreneurial flair who was adept at exploiting unusual external pressures would compel this structure to yield to and be realigned by radio technology. Reviewing this realignment in some detail is an essential part of the radio story, for the newly consolidated and centralized navy of 1917 would permanently alter the management of radio in the United States.

·   ·   ·

STANFORD C. HOOPER has been called the father of naval radio.[47] It is a title he enjoyed and believed he had earned. His version of the navy's ultimate adoption of radio has the self-serving tone not uncommon to the autobiographies of people who were pioneers in their field, yet the record supports Hooper's story of his efforts to integrate radio into naval operations. He achieved this integration at a propitious moment in naval history, but this does not detract from the adroitness and ultimate success of his strategy or methods. He was a shrewd man who read his organization—and the times—very well indeed.

The son of a California banker, Hooper grew up in an environment that encouraged individual enterprise and initiative. When he was eight, his father built him a telegraph sender and key and taught Hooper the Morse Code. By the age of ten he was working part time for the railroad as a relief ticket agent and then relief telegraph operator. His seven years of experience with telegraphy provided a necessary foundation for his later radio work. He saw at a young age how a transportation and a communication network were integrated and operated cooperatively. In 1901 his father arranged for Hooper to attend the Naval Academy, and at the age of fifteen he entered Annapolis and embarked on his career.

After graduating from the Academy in 1905, Hooper served in various ships of the Pacific Fleet. He began to read about and tinker with wireless. Sometime between 1907 and 1908, Hooper put in a request for postgraduate training at the Naval Academy, specializing in wireless. This request was denied by Lieutenant Commander S. S. Robison, who believed "wireless would never be enough to warrant an officer giving it his full attention."[48] Hooper continued to pursue this goal, trying various tactics and routes, and finally was sent to the Academy in 1910 as an instructor in electrical engineering, with wireless instruction added to his regular duties. Thus, not unlike many teachers, Hooper was to learn his subject matter shortly before teaching it to a class. But his assignment, as George Clark noted, marked a turning point in his career: "From then on he was in charge of a 'radio division' of the Navy, be it of the Department of Electrical Engineering at the Academy, or the Bureau of Steam Engineering in Washington, or of the Fleet."[49]

Lieutenant Hooper was one of the few officers in the navy who had experience as a wireless operator. By 1911, several commanders working in the Bureau of Steam Engineering had begun to consider more seriously the use of radio for communicating between vessels when in battle formation, but they could not adequately implement this plan without firsthand knowledge of radio communication. Consequently, Hooper was assigned to develop and write up instructions for tactical signaling between battleships. His plan would be tested during spring target practice of 1912, when radio signals would accompany all visual signals.

Another young officer who had had childhood experience with telegraphy had been assigned to report on the use of wireless during the autumn battle practice in 1911. Ensign C. H. Maddox assessed the technical merits of the apparatus and analyzed wireless's potential for tactical signaling. In his first report, he urged that wireless have its own set of tests rather than be tested in conjunction with target practice. Only then, he said, would wireless "get the full consideration that it deserves." During target practice, he reported, "a wireless test is too liable to be relegated to the list of those things that can be slighted for the sake of possible increase of 'hits per gun per minute.' "[50]

Maddox saw as the most immediate and pressing need "officers in the fleet who possess a thorough practical and theoretical knowledge of wireless and who are themselves expert operators." "At present," he reported, "the real head of the fleet's wireless system is the enlisted operating force of the flagship." He found these operators to be mediocre, in part because no specialization existed in the electrical force aboard ship. Wireless operators and dynamo tenders were often rotated between these two jobs and, thus, there was "small chance for improvement." He also recommended that the Atlantic Fleet have an officer in charge of the fleet's wireless who would "systematize and control this important factor in naval efficiency." He looked forward to the day when the navy would possess enough officers proficient in wireless for one to be assigned to "each division of the fleet, and eventually one to each battleship."

Reportedly, Hooper did not see Maddox's report or recommendations. In the spring of 1912, Hooper would reiterate these suggestions.

But Hooper, now twenty-six, was about four years older than Maddox, and was more experienced, higher in rank, and better connected. One important ally was Lieutenant Commander D. W. Todd, head of the Radio Division of the Bureau of Steam Engineering. Hooper's observations and report would obtain more sway as they passed through Todd.

Prior to the April 1912 tests, Hooper devised a tactical signaling code and made several other specific recommendations. He urged that each ship's wireless operator be brought up to the bridge from below decks during tactical signaling to reduce the delay between orders given by the captain and orders transmitted by the operator. Portable equipment would be installed quickly, the transmitting key on the bridge connected to the main transmitter below decks. This move demonstrated, physically and symbolically, the importance of tying radio directly to the chain of command.

Hooper's tactical signals and set of instructions on general signaling procedures for the maneuvers were printed up and included in a booklet of general instructions written by Commander T. T. Craven, who was in charge of fleet training in the Division of Operations, Bureau of Navigation. When the tests began, Hooper and Craven went to the flagship to observe results. Hooper monitored the radio signals, but he heard nothing all day. All the signaling was done by flag. At the end of the day, he visited several of the ships in the fleet to investigate what had happened. The only encouraging discovery Hooper made was that the Bureau of Steam Engineering had set up radio apparatus on the bridge of every ship. But no other steps in Hooper's plan had been followed. "The Navy, as usual up to that date, did not take radio seriously," he commented. "The commanding officers had handed the instruction pamphlet to the officer in charge of communications, but he, in every case not at all familiar with radio, did nothing more with it, probably not having faith in the ability of radio to do the work. In a few cases, the booklet had found its way down to the radio cabin, but the Chief had not had time to read and understand the scheme, nor did he have it expained to him."[51] Thus, responsibility for testing radio signaling had been passed down to those in the organization who had no authority or accountability, particularly in the sphere of tactics or strategy. It was 1912, and in the navy, radio was still being treated as an afterthought.

Hooper's report reflected his disappointment. He criticized the officers for not involving the operators in the tactical signaling process aboard ship, and urged that the skills of the operators be upgraded. Noting "Wireless is running away from us in certain regards," Hooper also recommended that there "be an officer in charge of radio matters in the Fleet who [was] an expert operator."[52]

Again, the spur for improvement came from the officers on shore, in Washington, not from ship's commanders. Commander Craven, persuaded by Hooper's report, urged the chief of the Bureau of Navigation to add to the staff of the fleet commander in chief the position of fleet radio officer. Both Craven and his friend Todd recommended Hooper for the position, and in August of 1912, he became the navy's first officer in charge of coordinating the use of radio at sea. Rear Admiral Hugo Osterhaus "objected strenuously" to having such an officer on his staff, and then arranged that Hooper's duties also include tactics and athletics.[53] Having to organize and supervise boat races and boxing matches helped delay Hooper in his work and undermined the importance and prestige of his main task. Thus, despite the efforts of the navy's equivalent of middle management, recalcitrant top executives were able to preserve the status quo. No fleet officer could institute full use of radio in naval operations alone. He needed allies on shore and within the fleet. Hooper and Maddox had recognized the need for a new organizational tier in the fleet: officers, not enlisted men, had to have control over radio. In the fall of 1912, Hooper recommended to Osterhaus that commanding officers of all major naval ships designate an ensign as radio officer and require him to become a proficient operator. Osterhaus followed Hooper's suggestion and issued the order.[54]

This was Hooper's first shrewd strategic move as fleet radio officer. Young officers would be less bound by naval tradition; unhindered by years of flag signaling, they would be more open to the new technology. They had, at this juncture in their career, little to lose and much to gain by becoming proficient in radio. Thus, there would shortly be dozens of officers just junior to Hooper in rank who were bound to his authority and competent with and sold on the new technology. He was arranging the beginnings of his network by creating an organizational cadre that would permit—and benefit from—full use of radio in navy operations.

In 1913, Rear Admiral Charles J. Badger, who had served on the 1902 Wireless Telegraph Board, became commander in chief of the Atlantic Fleet. His chief of staff, Commander Charles F. Hughes, was more sympathetic to Hooper's goals than Osterhaus had been and recommended that Hooper be relieved of his duties as fleet athletic and tactical officer. Hooper was now free to concentrate completely on radio, and his attempts to incorporate radio more fully into the daily functioning of the fleet would be more sympathetically and seriously considered.

While the ensigns were learning about radio, Hooper had another, more difficult task. He had to upgrade the performance of the operators while simultaneously wresting control of fleet radio from them. Because wireless had been kicked down the naval hierarchy from its introduction, it was now controlled almost entirely by the enlisted men.

From rear admiral to ensign, few officers had considered wireless important, and therefore few sought to learn how to use it. But the bureaus (first Equipment and then Engineering) kept sending the instruments to the ships, and someone had to oversee them. The "technological buck" stopped with the enlisted men who worked under the broad heading of electrical engineer. At this level of the navy, wireless did not mean subversion of autonomy or tradition. On the contrary, an enlisted man who knew about radio gained some small distinction. He enjoyed a certain degree of autonomy—he could transmit whatever he wanted—and he often possessed privileged information. Most of the messages sent out by naval operators before 1913 were personal messages such as this one cited by George Clark: "Longing for you darling, and waiting for the fog to lift. Lieutenant _____."[55] The ship's operator also conversed with other naval and commercial operators in the vicinity and could eavesdrop on various conversations.

As radio apparatus began to proliferate in the fleet, control of wireless was often maintained by the chief engineer of the flagship. Use of the airwaves came to be dispensed by the chief as a privilege, a perquisite. Radio technology provided the chief and the operators with control and diversion, often denied aboard ship, and these men were not about to relinquish these two things easily.

Hooper's goal was to compel the operators to "use their station for nothing but official business, to make use of a military routine whose first requirement was obedience to orders, and to improve their operating ability."[56] The fleet radio officer ordered that operators were only to send and receive official messages, only on official navy forms, and that personal conversations were to stop. Operators responded to this reform by both ignoring the orders and denigrating Hooper over the air. Hooper, as a result, spent his evenings learning to distinguish the sound of each ship's spark and the "fist" of each operator so he could ultimately determine which operators were violating his new regulations. He then devised a scheme he hoped would end all resistance. Admiral Badger authorized Hooper to send the following message to the commanding officer of any ship guilty of disobeying Hooper's rules: "Your attention is invited to Fleet Regulations. . . . Your radio operator is disregarding instructions and is using unofficial language. Badger, C-in-C."[57] By transmitting this message, which the operator had to deliver to his commanding officer, Hooper set the offending operator up for disciplinary action. The first operator who refused to acknowledge receipt of this message was court-martialed.[58] Through such methods, Hooper gradually began to enforce the discipline and obedience he needed. Now he had to build an efficient operating network from the bottom up which commanding officers at the top would eventually be convinced was indispensable.

While the enlisted men had enjoyed control over radio until Hooper's reforms, they had had no compelling need to be fast or efficient operators. Commanding officers had placed no premium on speed or accuracy, so why should the operators? They were not rewarded for making the technology work well for the organization or chastised for failing to do so. Hooper had to structure an incentive system that would give the operators a continuing interest in good signaling. He introduced a rating system whereby every operator would be labeled according to his level of proficiency. Linking performance to organizational rewards, Hooper also instigated sending and receiving competitions among the operators, with promotions as prizes. He began drilling the operators in learning and using his battle signal code. He standardized the number of operators on each ship—before this, one ship might have only one operator while another had six.

By 1913, Hooper had succeeded in having an officer on every ship in the fleet assigned to oversee the radio room. Removing control over radio from the enlisted men, who had no role in strategy and planning, and assigning control instead to a new managerial tier of young officers was probably Hooper's most important reorganizational ploy as fleet radio officer. Hooper had realigned and disciplined the lower levels of the fleet hierarchy. He needed this tier of enlisted men, ensigns and lieutenants, to be efficient, coordinated, and obedient to his authority in order to impress the men at the top levels that radio could be an invaluable tool for commanders. Hooper had recognized that before an organizational resource could exist, and be perceived as a resource, men and "machines" had to be fully integrated at the lower levels. Only then could the top brass legitimize the system through successful and continued use of it.

Hooper got his critical opportunity in 1913. Admiral Osterhaus "would not permit his ships to be maneuvered by radio and would only execute his signals by flaghoist."[59] Admiral Badger, a younger officer, was more inclined to try radio. Sometime in 1913, Badger ordered that during one day's exercises, all maneuvering would be handled by radio. For the first time, no flags were to be used at all. All of the commander in chief's instructions were accurately relayed and carried out, and, for radio, the maneuvers were a complete success. The next week a similar but unexpected test occurred. While steaming in Chesapeake Bay, the fleet hit a sudden squall, and visibility was reduced to zero: the flags were of no use. Radio had to transmit all instructions. The storm lasted for half an hour, and when it cleared all the ships could be seen in formation, exactly as they had been ordered. Tactical signaling by radio, done previously only on an experimental basis, increasingly became regular practice in the fleet. Luck had helped Hooper achieve an important breakthrough: commanders saw firsthand how radio could avert disaster. This was a major step toward gaining full acceptance of the new communications technology.

Hooper and his allies had partially integrated wireless into the operations of the fleet between 1912 and 1914, but the navy's network of shore stations lacked systematic coordination, a deficiency the department began to address after the war broke out in Europe. Early in the war, in August of 1914, Hooper was ordered to Europe to observe the use of radio. In early 1915, he returned to Washington, where he served for two weeks in February with three other officers, all with radio experience, on a Radio Reorganization Committee. The committee concentrated on the need to strengthen and better coordinate the navy's coastal chain. Until 1915, each shore station communicated with ships at sea, listened for distress calls, and worked with the two stations adjacent to it along the chain.[60] Most of these installations were at the navy yards. The stations were under the control of the commandant of the yard, an officer with a multiplicity of responsibilities and concerns and little incentive to seek improvement in the use of radio. Unlike the radio waves from his station, the commandant's influence was confined to the yard; he had no jurisdiction beyond it. Coastal radio could not be coordinated under this structure, whose fragmentation was reflected in the signaling setup of the chain. Messages were relayed from one station to the next along the north-south linkage. Thus, a message from Boston to Pensacola would be relayed many times. The stations were set up in series: if one link broke down, no transmissions were relayed beyond that point.

Hooper, who had welded the fleet into a signaling unit and was fresh from his European trip, found such a situation antiquated and potentially dangerous. His colleagues agreed. Again he proposed a marriage between technical improvement and organization building. No one oversaw the coordination of the shore stations because no slot existed in the organization for this purpose, and no one had conceived of the airwaves as appropriate jurisdictional turf for an officer. Through the committee, Hooper proposed a series of high-power stations, preferably with a range of a thousand miles or more; these large areas were to be called naval communication districts, and each was to be under the supervision of a district communications officer. The existing coastal stations would have their power and apparatus upgraded and serve as a secondary signaling tier. Again, Hooper had devised a more centralized network, with more clearly articulated lines of authority leading from the bottom to the top of the hierarchy, and from the field units to the central office. When Secretary Daniels approved these recommendations in February of 1915, some elements of the plan were already in place. The navy's high-power radio chain was under construction: Arlington was complete, the Canal Zone station at Darien was about to open, and construction of stations in Pearl Harbor, San Diego, Puerto Rico, the Philippines, Guam, and Samoa was underway.[61]

In 1915, Congress enacted a bill creating the post of chief of naval operations. This officer would serve as the much needed liaison between the secretary and the bureau chiefs, gathering information on material, operations, and personnel which would help the secretary develop more informed and long-range strategy. Although the chief of naval operations was the ranking active officer of the navy, he was not empowered with direct authority over the bureaus. Nonetheless, this influential advisory position just above the bureaus and just below the secretary provided the department with "professional coordination and operational direction."[62] During the same year, the Naval Radio Service was reorganized and became the Office of Communications, which supervised telegraph, telephone, cable, and radio communcations. The service was moved out of the Bureau of Navigation and became an important office, its director reporting not to a bureau chief but to the chief of naval operations.[63] These elevations in title, location, and organizational niche indicated how far up the hierarchy radio technology had traveled. The centralization and consolidation of radio operations, and their placement much closer to the center of power, ensured radio's progress under naval auspices. Radio's potential at the turn of the century had been fulfilled: changes had resulted in a more centralized structure at sea and on shore, and radio had become central to naval strategy.

As the navy's organizational structure came to resemble more closely the structure of successful civilian corporations, and as entrepreneurial officers like Hooper, Craven, and Todd came to preside over radio's development, there was more cooperation between the navy and the corporate world. The management of radio in both spheres had become much more bureaucratized, orderly, and determined. And there was a greater appreciation by each sphere of what the other had to offer. After

1912, more naval officers realized how important radio might be to furthering geopolitical ambitions, and after August 1914 there was no doubt about the interrelatedness of the two. The navy needed reliable, powerful apparatus as well as more harmonious relations with the companies that controlled the technology. Navy officers and members of the wireless industry became much more friendly than they had been before. AT&T and the navy cooperated fully on the transcontinental wireless telephone tests, and Federal Telegraph became a major supplier of equipment to the navy. The only company deliberately left out in the cold was American Marconi, which many naval officials continued to regard as being controlled by foreigners and therefore as suspect.

This military-corporate cooperation over radio, unthinkable as late as 1910, demonstrated how dramatically the management of radio had changed in five years. If one considers who controlled crucial patents in 1910 and 1915, the transition from individual to institutional control becomes starkly apparent. Cooperation turned into partnership when America entered the war, and the military management of radio provided powerful lessons in how the centralized control of radio might be achieved, and what, in both private and public spheres, such control might mean.

■   ■   ■

WORLD WAR I provided a most favorable political and ideological climate for the promotion of military wireless ambitions. When war broke out in Europe in August of 1914, U.S. reaction revealed how embedded wireless telegraphy had become in the fabric of American policy and how important it now was to the implementation of that policy. The European war brought to the center of public attention how wireless was being perceived and how it was being used to exert political influence. Wireless use was not confined to keeping ships' captains informed on navigational matters or to transmitting news stories. With the advent of war, wireless became both a potential weapon and a tool for spreading propaganda. Trying to prevent the invention from being used in either of these capacities was what guided American wireless policy between 1914 and 1917. The other agenda guiding this policy, an agenda vigorously endorsed by Secretary Daniels and certain navy officials, was that the war be used finally to gain for the navy control over America's wireless system.

Daniels's agenda was buttressed by public sentiment and presidential actions. In 1914, most Americans, reacting to the war news with a mixture of disbelief, sadness, and disgust, wanted U.S. neutrality firmly established. President Wilson, who was presiding over a country of fiercely contested loyalties, urged that Americans try to be "impartial in thought as well as in action."[64] On August 5, he issued his neutrality proclamation, which included specific regulations governing the operations of wireless stations on American soil. He forbade the transmission or delivery of any unneutral messages and prohibited any stations "from in any way rendering to any one of the belligerents any unneutral service during the continuance of hostilities." The secretary of the navy was to enforce these policies, and was authorized "to take such action on the premises as to him may appear necessary."[65]

In August of 1914, the government's major sources of concern regarding wireless were the British and German high-power, long-distance stations along the Atlantic coast. The Marconi Company had stations at Portland, Maine, and Siasconset, Massachusetts; the Belmar–New Brunswick stations, which were still under construction, also belonged to Marconi. The Germans had just completed two high-power stations in the United States which were capable of direct communication with Germany. Telefunken had established an American subsidiary, the Atlantic Communications Company, which had built a station at Sayville, Long Island, which was in constant contact with its sister station in Nauen. A German firm known as HOMAG had built an equally powerful station at Tuckerton, New Jersey, meant to communicate with a station at Eilvese, near Hanover.[66] At this station in late January 1914, a message traveling 4,062 miles was received for President Wilson from Kaiser Wilhelm, who expressed the hope that wireless would become a new link between the two countries.[67] Because the German cables had been cut by the British at the outset of hostilities, this new link became critical.

Through such powerful wireless stations, both the Germans and the British could monitor when and which ships were leaving the United States, send messages to their own military and merchant ships at sea, and stay in touch with their respective diplomatic and military leaders at home. As early as August 6, the *New York Times* printed allegations of breaches of neutrality. An amateur operator on Long Island monitored Sayville's transmissions and turned his findings over to the government. Military orders in cipher originating in Berlin were allegedly received at Sayville and then dispatched to German cruisers in the Atlantic.[68] The Tuckerton station was also accused of receiving unneutral messages. On August 7, the *Times* printed allegations that German diplomats had used the Sayville station to communicate with the cruisers *Dresden* and *Karlsruhe*. Relying on coded messages, they "guided the naval vessels, it is asserted, in their efforts to capture liners belonging to countries with which they are at war."[69] To ensure that America's neutrality was not violated by these stations, the navy dispatched censors to the foreign-controlled, long-distance stations; these censors were to be "in charge of the sending and receiving of messages."[70] Secretary Daniels ordered that no cipher or coded messages could be transmitted or received at these stations. The ensign who served as censor first reviewed every outgoing message to check for unneutral information; once the censor cleared the message, one of two naval operators listened in to the transmission of it to make sure it was sent in its censored form. The navy men also monitored and censored all incoming messages. That the censors at the German stations were allowed to demand translations if necessary suggests that there was considerable room for error—and evasion.

Neither the German nor the Marconi stations accepted the censorship with equanimity. The Germans protested that the prohibition on coded messages was itself an unneutral act, for the British could send coded messages via their cables, a route denied the Germans. The Marconi Company protested the entire censorship apparatus, charging that the secretary of the navy had absolutely no legal authority over wireless transmissions, that the only government agency with any authority over wireless was the Department of Commerce, and that even it could not, under the 1912 act, restrict the content or destination of messages. John Griggs, president of the company, stated that the Marconi Company of America was an American company, not a foreign one, and thus should not be under surveillance.[71]

The government found the Germans' position more compelling than the Marconi Company's. Censorship would indeed continue, but the government would now allow coded messages, provided that the American censors were given copies of the code books. No coded messages could deal with military matters, and the censors were under orders to keep the content of all messages confidential.[72]

At the same time, Daniels was sensitive to any opportunity that might consolidate the navy's position. The owners of the German station at Tuckerton, it turned out, had failed to apply for a license before the war broke out. Because the station was unlicensed, it could not legally operate, and the government could no longer license it, went the argument, because according to the 1907 Hague Convention rules governing wireless, to license a station owned by a belligerent would be a violation of neutrality. The government therefore closed down the station on August 24, 1914, an event announced on the front page of the *New York Times*. On September 5, Wilson announced through an executive order that the navy would take over one or more high-power stations on the Atlantic coast; four days later, the navy moved into Tuckerton, promising the Germans that their apparatus and the revenues collected during the navy's tenancy would be returned at the close of hostilities. The navy would handle news and diplomatic and commercial messages.[73] The station was closed until late October while the navy removed the German Goldschmidt alternator and replaced it with one of the Federal Company's arcs, by now the navy's standard long-distance transmitter.

Daniels apparently then set his sights on Sayville and the Marconi station at Siasconset. The Marconi Company, after filing a formal complaint with the government protesting the censorship procedures, took the case to court, charging that the navy had no legal authority to monitor messages. The government countered by charging, on September 10, that the Siasconset station, during the censor's temporary absence, had violated U.S. neutrality when it forwarded to a New York address an order from the British cruiser *Suffolk* for provisions and newspapers. The government threatened to close the station. Griggs, outraged, asserted that the message was not unneutral, that only the Commerce Department had the authority to close a station for just cause, and that no such cause existed.[74] President Wilson responded on September 19, asserting that Attorney General Thomas W. Gregory had determined that Wilson had full authority to close the station if the company refused to comply with the censorship code. Daniels forwarded the details of the opinion to Griggs, stating that unless the Marconi Company offered an explanation for the *Suffolk* message and indicated its intention to abide by the government's code, the Siasconset station would be shut down.[75] The Marconi Company did not respond, waiting instead for its day in court. On September 24, Daniels ordered the Marconi Company's most important American station closed. Two weeks later, the company's suit was thrown out of court when the judge ruled that the court lacked jurisdiction over the case.[76] Later in October, the government accused the Marconi Company's Honolulu station of handling unneutral messages. This time the company apologized and vowed this would not happen again.[77] Daniels did not allow the Siasconset station to reopen until January 17, 1915.

On April 23, 1915, headlines on the front page of the *New York Times* announced "Germans Treble Wireless Plant." A new, 100-kilowatt alternator replaced the 35-kilowatt quenched spark gap, and a new aerial consisting of three 500-foot towers was erected. Noting ominously that the change was made "quietly, almost overnight," the *Times* reported that all of the new apparatus had been manufactured in and shipped over from Germany, and that it required "six large freight cars" to bring the equipment to Sayville. Emphasis on the physical size and power of the machinery gave the article a tone of uneasiness, even paranoia. The story had been leaked to the paper by an unidentified naval officer. The new equipment, according to the *Times,* would transform Sayville into "one of the most powerful of the transatlantic communicating stations in this part of the world." The story observed that the new plant had not yet been licensed by the Department of Commerce, but that the license undoubtedly would be granted as soon as the station was ready for operation.[78]

However, this was the grisly spring of 1915, when, increasingly, civilian citizens of neutral countries were becoming victims of the war. On May 7, at 2:30 in the afternoon, while many of its passengers were finishing their lunch, the *Lusitania* was torpedoed and sunk without warning by a German U-boat. There were 1,959 people on board, including 188 Americans. The Germans, reported the press, made no effort to try to rescue survivors. Only 761 people survived; 1,198 died, 124 of them Americans. Fifty babies younger than one year and more than 100 others less than two years old were killed.[79] The outrage the incident provoked crystallized pro-Allied and even pro-interventionist sentiment in some sectors of the country and raised sharp suspicions about German activities in the United States. How did the U-boat know exactly where to find the ship? Some suspected wireless. The Berne correspondent of the *Morning Post* cabled that, in Germany, the sinking of the ship had been expected: "There is no doubt that the German Admiralty received a wireless message from New York, giving the date of the *Lusitania's* sailing, and that a number of submarines were told to torpedo her at any cost."[80] Others charged that the Sayville station must have been involved somehow and that an investigation was in order.[81] Given that most Americans, despite the *Lusitania,* still wanted to avoid entering the war, preserving U.S. neutrality and preventing belligerents from using the country as a base of military operations became even more important.

While Wilson pursued a diplomatic solution to the crisis, Sayville continued to operate as before. Then, on June 30, huge headlines announced "Twenty or More Americans Lost when Germans Sink the *Armenian*: Navy May Seize Sayville Wireless." The subheadlines read "Officers Think German Station May Send Messages to Submarines" and "Plant under Suspicion." Now the government suspected that seemingly innocent sounding messages, phrased in "plain English or German, were in reality in a cipher code and contained military information that would be useful to European belligerents."[82] Officials noted that, often, citing static interference, Sayville and Nauen requested that messages be repeated; now it was believed that these repetitions embodied some sort of code. The Providence *Journal* claimed it had been monitoring Sayville for months and was convinced that the station was "nothing but an adjunct to the German spy system."[83] Wasn't Professor Vennick, the German cipher code expert at Sayville, still an officer of the German Marine Reserve? Rumors and suspicion fed off each other. Reports circulated of secret wireless stations and secret submarine bases off the coast of Maine which would allow the Germans to attack ships closer to the American shore.[84] There was also supposedly a huge German spy ring, in which wireless played a critical role. The *Literary Digest,* citing several sources, reported that the Germans were "working toward the formation of a [wireless] network enclosing the whole world." This network consisted, in part, of "a German-controlled wireless station in each state of the Union, besides four stations in Mexico and sixteen distributed over South America, all under direct orders from Sayville."[85]

The navy could hardly have had a more favorable political climate in which to justify its takeover of Sayville. At this time, navy officials displayed a newfound sophistication in dealing with the press. Here they were leaking stories to the *Times* and exploiting international incidents to further the Navy Department's institutional ambitions. Front-page headlines on July 9 reassured readers that government control of Sayville began that day. The official explanation was that the German upgrading of the station made Sayville, in reality, a new station, and since it was controlled by one of the belligerents, the station could not be issued a new license without violating neutrality. Contending that Sayville had been operated in part by German military forces and that the American money involved in the Atlantic Communications Company was "infinitesimal," the navy maintained it had little choice but to assume control. At first all the German employees were to be replaced immediately by navy men, but since the Germans were familiar with the apparatus, they were temporarily retained.[86] However, the German employees' duties were restricted to those necessary to maintaining the apparatus; the Germans were no longer allowed to handle messages.

In September the *Literary Digest,* citing an article printed in the *Electrical Experimenter,* revealed that Charles Apgar, an amateur operator from Westfield, New Jersey, had helped the Secret Service obtain evidence of neutrality violations at Sayville. The article failed to mention that the patriotic Apgar was also an employee of the Marconi Company. Apgar, who used an audion and the regenerative circuit, monitored the Sayville transmissions and often heard Nauen, as well. In addition, Apgar had devised a method of recording incoming wireless messages on phonographic wax records; these records were later replayed and their contents carefully transcribed. The results were sent off to the Secret Service. The records allegedly confirmed that "secret messages, in code and otherwise, were . . . sent along with regular censored messages" by the insertion of extra dots or by transmitting messages at an abnormal pace.[87] Secretary Daniels held up the Sayville incident as a dangerous example of the perils of private ownership of wireless. In his *Annual Report* for 1916, Daniels asserted: "It is becoming increasingly evident that no censorship of radio stations can be absolutely effective outside of complete government operation and control." He continued: "The government must in the end follow the lead of almost all other governments and obtain control of all coast radio stations and operate them, in conjunction with naval stations, for commercial work in times of peace."[88]

Navy management of the two German transatlantic stations may have curtailed the transmission of coded, unneutral messages, but it did not help determine which kinds of messages were unneutral and which were not. By the fall of 1916, the United States had weathered several diplomatic crises over Germany's use of submarine warfare, and tension over the issue remained high. In early October, a German U-boat docked briefly at Newport, Rhode Island, and then put out to sea. On October 9, the submarine sank six ships off the coast of New England, near Nantucket. It turned out that the *New York Herald* station, as part of its news bulletin service, sent out a wireless message warning all ships in the vicinity that the German submarine was operating off the Nantucket lightship. The navy maintained that sending such a message constituted an unneutral act, because a British cruiser could have picked up the message and used the information to locate and sink the submarine. The navy installed censors at the *Herald* station. Navy officers visited other major commercial stations along the north Atlantic coast to "remind" the operators to preserve neutrality. The *New York Times* despaired in an editorial that such stations could not "be neutral without being unneutral."[89] Sending the message helped the British; not sending it would have helped the Germans. Either way, the *Herald* station could not be truly neutral.

On February 4, 1917, when Wilson broke diplomatic relations with Germany, all employees of German extraction, including those who could not be vouched for with certainty, were dismissed from Sayville. All the Germans at Tuckerton were dismissed, as well; a few months later they were arrested, as was Dr. Karl Frank, head of Atlantic Communications.[90] The detention of these men reflected the widespread fear of German spy rings which had been raised to near hysterical levels after publication of the Zimmerman Telegram on March 1, 1917. Their arrests anticipated the repressive measures that the war would spawn.

Through censorship and then takeover, the government sought to prevent U.S.-based radio stations from being used as weapons of war. More difficult to control was use of the airways for propaganda purposes. Reported the press in November 1914: "The spread of official news by wireless broadcast over the globe for anyone who can pick it up is a new method of political propaganda which has been introduced through the European War."[91] With the German cables cut, the German chancellor had begun to use wireless during the second week of the war to present Germany's "side of the case" to the American people. Such messages were, first and foremost, intended for the American press, but they also reached British, Belgian, and French stations and thus served well as psychological warfare. They also reached all those American amateurs. The French decided to send "military and political bulletins on the progress of events from the French point of view from their station on the Eiffel Tower." The British Marconi Company followed suit, "disseminating British bulletins composed by the Foreign Office."[92] The airwaves were now filled with political debate and with conflicting views of the war, all of them designed to sway public opinion. One British citizen saw a real danger in the broadcasting of Germany's "false versions of war occurrences" and wondered whether Britain shouldn't try to jam these transmissions. At the very least, he thought, the Allies should "counteract the effect of this 'news' " so neutral countries would recognize that "Germany was a common danger and a common enemy to the whole civilized world."[93]

Once the United States entered the war, the American press, especially newspapers and magazines published in New York and elsewhere in the Northeast, where pro-British sentiment was strongest, began to view the ether less as an exciting, mysterious realm or as a resource whose use had to be regulated to ensure safety at sea. Now it was clear that the ether constituted national territory, in which Americans had political and international rights and prerogatives, and which had to be defended as staunchly as the shores. As the ether became militarized by the war, press coverage increasingly reflected military concerns. Stories revealed fear and resentment: the Germans, from their secret stations, were using our ether to further their military ends. The stations were hidden, hard to locate, and highly dangerous because, it was alleged, they sent orders to U-boats and coordinated spy plots on land. How dare the Germans use our ether to subvert our national goals and advance their own? Implicit in the press accounts was the fear that if foreigners with political and, particularly, military agendas gained access to the ether, they could use this power to ensnare and lead public opinion. The military and the press got a taste of what it was like when the ether was used by one country to convey its beliefs and political goals to another. While American government officials may have found the taste bitter, they also thought the recipe worth duplicating.

On April 6, 1917, the United States entered the Great War. Under section three of the 1912 Radio Act, President Wilson was authorized "in time of war or public peril or disaster" either to close down private radio stations or to place these stations under the control of "any department of the government, upon just compensation to the owners." Consequently, on April 7, all radio stations in the United States, except those already under army control, were taken over by the U.S. Navy. Fifty-three commercial stations, most of them formerly operated by American Marconi, were added to the naval communications network, and another twenty-eight commercial stations were closed.[94] Now the navy presided over a national and international radio network.

The navy takeover of America's wireless network gave Daniels and other like-minded officials the control over radio they had so fervently sought. Navy intervention in the marketplace after April 1917 was swift and decisive. The Navy Department's technical needs and organizational imperatives were forcefully imprinted on civilian suppliers. The navy required increased numbers of efficient and sturdy transmitters and receivers, as well as portable sets. Although radio was a "two-edged weapon," because messages sent by radio were not secret and could be monitored, it was nonetheless widely used during the war. Radio communication was essential to naval operations, and portable sets were used in the trenches. Radio also provided communication between airships and ground bases. To provide all this apparatus in time, rapid production, rapid integration, and centralized coordination were essential. During the war, the navy controlled the design, purchase, installation, and upkeep of all government radio except the army's. This centralization led to standardization of apparatus, the navy's long-sought goal, and better control over suppliers, rate of production and delivery, and "competition" from other agencies needing radio. For example, while there had always been several suppliers of most radio components, the Crocker-Wheeler Company enjoyed a near-monopoly in the production of motor-generators. When demand increased during the war, Crocker-Wheeler was deluged with orders from various companies demanding to be supplied immediately because of war contracts. Stanford Hooper

**U.S. Signal Corps recruits training to become wireless operators
during World War I.**

intervened and set up a schedule for production and delivery based on
the navy's needs. This incident prompted Hooper to view production
more as a corporate executive would: he wanted to ensure that he had at
least two sources of supply for whatever he might need. He demanded
that Crocker-Wheeler turn its blueprints over to General Electric. Failure
to do so could mean government takeover of Crocker-Wheeler. In the
face of this threat, Crocker-Wheeler naturally compromised, suggesting
the Triumph Electric Company as second supplier.[95]

Naval officials also contrived to eliminate competition over the limited output of radio equipment. Hooper learned of a plan to build a new
merchant fleet, to be called the Emergency Fleet Corporation. He saw
this fleet as a potential competitor in purchasing, and he feared that
civilian (and in his view inexperienced) purchasers would be willing to
pay more for radio apparatus and thus both raise prices and deprive the
navy of needed equipment. He intervened even before construction began and persuaded the chief engineer of the Emergency Fleet Corporation to let the navy supply it with radio.[96]

Because of the great demand for radio, American companies producing radio apparatus, such as General Electric, Western Electric, De Forest, and AT&T, began to enjoy navy patronage. One of the American companies' major competitors, the Germany company Telefunken, would obviously no longer be supplying the navy. A government-imposed patent moratorium instructed all suppliers to make use of the best components, no matter who owned the patent. The government guaranteed to protect all suppliers against infringement claims and encouraged the inventors not to be oversensitive to relatively free use of their apparatus during the national emergency. Under this arrangement, with the inventors and radio companies concentrating less on marketing strategies and litigation, and more on research and development, significant advances in continuous wave technology were achieved. Civilian-military cooperation produced apparatus more ideally suited to the navy's special needs.

The symbol and centerpiece of this cooperation was the 200-kilowatt Alexanderson alternator at the Marconi station in New Brunswick. Negotiations between the Marconi Company of America and General Electric had begun in 1915, and by early 1917, G.E. had agreed to install an alternator at the Marconi Company's New Brunswick station.[97] By March of 1917, a 50-kilowatt machine was in place, just in time for the navy takeover. Navy officials were highly impressed with the alternator, which surpassed the synchronous spark and even the arc transmitter in its ability to maintain regular transatlantic communication with Europe. Although the alternator failed to provide constant twenty-four-hour communication, it nonetheless represented a major technological leap in clarity of signal and distance achieved. Its high, piercing signal was heard throughout Europe. But the navy wanted continuous twenty-four-hour service and urged Alexanderson to complete as soon as possible the 200-kilowatt machine he was working on. That alternator, which was installed in September of 1918, transformed New Brunswick into the government's flagship radio station. New Brunswick quickly was designated to handle most of America's radio communication with Europe.

Several other components were dramatically improved during the war years. Without a doubt, the most important of these was the vacuum tube. Although the first critical steps in refining the tube as a detector, and then using it as an oscillator, had been taken prior to the war, it was in the urgent, heady, and litigation-free environment of 1917 and 1918 that the tube's full potential was realized. By the end of the war, the vacuum tube had become a much more sensitive, rugged, reliable, and long-lasting detector of radio waves. In addition, it was being used as a small, relatively inexpensive, but not yet powerful generator of radio waves.

Aerial construction was revolutionized, as well; for radio reception, high poles supporting seemingly endless lengths of wire were replaced by what was called the loop. The loop resulted from tests conducted by Roy Weagant, chief engineer for the American Marconi Company, who was working to reduce static interference. Because the loop was small and adjustable, it could be turned to the best direction for receiving and, in the process, overcome static interference.[98]

By 1918, Daniels wrote, almost complacently, "The Navy occupies a strong position in the commercial radio field on account of efficient service rendered, and I think presages the way for making this service entirely governmental."[99] The year had proved critical to the strategy for postwar naval control of radio. The U.S. Navy had never enjoyed cordial relations with the Marconi Company; the war years turned friction into deep suspicion and hostility. The Marconi Company would always, in the eyes of navy officials, be a British company. With nationalism, even xenophobia, coursing through official arteries, the thought of foreigners enjoying a virtual monopoly of something so critical as the nation's wireless communication network was intolerable to the navy men. At times it seemed the navy's obsession with driving the Marconi Company out of America was even stronger than the navy's desire to control America's wireless system. Navy officials now appreciated the distinctiveness and superiority of American radio technology, and any prospect of the Marconi Company gaining access to that technology produced an immediate response. Lieutenant Commander George C. Sweet, head of the Shore Section of the Radio Division in the Bureau of Steam Engineering, learned that the Marconi Company had offered to purchase Federal Telegraph's patents and physical assets for $1.6 million.[100] He brought this to Secretary Daniels's attention and urged that the navy make Federal a comparable offer. This the navy did, and on May 15, 1918, the Navy Department bought out Federal Telegraph. Six months later, the department purchased from American Marconi its low-power ship and shore stations for $1.4 million. Thus, by November 11, 1918, when the war ended, there were only three important features of American wireless the navy did not control: the Marconi long-distance stations, the patents on vacuum tubes and G.E.'s alternator. There were other important patents, privately held, that the navy did not own. Nonetheless, Secretary Daniels had reason to believe that postwar naval control was inevitable. From his vantage point, the navy had presided over America's wireless communications with no small distinction. In the navy's custody, radio apparatus had become more standardized, powerful, and efficient. The production and distribution of radio apparatus had been highly coordinated. The navy had prevented a foreign-contolled firm from acquiring the Federal Telegraph Company; government ownership would keep other critical American technology out of foreign hands. Basically, the navy's wireless strategy both just before and during the war had given it de facto control over radio, which, officials hoped, would argue in favor of extending de jure control. To reverse the trend seemed to Daniels foolish and unnecessary.

Another argument for military control was the extent to which such control had furthered national political and diplomatic goals during the war. Daniels was proud to report in 1918 that the Naval Communication Service was "cooperating with the State Department and the Committee on Public Information in the broadcasting of information of advantage to the United States to all parts of the world by high-power radio." Each night the navy broadcast a program titled "Home Stuff" which included news items from many American cities. Daniels reported, "This dispatch is received simultaneously in France and England, and is posted in all Y.M.C.A. huts and other places where our men in foreign service congregate. It is their daily home paper."[101] Woodrow Wilson's Fourteen Points speech of January 8, 1918, and his message to the German people the following September were both broadcast to Germany from the New Brunswick station. Both of these speeches included a rejection of the German form of government, which was "autocratic" and did not truly "represent the German people," as well as the suggestion that Germany adopt a political system more like that of the United States. The president, the CPI, and the navy—all had taken advantage of government control of radio to use the airwaves to extend America's ideological influence. Those advocating permanent naval control of radio argued that government control over the flow of information had made it possible for the United States to be regarded, in 1918, as the moral leader of the West.

Just after the armistice, Daniels began coordinating his plan to introduce legislation that would place America's wireless systems permanently under the navy's control. He was motivated by his strong belief that such a strategically important method of communications—it linked the U.S. government with other governments and provided the only communication between the naval shore command and the fleet—had to be controlled by the navy. Reviewing the navy's impressive management of radio during the war years, he was convinced that naval control would result in the most efficient and technically progressive wireless system in the world. Daniels was right to take pride in the navy's wartime supervision of radio: considerable organizational and technological progress had occurred. But he was thinking about the issue too narrowly.

After the war, Congress and the public were not considering government control of wireless; they were assessing government control of private industry in general. Thus, they focused not on the happy results surrounding wireless, but on the rather unfortunate outcome of government management of other public utilities.

During the war, the government had taken over the country's telegraph, telephone, and cable systems. A brief review of how the Bell system and its customers fared under such control sheds light on postwar animosity toward government ownership of public utilities. Prewar agitation for nationalizing the country's telephone system was based on the belief that government control would bring a reduction in rates. Yet shortly after federal takeover of the Bell system, which occured in July of 1918, the government found itself compelled to raise rates, a move that came as a most unpleasant surprise to consumers. Long-distance rates increased by 20%. Even given wartime inflation, the highest inflation since the Civil War, the increases impressed the public as much too high. In addition, the government instituted a service-connection charge that new subscribers had to pay for the physical installation of their phones. The Bell companies had been trying to impose such a charge for years but had been turned down repeatedly by local utilities commissions. To many, then, government control of the Bell system was considerably worse than private management. It produced an unregulated monopoly that could completely disregard its customers and impose whatever rates it chose. Adding insult to injury was the revelation that these rates had not generated profits; under government ownership, Bell operations showed a deficit of more than thirteen million dollars, nine million of which was repaid by the U.S. Treasury.[102]

Other public utilities fared no better. The government had also taken over America's railroads, with similarly disastrous results. Secretary of the Treasury William Gibbs McAdoo became director-general of the federal Railroad Administration, and to him no price was too great to pay to keep the railroads running. He authorized a major pay raise for the workers and paid for it through dramatic rate increases imposed on the public.[103] Even progressive leaders, who had thought government control of the railroads would be a great reform, were disillusioned by the government's mismanagement. The railroad controversy was widely discussed in the popular press and fueled a general and heated debate about the merits of government control of any private industry. This debate affected the future management of radio, as well.

On November 24, 1918, Congressman J. W. Alexander introduced a bill endorsed by both Daniels and Wilson which would give the navy permanent control over America's wireless system. Alexander was chairman of the committee to which the bill would be referred, the House Committee on Merchant Marines and Fisheries. Under the plan, the navy would buy the remaining sixteen American Marconi stations it had failed to acquire during the war. "Just compensation" would be made for these stations. Daniels issued a statement justifying the proposal. He maintained that prior to the war, none of the private companies had been able to "make an adequate return and in most cases no profit has been made except through the sale of stock." No profit was made, argued Daniels, because radio required a complete monopoly to be remunerative. He maintained that it was interference that made monopoly a necessity. He exploited Wilsonian rhetoric to package the proposal in the most persuasive language: "At this time, when most of the world is to be made over, when the United States is fostering the beginnings of a great merchant marine, whose servant radio-telegraphy is, and when the American news and American viewpoint are to be disseminated throughout the nations, . . . the greatest good to the people of the United States as a whole will accrue to them from well-regulated communications . . . at reasonable rates and without interference."[104]

Daniels, however, had misread public opinion and failed to appreciate that under the prevailing political climate, Americans would react against government ownership as a policy, and would not make distinctions between specific communications systems when it came to government control. He also failed to grasp the significance of the recent congressional elections. In November 1918, the Democrats lost control of Congress. The Republicans, eager to break free of the harness that had bound them together with the Democrats during the war, exploited the frustrations many Americans had felt with wartime control of private industry. Daniels was an excellent target for opposition attacks, as the Navy Department had endorsed postwar government control of telephone, telegraph, and cables, as well as wireless.

On December 12, when Daniels appeared before Merchant Marines and Fisheries to testify on behalf of the bill, he found "an atmosphere of antagonism" toward him and his proposal. Daniels, it turned out, had proposed a similar bill in 1917 which would have provided the navy with the funds to purchase private wireless stations outright. The committee at that time refused to report the bill, considering purchase unnecessary since the navy already enjoyed complete control over the stations during the war. It is not difficult to imagine the committee members' indignation when they learned that Daniels had then taken money from general navy appropriations to buy the Federal Telegraph Company and the Marconi shore stations. The *Times* reported that Congressmen Edmunds and Greene assailed Daniels and charged that he had exceeded his authority in making the purchases. Congressmen never like having their authority flouted, and Daniels's failure to appreciate this doomed an already difficult case.

E. J. Nally, vice-president and general manager of American Marconi, and John Griggs, the company's president, testified before the committee on December 17. Not surprisingly, the firm was "unalterably opposed" to the bill, and Griggs confidently told the press he doubted the bill would make it out of committee. Nally and Griggs released figures for the years 1914–17 showing the company's profits, to counter the navy's charge that no private firm had been financially successful. During their testimony, they emphasized the risks that had accompanied the firm's research and development, risks assumed by the company's twenty-three thousand stockholders, who had received only two dividends in sixteen years. The payment of further dividends had been interrupted by the war. Now, asked Griggs and Nally, "when success was in sight," was Congress going to sanction government confiscation? Such an act, according to Griggs, would "be clearly against the interests of the public" and would also "destroy all the hopes" of the company's stockholders. The Marconi men also charged that with the navy in control of wireless, "a permanent censorship would be established on commercial and newspaper messages."[105] Government censorship during the war, especially by Postmaster General Albert Sidney Burleson, had been highly capricious and heavy-handed, and had become increasingly controversial. Suggesting that such control over information might persist after the war added a valuable inflammatory element to the Marconi company's position. Nally and Griggs were joined by representatives of the amateur operators, including Hiram Percy Maxim, who deluged congressmen with letters of opposition and lobbied vigorously against the bill.

On January 16, 1919, the committee tabled the bill, and its members began circulating their antagonism toward Daniels and his proposals throughout the House of Representatives. One committee member, Congressman Fredrick W. Rowe of New York, urged that the House instruct the attorney general to force the navy to sell the Federal Telegraph and Marconi properties back to their owners.[106] Two weeks later, Congressman James Mann attacked Daniels in the House, "saying the Secretary had violated the law and 'ought to be impeached.'"[107] One week after this suggestion, Congress amended the general Naval Appropriations Bill specifically to prevent Daniels from using any of the appropriations to acquire radio systems. The sponsor of the amendment said the purchase of Federal and the Marconi stations had "brought down a storm upon the department."[108]

Publicly, Daniels remained undaunted. In July of 1919 he approached Congress again, this time proposing that all navy stations be allowed to handle commercial messages. He sought to portray the navy as the selfless servant of American business. "The ability to transact business depends on communications," he wrote, adding that the navy had the transatlantic and transpacific stations to offer a reliable alternative to the cables. He was, of course, trying to keep the navy's long-distance stations operating as commercial stations to forestall, and, he hoped, prevent, the Marconi Company from regaining its foothold in transoceanic radio.[109]

A New York Times editorial offered succinct and pointed reaction to Daniels's latest ploy: "Not for any temporary and not for any permanent cause, or merely assumed cause, should the government be allowed to put its bungling and paralyzing hand upon private business." The paper believed that under the proposal, the government would become "a potential censor and muddler of business messages." Declaring that "the country does not pine for nationalization," the editors noted sarcastically, "The people know how beneficiently the government inserted itself into telegraph and telephone and cable management. They await anxiously the hour when the railroads will be free from bureaucratic control."[110] The Independent echoed this position: "The public . . . is generally ready to shake its head and emphatically deny the Secretary's request, because of its experience with government-controlled telegraph and telephone wires in wartime."[111] By the summer of 1919, it had to be clear to Daniels, Hooper, and other naval officials that not Congress, the press, or the public would support continued naval control of wireless.

Yet, the alternative was returning the high-power stations to the Marconi Company, letting the firm gain access to American technology, especially the alternator, and, through that technology, establish a virtual international monopoly over wireless communications. To the high-ranking naval officials involved with wireless, such a prospect was unthinkable. In their efforts to seek a solution to the situation, naval officials displayed considerable entrepreneurship. They began orchestrating the formation of an all-American company that would buy out American Marconi and remove, once and for all, foreign interests from America's wireless communications networks.

The story of the formation of the Radio Corporation of America has been told many times, most recently and with the most detail and sophistication by Hugh G. J. Aitken. That story will not be told in detail here.

But several key points and events require our attention. The formation of RCA marked the culmination of the private, behind-the-scenes, institutional activities surrounding wireless. It marked the end of the wireless era and the beginning of the radio age. It represented the end of the militarization of the ether and the beginning of effective monopolistic control over access to the airwaves. Thus, it signaled the death knell for certain visions surrounding wireless and prompted a new, redefined set of visions in the popular press about how radio fit into American culture.

• • •

THE AMERICAN MARCONI COMPANY had begun negotiating with G.E. in 1915 to purchase Alexanderson alternators. The negotiations were disrupted by the war, but they resumed in February of 1919 after Congress made it clear that the navy would not be awarded postwar control of radio. The Marconi Company had wanted exclusive rights to the alternator, which G.E. declined to assign. Under the terms of a tentative contract, however, a de facto exclusivity would result, for the Marconi Company planned to order at least twenty-four alternators. This order would occupy G.E. for several years and prevent it from supplying anyone else with the machines until long after the Marconi Company had their alternators in place and operating. The real dispute emerged over the royalty rates to be paid G.E. for the right to use the machine, and by late March the two firms faced an impasse on this issue.[112]

Meanwhile, Hooper and Franklin D. Roosevelt, assistant secretary of the navy, and acting secretary while Daniels was in Europe for the peace talks, asked G.E. officials to hold off on finalizing any deal with the Marconi Company until the navy could discuss the wireless situation with G.E. representatives. This discussion took place in New York on April 8, 1919. Hooper and Admiral W. H. G. Bullard, who had been superintendent of the Naval Radio Service from 1912 to 1916, presented the navy's case to G.E. officials. They emphasized what they considered to be the essential feature of any postwar wireless communications network: it had to be controlled by Americans. Since it was by now clear that the navy would not retain custody of the system, another alternative had to be found immediately, an alternative that would keep American technology and American-based long-distance wireless stations out of foreign hands. The alternative was to form a new company, an American company, which would buy out American Marconi and become a new wireless-operating company controlling the country's long-distance, point-to-point wireless systems. Bullard and Hooper assured G.E. complete naval cooperation with such a venture, and also told G.E. officials that President Wilson himself had sent from Europe a secret message urging G.E. not to sell its alternators to Marconi. Bullard and Hooper were highly persuasive: on April 9, G.E. informed the Marconi Company that it considered the negotiations concerning the alternator sales terminated. Several weeks later, the navy and G.E. reached a tentative agreement on the formation of the new American firm.[113]

Owen D. Young, vice-president at G.E., met with E. J. Nally, president of American Marconi, in May. They discussed the creation of a new, all-American firm, and through the summer months negotiated an agreement. Further negotiations, beginning in August, were necessary between G.E. and British Marconi, and the two reached a preliminary agreement on September 5. In October of 1919, the Radio Corporation of America was incorporated in Delaware. Nally became president, Young chairman of the board. RCA took over all the stations, offices, and factories of American Marconi and retained Marconi personnel. The Marconi Wireless Telegraph Company of America ceased to exist. The navy finally had its all-American company, and G.E. had a client for its alternator and other radio apparatus.

The most surprising aspect of the formation of RCA is how little press attention the new company received. Here was a corporate realignment that in only a few years would have a profound impact on the American economy, on the control of radio technology and access to the ether, and on the transmission of cultural values, yet there is not one article on the subject listed in the *Reader's Guide* for 1919 or 1920. The *New York Times* printed a front-page story on January 5, 1920, headlined "American Radio to Span the Globe," which described RCA's purpose as managing an international wireless-operating system that would compete with the cables. It also noted that the ownership and control of RCA was "vested exclusively in Americans." The editorial pages of the *Times* during early January were preoccupied with the threat of Bolshevism here and abroad, and commentary was devoted to endorsing Attorney General A. Mitchell Palmer's raids against alleged communists. There was no editorial on the demise of American Marconi and the formation of RCA. A watershed institutional event in the history of radio, and America, was ignored by the popular press.

The journalistic silence about the formation of RCA was no anomaly; the press, during World War I, had neglected most of what was going on in wireless circles. AT&T's successful wireless telephony tests received more coverage than any other wireless story between 1914 and 1920. There were no reports about the transfer of patent rights from individuals to corporations. Although the *New York Times* and the *Provi-*

dence Journal printed charges that the German station at Sayville had played a role in the sinking of the *Lusitania,* magazines such as *Outlook, Literary Digest,* and the *Independent* did not once mention Sayville in their coverage of the disaster. When Congressman Alexander first introduced legislation in 1917 authorizing government ownership of radio, only *Scientific American* printed an editorial on the subject. When the bill was reintroduced in 1918, the magazines ignored it. Radio historians have made much of the broadcasting of Wilson's Fourteen Points to Europe, and of his broadcast to the German people in September 1918, yet these highly symbolic events went completely unreported. The *New York Times,* in its coverage of Wilson's Fourteen Points address to Congress, noted that the message was cabled abroad and printed on pamphlets that were dropped from airplanes throughout Europe. The article did not mention radio.[114]

How are we to account for this silence? Certainly the press was preoccupied with the war, and war news crowded out other kinds of stories. But there were other factors, as well. Wireless was not new anymore. It had become an established part of shipping and of long-distance communication. More importantly, the critical wireless developments in the teens were not the sort of events that got into newspapers. The technological transformation of wireless into radio occurred, with the exception of the AT&T-navy demonstration in 1915, in private, primarily in industrial research laboratories. The changes were incremental; some of them had to remain secret until patents were obtained. The improvements were in sophisticated, discrete components; to explain and understand them required scientific literacy. The changes were also bureaucratic and institutional. They were part of a larger, ongoing national process still only partially apprehended by reporters—in fact, by many Americans.

All of these factors were at odds with journalistic standards of newsworthiness, standards that favored technological display, the large public test marking the first, the newest, the biggest, the farthest, and fastest. These standards also compelled reporters to relate technical change to their readers' lives, to describe how an invention would affect the activities and aspirations of ordinary middle-class Americans. This could be done when discussing the wireless telephone, even if the predictions were unrealistic, but reporting on vacuum tubes, feedback circuits, or alternators was another matter. Also important during the teens was the rise of the public relations agent in corporations and in the government: he told the press what he wanted published and was silent about the rest. The press, then, during the teens, by covering certain stories (and from a particular viewpoint) and ignoring others, legitimated the transfer of control over radio from individuals to institutions. Stories such as those celebrating AT&T's successful tests, or condoning navy censorship and takeover of the high-power stations, validated the centralized management of radio technology in America. Nowhere in the popular press was the marriage between radio technology and exertion of American political will questioned, critiqued, or explored.

· · ·

REVIEWING THE IMPACT of World War I on the development of radio in America, one critical influence stands out: military control and management. The navy brought enormous resources—money, manpower, an integrated and far-flung organizational structure—to bear on wireless development. Radio was, for the first time, under rigid monopolistic control. All stations, on ship and on shore, high power and medium power, were controlled by one organization. The content of messages, and who was allowed to send them, was controlled, as well. The navy imposed standardization on technical design, and centralized coordination on the production and distribution of radio apparatus. The airwaves were used primarily to transmit point-to-point messages, and the navy placed strong emphasis on perfecting long-distance transmission and reception. When the navy's high-power stations, especially the one at New Brunswick, were used for broadcasting, the messages were meant to exert diplomatic, political, and cultural influence. The messages were designed to produce consensus and compliance with American ideological goals.

When RCA was formed in 1919 as a government-sanctioned monopoly, all of this experience provided powerful lessons on how to manage civilian American radio in the future. Indeed, the press release announcing the formation of the company stated the firm's goal as this: "To link the countries of the world in exchanging commercial messages."[115] The company was to provide long-distance, point-to-point communications and to compete with the cables, but to take advantage of the technological breakthroughs made during the war, RCA would have to perpetuate a version of the wartime patent moratorium imposed by the navy between 1917 and 1919. For, by the end of the war, AT&T and G.E., as well as United Fruit and Westinghouse, controlled the rights to critical patents. No firm controlled a complete technological system, and no firm enjoyed undisputed rights to manufacture vacuum tubes. Hooper, again modeling a civilian solution after a military precedent, urged that the patent moratorium be extended through postwar cross-licensing agreements.

The first such agreement was between RCA and G.E., each licensing the other to use its patents. The next firm RCA had to reckon with was AT&T. Irving Langmuir at G.E. and Harold Arnold at AT&T had transformed De Forest's audion into the vacuum tube, and while AT&T controlled the audion rights, RCA, through its absorption of American Marconi, controlled the rights to the Fleming valve. In a 1916 patent suit between De Forest and Fleming, the court established the priority of Fleming's valve and ruled that De Forest could not manufacture the three-electrode audion without the consent of the Marconi Company. The court also ruled, however, that the Marconi Company could not use the triode without De Forest's permission.[116] Getting beyond this critical technological impasse through adjudication promised to be costly and prolonged. Early in 1920, G.E. and AT&T began negotiating, with the understanding that any agreement reached would also include RCA and AT&T's manufacturing arm, Western Electric. The two firms quickly worked out cross-licensing agreements on the vacuum tube, and then moved to discussions surrounding corporate turf: which companies would control which aspects of wireless communications. RCA established its exclusive rights to use the pooled patents for transoceanic wireless telegraphy and for ship-to-shore communication. Transoceanic wireless telephony was also within RCA's province, although these rights were nonexclusive. AT&T retained its rights over wireless telephony, especially over land, and gained exclusive rights to "all land radio telephony for toll purposes."[117] One of G.E.'s exclusive fields was the manufacture of amateur apparatus, especially vacuum tubes. G.E. also gained exclusive rights to manufacture radio receivers, while AT&T had the exclusive right to manufacture wireless telephone transmitters. Agreement, reached in July of 1920, resolved the vacuum tube dilemma. It also significantly enhanced RCA's strength and ability to maneuver.

E. J. Nally, president of RCA, spent the spring and summer of 1920 in Europe, negotiating transatlantic traffic agreements with other countries. Nally's previous affiliation with the Marconi Company helped open doors for him, and although not all of his negotiations were easy, by the time he returned to the United States in September 1920, he had successfully preempted the transoceanic wireless business. This proved to be a critical move, for Westinghouse, which had manufactured wireless apparatus during the war, had decided to try to compete with RCA. Its first step was to form a partnership with the International Radio Telegraph Company, owner of the Fessenden patents, but when Westinghouse tried to establish traffic agreements with European governments or communications companies, it found that RCA had already beat it to the punch. Westinghouse also consolidated its technological position by acquiring

Edwin Armstrong's feedback circuit patents and his patents on the super-heterodyne, a method of reception which amplified and filtered weak incoming signals. These patents were critical to a complete technological system of continuous wave transmission and reception.

Although Westinghouse executives initially rebuffed an offer by Owen Young to reduce competition between the two, by late 1920 they were ready to negotiate. In March of 1921, Westinghouse officially became part of the Radio Group, a corporate alliance among RCA, GE, and United Fruit designed to control radio technology. RCA agreed to divide its manufacturing orders between G.E., which would produce 60 percent of RCA's radio apparatus, and Westinghouse, which would produce 40 percent. RCA purchased the International Radio Telegraph Company, and Westinghouse and RCA agreed on cross-licensing agreements. Less than two years after the formation of RCA, the consolidated, monopolistic structure the navy had imposed on radio during the war, which Secretary Daniels had argued was the only possible way to manage radio technology, had been successfully emulated and embellished by what later came to be known as the radio trust. The trust was set up to control access to the technology and to the ether itself.

Radio technology was now embedded in interlocking corporate grids, and RCA had become a civilian version of the military monopoly that had controlled radio during the war. Military-industrial alliances in this country have been strengthened with each successive war; between 1917 and 1919, we see the unabashed beginnings of that alliance in communications technology. The results were improved radio components, a trend toward consolidation and centralization of the industry, and a legitimation in the press of monopolistic control. So, although Secretary Daniels had failed in his campaign to place radio under permanent government control, the postwar military imprint on radio was nonetheless large and lasting. The most important naval legacy, possibly, was conceptual: the conviction, repeatedly articulated by Daniels, that radio was a natural monopoly, and that only as a monopoly could radio function efficiently in the United States.

The other preconception RCA inherited from the navy, and from the Marconi Company, was that radio's primary purpose was to establish long-distance, point-to-point communication between specific senders and receivers. For both the navy and the Marconi Company, the attention of their respective interlocking networks was directed outward, abroad, in a competitive quest for international hegemony. RCA adopted this orientation from the two organizations that were most responsible for its existence, and consequently its executives also looked outward, across oceans. In doing so, they remained blind to alternative ways of thinking about and using radio technology. Thus, while corporate negotiations parceled out control over patents and turf, the American amateurs unwittingly revealed where the real radio profits lay.

CHAPTER NINE

THE SOCIAL CONSTRUCTION OF AMERICAN BROADCASTING

1912–1922

ALTHOUGH THE MILITARY and major corporations, in turn, monopolized radio technology between 1912 and 1917, they failed to control, or
to take seriously, how many individuals outside these bureaucracies
thought about using the apparatus. These individuals, the amateurs consigned in 1912 to their ethereal reservation of 200 meters and less, had
not stopped exploring the "folds in the night." Their numbers had increased dramatically after 1912, and their influence on how radio would
eventually fit into American society was enormous. Their ideas and their
activities provided an important countervailing force to the bureaucratic
management of, and mindset about, radio.

Despite the 1912 Radio Act—and partially because of it—amateur
radio gained many new enthusiasts during the teens. Amateurs soon
learned that there was considerable latitude in how the law was enforced. First, the appropriations for implementing the law were extremely modest: $37,880 in 1913 to inspect all commercial ship and
shore stations and to test and license all commercial and interested amateur operators in the United States.[1] Many amateurs simply ignored the
law and continued their wireless activities, although those near naval
stations exercised more discretion than they had in the past. Even those
who were licensed realized that 200 meters was not an ironclad wavelength assignment and that they could move up to 375 or 400 meters
without getting into trouble.[2] In addition, the Department of Commerce
published a call book that contained the names of all the amateurs who
had successfully passed the government tests and secured transmitting
licenses. According to one amateur, "the astounding number listed in this
book was a revelation"; the call book also documented that the amateur

hobby was national in scope.[3] The government, by publishing and circulating the call book, inadvertently encouraged amateurs to try to achieve greater distances so they could communicate with their compatriots across the country. As the amateurs' experimentation intensified and their enthusiasm grew, they recruited even more boys and young men to explore the ether.

The number of licensed amateurs and amateur stations increased sharply in the three years following enactment of the Radio Act: from 322 licensed amateurs in 1913 to 10,279 in 1916. Between 1915 and 1916 alone the department licensed 8,489 amateur stations. During the same period, only 5,202 commercial operators and fewer than 200 shore stations were licensed.[4] In 1917, the number of licensed amateur operators totaled 13,581.[5] Estimates placed the number of unlicensed receiving stations at 150,000.[6] Thus, even after federal legislation, the amateurs continued to dominate the airwaves. They were most numerous in the Midwest, in Great Lakes cities such as Cleveland and Chicago, and in seaports such as Baltimore, Boston, New York, San Francisco, and Seattle, each of which had hundreds of licensed amateurs and thousands more who were unlicensed.[7]

More amateurs began to use their apparatus to broadcast voice and music. One famous early broadcaster was "Doc" Herrold, who broadcast music and advertising as early as 1914. He used an arc transmitter and powered his San Jose station by illegally hooking into the streetcar lines of the Santa Fe Railway.[8] High school and college radio clubs, whose transmissions usually covered a radius of twenty-five to fifty miles, also began broadcasting more regularly. Amateurs recalled that between 1913 and 1915, voice and music broadcasting increased markedly.[9]

Radio broadcasting was enthusiastically promoted by its earliest pioneer, Lee De Forest. When De Forest negotiated with AT&T over the audion rights, he agreed to stay out of the radio-telephony business as AT&T conceived of it: point-to-point voice transmission from a specific sender to a specific receiver. He insisted, though, on being allowed to continue using his apparatus for the distribution of news and music, and on being allowed to manufacture and sell apparatus capable of receiving these broadcasts. To AT&T, such broadcasting was frivolous, a hobby, and certainly not a pastime that related in any way to its corporate goals, but to De Forest, broadcasting was a very serious business indeed. Since at least 1907, when he began performing experiments with his radiophone, De Forest had dreamed of bringing music, especially opera, into the homes of those unable to attend in person. De Forest did not abandon

An amateur station assembled at home: spark gap and sending key
to the left, crystal detectors and tuning coil at right.

this dream, and once his trial was over and his company revived, he resumed his broadcasting experiments. His pioneering work on voice and music transmissions had made him a great hero "to amateurs who were eager to hear him on the air again."[10] It is impossible to ascertain whether De Forest expected, as early as 1914, that broadcasting would turn into such a huge business. What he did know was that there was a still-growing market for radio apparatus and a desire to use that apparatus to pick up faraway stations and unusual transmissions. De Forest, who had always hungered for attention and recognition, was a born ham, in both senses of the term. In 1915, he erected a 125-foot tower on the roof of his Highbridge factory and inaugurated regularly scheduled half-hour "nightly concerts" of phonograph music.[11] In October 1916, he transmitted music to the Hotel Astor.[12] That same fall, he broadcast the Yale-Harvard football game, and on election night, he provided six-hour coverage of the neck-and-neck presidential race, signing off at 11:00 P.M. with the announcement that Charles Evans Hughes had been elected president.[13] On New Year's Eve, he broadcast music to the Morristown, New Jersey, home of Theodore E. Gaty, vice-president of the Fidelity and

Casualty Insurance Company of New York. Billed as "the first wireless dance," the broadcast included a variety of musical selections, each introduced by the radio operator at Highbridge. Megaphones amplified the music in Gaty's home.[14] De Forest also took advantage of his broadcasts to promote his own apparatus, and later claimed he was the first to advertise over the airwaves.[15] De Forest well anticipated how radio would be used: to broadcast music and entertainment, sports events, news, and advertising into people's homes.

De Forest, whom historians don't usually characterize as a shrewd businessman, capitalized quite well on the transition from individual to institutional control over radio technology. He had, by 1914, developed a strong sense of who his audience was and what they wanted. He did not try to compete with the powerful communications corporations; rather, he catered to the market they dismissed. He took the amateur operators seriously, both as an audience and as a market; he did not ignore their importance. And during the war years, this audience continued to grow.

The number of radio clubs, and their members, increased as well. The Radio Club of America, headquartered in New York, installed a station in the Hotel Ansonia in 1915. The American Marconi Company, which in the first decade of the century began publishing a publicity magazine, changed the magazine's name in 1913 from the *Marconigram* to the *Wireless Age*. Two years later the magazine announced the formation of a club sponsored by the magazine, the National Amateur Wireless Association. The same year, Hugo Gernsback announced the organization of the Radio League of America.[16] These clubs came to boast thousands of members, but the clubs' political and organizational activities remained modest.

Hiram Percy Maxim, who had organized the Hartford Radio Club and then, in 1914, the American Radio Relay League, thought radio clubs should be more politically active and publicly visible. He sought to have the league expedite the exchange of information among amateurs and to develop the league into an efficient, well-coordinated organization. Maxim, a graduate of MIT and a practicing engineer for decades, had considerable business experience before starting the Maxim Company of Hartford in 1908. He hoped to apply the methods big businesses used in coordinating their national operations to the management of a national system of amateur radio. The organizational ethic and the desire for legitimacy was directing the efforts of the leaders of amateur radio, too. Maxim and other league members dreamed of a coast-to-coast relay through which amateurs would demonstrate their seriousness of purpose and their technical expertise. Maxim persuaded the Department of

Commerce to grant special licenses for transmission at 425 meters to stations at strategic points along the relay chain. He divided the league into districts and established wireless trunk lines between major relay points. The league published its own call book annually, which contained the names, addresses, call numbers, power, range, receiving speed, and operating hours of stations across the country. Accompanying this book was a map showing the location of these stations. In December 1915, the league published the first issue of its magazine *QST,* whose purpose was "to maintain the organization of the American Radio Relay League and to keep the amateur wireless operators of the country in constant touch with each other."[17] Maxim organized drills to increase the efficiency of the trunk lines and to eliminate operators who were too slow or whose apparatus was inadequate.

To demonstrate that the amateurs constituted a viable alternative communications network, Maxim planned the first countrywide message relay, which took place on Washington's birthday in 1916. At 11:00 P.M., a message from Colonel Nicholson of the Rock Island, Illinois, Arsenal was broadcast from station 9XE in Davenport, Iowa. The message read "A democracy requires that a people who govern and educate themselves should be so armed and disciplined that they can protect themselves." Amateurs relayed the message from station to station throughout the country; it reached New York at 1:30 A.M. The ARRL had arranged for the message to be delivered to the governor of each state and the mayors of major cities. In addition, the message was "read by Boy Scouts at Mount Vernon and on the battlefield of Bunker Hill."[18] The success of the relay, and the favorable publicity it received, added to the growing prestige of amateur wireless. Maxim had been shrewd in staging the relay so that the ARRL appeared willing and quite ready to cooperate with the government if and when that might be necessary. On his next trip to Washington, Maxim succeeded in getting the special license allocation upped from 425 to 475 meters. He now dreamed of a transcontinental relay with only two intermediate links.

One year later, the ARRL announced the success of its second national relay. The announcement came on March 8, 1917, seven days after news of the Zimmerman Telegram dominated the front pages. The press reported that hundreds of thousands of Germans were already in Mexico preparing to invade, and concern over American preparedness turned into panic in many quarters. Maxim framed the amateurs' success accordingly. Members of the league had relayed a message from New York to Los Angeles, and a reply back to New York, in less than two hours. Maxim said the ARRL was now prepared to provide "transcontinental service through amateur plants, which, in case of war, would enable communication to be maintained, even if telegraph and telephone wires were cut." The *New York Times* reported the league "Ready for War Service."[19] Maxim clearly sought to discipline America's amateurs and to establish distinctions between those who were skilled operators with efficient apparatus and those who were hacks. He wanted to make the amateurs, both in reality and in image, more docile and cooperative, more in harmony with the prevailing social order.

Press coverage of the amateurs since the 1912 Radio Act emphasized their new utility to the government, especially in the event of war. Gone were articles titled "Curbing the Wireless Meddler." They were replaced by articles such as "The Good of Amateur Wireless" and "Wireless Amateurs to the Rescue." The romantic glow of amateur work continued to shine in "The Romance of Wireless" and "A New Style of Adventures." The *American Magazine* ran a competition in which readers were to submit essays on "My Hobby and Why I Enjoy It." Third prize in the July 1916 issue went to a twenty-two-year-old amateur who wrote, "Our hobby is wireless and we talk, read and think wireless continually." He asserted: "One who has never operated can nowhere begin to feel the thrill of satisfaction that we operators enjoy when working over great distances of space."[20] During 1916 and early 1917, stories celebrating the romance of amateur radio were eclipsed by stories that cast the amateur network as a potential military resource. Typical was an article in the *Woman's Home Companion* titled "Almost a Soldier . . . Is the Boy Who Understands Wireless Nowadays." The piece offered a brief description of how a boy could learn the hobby and announced the formation of a new organization, the Junior American Guard, sponsored by the Radio Signal Service of the U.S. Army. The boys and young men in this organization constituted "a third arm in the defense of the country, should occasion ever arise."[21] Yet, for the most part, the press failed to take note of or see the significance in the explosive growth of amateur radio during the teens.

When the United States declared war on Germany in April of 1917, all amateurs were ordered to close down and dismantle their stations. To accelerate the process, local police searched for and seized independent stations; by April 10, according to the *New York Times*, New York police had closed down more than eight hundred stations. German-Americans who had their own wireless stations were subject to charges of espionage. Some were called in for questioning; others had their stations destroyed and their homes ransacked by police or special agents.[22] The amateurs, who had just one month earlier relayed a message across the

United States and back, were completely shut down. Press reports indicate that many amateurs were disgruntled by the order and in many areas of the country failed to see the need for the measure. An *Electrical Experimenter* article, reprinted in the *Literary Digest,* advised the amateurs: "The government is with us, not against us."

This was quite true. With the declaration of war, the military's need for skilled operators skyrocketed, and they were "scarcer than hen's teeth."[23] Training people would take months. Uncle Sam, it turned out, needed the amateurs, and launched a campaign through radio clubs and the press to persuade qualified amateurs to enlist as soon as possible. Telegrams went out to the ARRL, to Hugo Gernsback, to editors of technical journals, urging the amateurs to join up. Articles in magazines as diverse as *St. Nicholas, Scientific American,* and *Women's Home Companion* advised young men to turn their hobby to national service. "It's up to you if you prefer the trench to the radio tent behind the lines," offered one writer. "You can be heroic and manly in either."[24]

Thousands of amateurs responded to the call. The government set up radio schools for men and women at local YMCAs and at colleges and universities. The women were trained to serve as teachers of new recruits. But many of the amateurs needed only minimal training; they were already more than adequately skilled for military service and quickly passed the government's radio exam. In early January 1917, there were 979 navy radiomen; by November of 1918, that number had jumped to approximately 6,700, a large proportion of them from the ranks of the amateurs. The amateurs were no longer a source of competition and interference in the airwaves. Instead, the subculture of American men and boys who had previously fought with the navy over who owned the ether now supplied the armed services with thousands of willing, cooperative recruits. They were no longer outside the system, they were part of it.

When the war ended, however, the amateurs were eager to get back to their hobby. They had expected that the wartime restrictions on amateur radio would be lifted after the armistice in November 1918. But Secretary Daniels, who was stalling while he tried to reshape postwar wireless, maintained that the restrictions would not be lifted until the final peace treaty was signed. On April 12, 1919, the Navy Department lifted the ban on amateur receiving; the ban on transmitting remained until September 26. During the wait, the American Radio Relay League reorganized, reestablished its mailing list, resumed publication of *QST,* and reconstructed itself as a national organization. After the transmitting restrictions were lifted, the amateurs, whose licenses had expired during the war, had to reapply for new ones. This requirement produced additional delays, but by late fall of 1919, shortly after RCA was incorporated, the amateurs began to return to the air. On December 4, 1919, the ARRL coordinated another transcontinental relay. The number of licensed amateur stations continued to climb; by June of 1920, there were already fifteen times as many amateur stations in America as there were other types of stations combined. The number continued to grow: the Department of Commerce counted 6,103 licensed amateurs in 1920; one year later, the figure was 10,809.[25]

Because many of these amateurs had served as radio operators during the war, they were familiar with the advances in circuitry, tube technology, and aerials. Many were eager to switch from spark technology to continuous wave, and they especially wanted to adopt the new transmitting tubes. Yet in 1920 these tubes were not on the market; they had to be obtained in other ways. Those who found work after the war with one of the electrical concerns involved with radio were able to get transmitting tubes for their own uses. Amateurs who obtained such tubes began to broadcast not dots and dashes, but speech and music to their fellow operators. The most famous of these amateurs was Frank Conrad.

Conrad, like many of his fellow enthusiasts, had been an amateur operator before the war. He also worked for Westinghouse and was a self-taught and gifted engineer. During the war, Conrad supervised Westinghouse's manufacture of portable transmitters and receivers for the Signal Corps. After the war, Conrad resumed his amateur work and, using transmitting tubes (to which he had access), began talking to other amateurs and playing phonograph music over the air from his station, 8XK. Soon he was receiving letters from other amateurs praising his broadcasts and making requests for particular songs. He scheduled his concerts on a regular basis: at first, every Saturday evening, and soon after, weeknight performances, as well. By May 1920, the Pittsburgh newspaper reported on the concerts, which included live performances, such as piano or saxophone solos, and phonograph music.[26] The amateurs, whose numbers continued to swell, were no longer receiving the dots and dashes that only a practiced initiate could decipher; now they were picking up speech and music. As a result, the amateurs could introduce their parents, friends, and siblings to the excitement of tapping the ether. The audience for broadcasting grew even larger.

Seeing an opportunity to increase its sale of amateur apparatus, the Joseph Horne department store in September 1920 ran an ad in the *Pittsburgh Sun* describing the Conrad wireless concerts and informing the public that sets capable of picking up these concerts were on sale at Horne's for ten dollars. When Harry P. Davis, a Westinghouse vice-president, saw the ad, he suddenly grasped that the company's conception of the wireless market had been much too limited in scope. He realized that "the efforts that were then being made to develop radio telephony as a confidential means of communication were wrong, and that instead its field was really one of wide publicity, in fact, the only means of instantaneous communication ever devised." He now comprehended that the amateurs did not represent a discrete market limited to technically inclined boys and men; rather, the amateurs were simply the forerunners of a much larger market for radio receivers. As Davis later remarked, "Here was an idea of limitless opportunity."[27]

Davis urged that Westinghouse authorize Conrad to build a more powerful transmitting station at the Westinghouse plant and that Conrad broadcast on an even more regular basis. These broadcasts, according to Davis's plan, would stimulate sales of radio receivers, and the profits from the sales would defray the cost of the station. Davis wanted the station completed by November 2, so Conrad could broadcast the presidential election returns. At 8:00 P.M. on November 2, 1920, the newly licensed station KDKA, operating at 360 meters, broadcast the election results. Amateurs listened enthusiastically, sometimes rigging up loudspeakers so friends and family members could listen, as well. To ensure that the broadcast had the right effect, both within and outside of the company, Davis provided Westinghouse officers with receiving sets, and also helped arrange for local department stores to have their radios tuned to Conrad's station. Newspapers in Pittsburgh and elsewhere took note of the event, but most newspapers and magazines ignored the broadcast. News of it was spread most rapidly and enthusiastically by word of mouth among amateurs and their families and friends. Over the next year and a half, the "broadcasting boom" swept the United States, beginning in the Northeast and moving south and west, reaching unprecedented levels of intensity by the spring of 1922.

Amateur operators and commercial establishments—primarily department stores, newspapers, and Westinghouse, G.E., and then AT&T—set up broadcasting stations. RCA had begun selling transmitting tubes in the spring of 1921, enabling amateurs to transmit speech and music. Although the Radio Act of 1912 had restricted amateur transmission to 200 meters or lower, during the teens Hiram Percy Maxim had succeeded in obtaining authorization for some stations to broadcast in the 350- to 400-meter range. Other amateur stations had also moved above 200 meters to longer wavelengths. With the advent of commercial broadcast-

Westinghouse photograph promoting the radio hobby to girls.

ing, however, the Department of Commerce warned the amateurs that they must retreat to and remain at 200 meters. The corporate stations were assigned the 360-meter slot. An amateur station had to be re-licensed as a commercial station to use 360 meters, and various of the more powerful and sophisticated amateur stations made the shift, partic-ularly those which were supported by newspapers or department stores. Other amateurs were recruited by colleges or universities to help start educational stations.

The amateurs revealed that many middle-class Americans were hungering for a sense of what people in different cities or states were like, what they thought and how they lived. The amount of listening in to far-off messages that took place, and the delight the amateurs took in this eavesdropping, suggests that these Americans had a feeling that there was more information available to them than they routinely received. They wanted this information in a less distilled, more immediate form: the popularity of motion pictures confirmed this. Amid the growing hubbub of public amusements, of cabarets, chautauquas, movies, and vaudeville, however, it must have been soothing to sit quietly at home

Publicity photograph of amateurs listening to a ballgame, ca. 1922.

and yet be transported to distant places and be privy to all sorts of messages, personal and official. It is true that during the early twentieth century, many men and women of the middle and working classes pursued the various new amusements in the public sphere on an unprecedented scale. But at the same time, there was in many the reticence, the love of comfort and quiet, the distaste for crowds, and the resentment at having to compete with others over space, time, and status during leisure hours which produced a cultivated, well-nurtured domestic inertia. These people wanted new experiences, information, and entertainment, too, but they were not always willing to go out for this stimulation. The amateurs were the first subculture of Americans, during the explosive rise in public entertainment, to spend much of their leisure time at home, using a new communications technology to entertain themselves and others.

The amateurs and their converts had constructed the beginnings of a broadcasting network and audience. They had embedded radio in a set of practices and meanings vastly different from those dominating the offices at RCA. Consequently, the radio trust had to reorient its manufacturing priorities, its corporate strategies, indeed, its entire way of thinking about the technology under its control. The amateurs had also taken the press quite by surprise. Newspapers and magazines, which had paid scant attention to the formation of RCA or to the burgeoning amateur population, were confronted with a new communications system of major significance. As they sought to grasp its meaning, they also worked to resolve, through symbolism and romantic language, the desires of corporate monopolies, and of "the people," to get what they wanted from the ether.

• • •

IN THE WINTER and spring of 1922, magazines and newspapers rediscovered radio. To the press, the fad seemed to come from nowhere. "Little more than a year ago," observed a writer in *Current Opinion*, "the public regarded radiotelephony as a great mystery."[28] Now, millions were "listening in." Official announcement of the boom came from Herbert Hoover, then secretary of commerce, who described the "wireless fever" as "one of the most astounding things that [has] come under my observation of American life."[29] This proclamation from an official of Hoover's stature alerted the press that it had better take note of a pastime quickly assuming major cultural and economic significance. "The rapidity with which the thing has spread has possibly not been equalled in all the centuries of human progress," noted the *Review of Reviews*. "Never in the history of electricity has an invention so gripped the popular fancy."[30] Radio emerged "with almost stunning suddenness," becoming "within a few weeks . . . a force in public opinion and public taste fitly comparable to the press."[31] In March of 1922, the *New York Times* observed, "In twelve months radio phoning has become the most popular amusement in America. If every boy does not possess a receiving outfit, it is because he lacks either imagination or money. . . . In every neighborhood people are stringing wires to catch the ether wave currents."[32] The public demand for receiving apparatus seemed insatiable, and RCA, Westinghouse, and many smaller firms went into overdrive to supply customers. The first issue of *Radio Broadcast,* in May 1922, described people standing "in the fourth or fifth row at the radio counter waiting their turn only to be told when they finally reached the counter that they might place an order and it would be filled when possible."[33] In 1922, sales of radio sets and parts totaled $60 million; in 1923, $136 million; by 1924, $358 million.[34]

Now the press, responding to the "tidal wave of interest in the subject," overflowed with interpretive articles on the social destiny of radio. Magazines such as *Collier's* and *Literary Digest* inaugurated radio sections in 1922, and new magazines such as *Radio Broadcast* were devoted entirely to the new craze. How would radio change America? What did the spread of broadcasting mean for Americans? These were the questions the popular magazines addressed.

The radio boom seemed all the more sudden because radio had been badly neglected by newspapers and magazines between 1915 and 1922. The press, through the content and tone of its articles, constantly emphasized the newness of the phenomenon. Little attention was paid to broadcasting's twenty-year gestation period. In this way, in its coverage of radio, the press helped to reinforce the media's tendency to ignore and thus deny their own history. This ahistorical stance made radio seem an autonomous force, so grand, complex, and potentially unwieldy that only large corporations with their vast resources and experience in efficiency and management could possibly tame it.

The sense of awe that had permeated the early articles on wireless telegraphy also colored the early articles on radio. To many writers, it was as if a fantastic dimension that people had suspected and hoped existed had finally been penetrated. People responded as if radio put them in touch with primordial forces. In "Broadcasting to Millions," A. Leonard Smith described hearing the sounds of static through his headphones: "You are fascinated, though a trifle awestruck, to realize that you are listening to sounds that, surely, were never intended to be heard by a human being. The delicate mechanism of the radio has caught and brought to the ears of us earth dwellers the noises that roar in the space between the worlds."[35] Joseph K. Hart wrote in the *Survey:* "We are playing on the shores of the infinite." He found this probing of the cosmos thrilling; he also sensed that the hubris that had made such exploration possible had a potentially dangerous underside: "The most occult goings-on are about us. Man has his fingers on the triggers of the universe. He doesn't understand all he is doing. He can turn strange energies loose. He may turn loose more than he figured on; more than he can control."[36] Grappling with the concept that something seemingly dark, quiet, and empty actually contained invisible life, another writer observed, "You look at the cold stars overhead, at the infinite void around you. It is almost incredible that all this emptiness is vibrant with human thought and emotion."[37] The air had been cracked open, revealing a realm in which the human voice and the sounds of the cosmos commingled.

Could this great void be filled not just with our voices, but with the voices of others, farther out in the cosmos? What were the sounds we called static, anyway? And could those in other spheres be listening in on us? Such questions were irresistible, especially when provoked by legitimate scientific observation. In the spring of 1919, Marconi announced that several of his wireless stations were picking up very strong signals "seeming to come from beyond the earth."[38] Nikola Tesla, another prominent inventor, believed these signals were coming from Mars. Marconi, too, considered Mars a not unlikely possibility. While *Scientific American* urged skepticism, *Current Opinion* quoted Tesla extensively in support of the Mars hypothesis. *Illustrated World,* a magazine that popularized recent technical developments, ardently embraced the prospect of interplanetary communication. In its article "Can We Radio a Message to Mars?" the magazine urged Americans to try to respond to the signals from beyond. This would no doubt require scientists to mobilize "all the electrical energy of the nation" to transmit signals of sufficient power. But the effort had to be made, for only then would the Martians know that "their signals were being responded to, and that intelligent beings actually inhabit the earth." The article enthused: "We can imagine what excitement this would cause on Mars." The most important reason for trying to contact Mars was to learn what the magazine assumed the Martians must know about improving, even perfecting, the quality of earthlings' lives. "It is not unreasonable to believe," predicted *Illustrated World,* "that the whole trend of our thoughts and civilization might change for the better."[39] These Martians could not only view our civilization with considerable detachment, they could also, presumably, give us all the secret answers, at last.

Illustrated World was a publication in which the distinctions between science and science fiction were minimized; its articles were written with an unsophisticated or credulous audience in mind. Its predictions about signaling to Mars would not have been taken seriously by some sectors of American society. Yet the underlying longings this article exposed are revealing, and they could hardly have been confined to readers of science fiction. In fact, the article contained themes that would be embellished in less fantastic, more earthbound articles about radio's potential. There was a hunger for contact over great distances and with beings who presumably knew more, and were wiser, than most contemporary Americans. Such contact would temper our deep and long held fears about being alone in the universe. Such contact would bring wisdom; it would be reassuring; it would be religious. Thus did the rhetoric surrounding radio draw from the past while it looked to the future.

The aspect of radio most universally praised in the press was its ability to promote cultural unity in the United States. "The day of universal culture has dawned," proclaimed The *Survey*.[40] The author of an article titled "The Social Destiny of Radio" maintained that prior to broadcasting, a sense of nationhood, a conception that Americans were all part of one country, was only an abstract idea, often without much force. The millions of towns and houses across America were unrelated and disconnected. But now that atomized state of affairs was changing: "If these little towns and villages so remote from one another, so nationally related and yet physically so unrelated, could be made to acquire a sense of intimacy, if they could be brought into direct contact with one another! This is exactly what radio is bringing about. . . . How fine is the texture of the web that radio is even now spinning! It is achieving the task of making us feel together, think together, live together."[41]

Stanley Frost, in his *Collier's* article "Radio Dreams That Can Come True," saw radio "spreading mutual understanding to all sections of the country, unifying our thoughts, ideals, and purposes, making us a strong and well-knit people."[42] Those isolated from the mainstream of American culture would now be brought into the fold. Farmers, the poor, the housebound, and the uneducated were repeatedly mentioned as the main beneficiaries of the culture surrounding people "in the flexible, tenuous ether."[43] Frost reprinted in his article two letters of thanks written to Newark station WJZ by culturally dispossessed listeners. To set the stage for the first letter, Frost wrote, "There is a dingy house in a dreary street in a little factory town, where the miracle is working. A worried mother frets through the day to achieve a passable cleanliness for her flock, without power to give them the 'better start' and wider happiness she had dreamed. [A] little flurry of prosperity" allowed her to get a radio. The letter followed: "My husban and I thanks yous all fore the gratiss programas we received every night and day from WJZ. . . . The Broklin teachers was grand the lecturs was so intresing . . . [the] annonnser must be One grand man the way he tell the stores to the chilren." Frost stated: "There are others, hundreds of letters a day of appreciation and delight from illiterate or broken people who are for the first time in touch with the world about them."[44]

A writer predicted in *Century Magazine* that radio would "do much to create a sense of national solidarity in all parts of the country, and particularly in remote settlements and on the farm."[45] The farmer's loneliness would be abolished, radio making him a real "member of the community." The writer continued, "If I am right, the 'backwoods,' and all that the word connotes, will undoubtedly dwindle if it does not entirely disappear as an element in our civilization."[46] Repeatedly, the achievement of cultural unity and homogeneity was held up, implicitly and explicitly, as a goal of the highest importance. One writer went so far as to complain, "At present, broadcasting stations are far too eclectic." The ultimate ethnocentric extension of the impulse toward cultural unity was the prediction that English would become the universal language. Argued one writer, "It so happens that the United States and Great Britain have taken the lead in broadcasting. If that lead is maintained it follows that English must become the dominant tongue."[47]

Yet, this desire for unity, for sameness, was not without its opposite, the pleasure taken in discovering cultural diversity. In the first years of the broadcasting boom, listeners delighted in picking up as many stations as possible. Dedicated enthusiasts posted a special map of the United States on a wall near the radio. Red dots on the map designated the location of operating broadcast stations across the country; the call letters of each station were also listed. Listeners would spend the evening tuning their radios in the hopes of hearing stations thousands of miles away. One self-described radio maniac referred to the actual radio programs as "the tedium between call letters." He maintained, "It is not the *substance* of communication without wires, but the *fact* of it that enthralls. . . . To me no sounds are sweeter than 'this is station soandso.' "[48] He described his delight in hearing "the soft Southern voice of Atlanta," while another enthusiast relished picking up the Spanish emanating from the station in Havana. Many of these stations adopted slogans that highlighted their special regionalism. Atlanta was "The Voice of the South," Minneapolis "The Call of the North," Davenport "Where the West Begins."[49] Radio allowed people to skip across the country, to go to never-seen and exotic places, all by turning a dial. Like the movies, radio blended the urge for adventure with the love of sanctuary in an ideal suspension. The difference with radio, at least in these early years, was the greater sense of control the listener enjoyed.

This feeling of mastery, coupled with the sense of adventure, kept radio enthusiasts at their sets night after night. Picking up far-away stations was frequently likened to other sports, especially fishing. "There are times when it is as difficult to land a given station—making the same demands upon patience, ingenuity, and even skill—as to bring to boat that elusive creature, the sailfish."[50] Another writer used the same metaphor: "This fishing in the far away with the radio hook and line is rare sport. The line is long, the fishing is getting better all the time, and it usually does not take many minutes to find out what you have on the hook."[51] As such a metaphor suggests, this active type of listening, which involved some technical expertise in adjusting the apparatus and bringing it to its maximum efficiency, was confined almost entirely to men and boys. Those who wrote about their ethereal adventures celebrated the manly challenges radio posed: "Your wits, learning and resourcefulness are matched against the endless perversity of the elements."[52] Within the safety of one's home, and out of public view, one's masculinity could be tested and reaffirmed.

Even after the desired station was reeled in, the essentially passive act of listening to radio programming was imbued in magazine articles with a sense of empowerment. Listeners had a choice: they could turn the dial until they got exactly what they wanted to listen to; if they didn't like what they heard, they could shut the radio off. More people, whatever their circumstances, had access to cultural events than ever before. "We have all free tickets to the greatest radio show on earth," noted one writer. As Stanley Frost put it, "With radio we, the listeners, will have an advantage we have never had before. We do not even have to get up and leave the place. All we have to do is press a button, and the speaker is silenced." Therefore, predicted Frost, "We will get what we want."[53] This sense of control over cultural content, combined with increased access to cultural events, cultivated a sense of cushioned privilege. One "music loving gentleman" decided to turn in his ticket to hear the Philharmonic Orchestra and to listen to the performance on radio instead. "I can only afford a top gallery ticket," the man explained, "but the radio microphone always gets a good seat down-stairs. I enjoy the music just as well here by my fireside and I save a lot of climbing."[54]

Another writer hinted at how monetary and class differences had, in the past, determined who got the good seats at a concert. Those with the cheapest seats usually could not hear the music very well. With radio, though, everyone hears the music "as plainly as if he had the best seat in the auditorium."[55] Everyone who previously could not attend such concerts now could. Thus was radio seen as democratizing some of the advantages previously enjoyed by the well-to-do, and bringing all the benefits of high culture to the masses. At the same time, radio helped insulate its listeners from heterogeneous crowds of unknown, different, and potentially unrestrained individuals. One writer absolutely reveled in the marriage between entertainment and solitude: "This vast company of listeners . . . do not sit packed closely, row on row, in stuffy discomfort endured for the delight of the music. The good wife and I sat there quietly and comfortably alone in the little back room of our own home that Sunday night and drank in the harmony coming three hundred miles to us through the air." He imagined other listeners in their back rooms, garages, dining rooms, attics, or cabins, "each and all sitting and hearing with the same comfort just where they happen to be."[56] The listeners sat suspended, in delicious tension, between their hunger for contact with the outside world and their craving for the comforts of home. With radio, both appetites were satisfied at once.

Although radio had indeed become embedded within the larger network of commercial entertainment in America, for those who wrote about radio and its role in American life, radio represented an antidote to what critics considered the more debasing effects of mass culture. Reformers who fancied themselves the true custodians of American culture believed that leisure activities should be educational and morally uplifting, and should not overly stimulate the senses. These reformers were witness to the rise of public amusements and commercialized leisure activities that often deliberately flouted such genteel precepts. Dime novels relied on hackneyed writing and action-packed stories, and at times they even glorified their criminal protagonists. Comic strips told their stories with pictures. The dark, crowded nickelodeons in working-class neighborhoods seemed to reformers to be dens of iniquity. Amusement parks were specifically designed to overstimulate the senses. Leisure had not taken the course many reformers had hoped. Radio seemed to hold out a remedy, or at least an alluring alternative, to all this. Like the first press coverage of wireless in 1899, the new hopes invested in radio were shaped by a faith in technological determinism, a belief that certain machines could make history. The educated bourgeoisie who believed their conception of culture to be at risk became newly optimistic with the advent of broadcasting. Here, at last, was a mass medium that could instill the right values in people.

The educational possibilities seemed unlimited. *Collier's* "radio maniac" claimed that radio provided "an education both precise and varied." Through the radio hobby, his son had become more technically informed and manually dexterous while mastering American geography. In listening to the programs, he had learned about politics, music, agriculture, and sports.[57] Magazines also offered grander visions. Radio could "give everyone the chance and the impulse to learn to use his brains." In doing so, radio would "tend strongly to level the class distinctions, which depend so largely on the difference in opportunity for information and culture."[58] "Who can help conjuring up a vision of a super radio university educating the world?" asked one writer. With radio, minds could "be detonated like explosives."[59] In his essay "Radiating Culture," Joseph Hart envisioned previously bored students now being instructed "by a single, inspiring teacher who speaks to the thousands of revived students through a central radio-phone. A whole nation of students might thus come under the stimulating touch of some great teacher."[60]

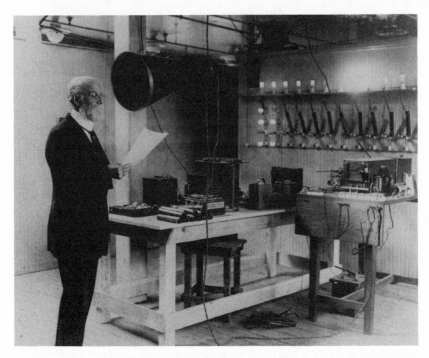

The first lecture being delivered by radio from Tufts University, 1922.

Colleges and universities set up radio extension schools, and anyone could listen in. As one writer stated, every home had "the potentiality of becoming an extension of . . . Harvard University."[61]

Anxieties about musical tastes surfaced in these articles, although there was optimism that as radio matured, the quality of music played would improve. Several writers made explicit distinctions between "good music" and jazz, which was more popular and more frequently played. One complained, "Most of the musical talent that is now attracted by the broadcasting stations is of mediocre nature." He also contended, however, that there were "thousands upon thousands of people whose musical tastes [ran] high above the average received from the air." To prove that people preferred a "higher class of talent," he cited the popularity of AT&T's WEAF, which could afford to recruit such talent because it accepted advertising.[62] Thus might radio, by bringing opera and other "good music" to the millions, upgrade American musical standards.

Another area of American life radio might improve was politics. "We may even become more thoughtful about the selection of our presidents," noted one observer sarcastically, "if we have to run the risk of hearing them speaking directly to us, however far from them we may try to keep ourselves."[63] Another commentator believed radio would make politicians more sensible and accountable to their constituencies: "Let a legislator now commit himself to some policy that is obviously senseless, and the editorial writers must first proclaim his imbecility to the community. But let the radiophone in the legislative halls of the future flash his absurdities into space and a whole state hears them at once." Citizens would be better able to judge a president who was "a real personality" instead of "a political abstraction."[64]

Bruce Bliven, writing for *Century Magazine*, gave voice to a progressive hope for radio's salubrious effects on politics. Crowds listening to a politician's speech in a large public setting were subject to "the mob spirit, with its factitious enthusiasm." The astute politician sought to take advantage of such mob psychology, and thus would cater more to the emotions than to the intellect. But with radio, argued Bliven, people would listen to the speech not as members of a crowd, but as individuals. The politician's ideas therefore would "have a better chance of being weighed for what they are really worth." Thus, radio might even produce a new kind of politician, a "man without the ordinary tricks of delivery, but possessed of a quiet, logical persuasiveness." Bliven allowed that such a man would have to have a "deep resonant voice such as will carry well in the microphone."[65] A major benefit was that more people than ever before would be able to hear their political leaders simultaneously. Political speeches reprinted in the newspapers often went unread, according to Bliven. With radio, more people could become politically informed than ever before, and they would have a sense of immediacy about the information they received. Politicians would seem less remote, more accountable, while the audience would gain a new sense of cohesiveness, even political empowerment, through the knowledge that everyone in a city, state, or region, or even everyone in the country, had heard the same speech at the same time without the distorting effects of mob response. This knowledge would further the sense of cultural unity: millions of people across the country, hearing together, reacting together, thinking together, as informed, politically aware citizens.

Religion was another area of life destined for change through radio. Sermons were an early staple of broadcasting; by 1921 KDKA transmitted the complete church service of the Calvary Episcopal Church of Pittsburgh every Sunday night. "Think of what this means to many people," urged the reporter for *Scientific American*.[66] He, too, mentioned the farmer, the invalid, the housebound who, prior to radio, had been cut off from religious services. All of these people could now "almost imagine being in church." "The preacher who has a little black box mounted on the pulpit," wrote another commentator, "comes very soon to know that the congregation seated before him is to the great invisible listening throng but as the sprinkle of a few drops over the baptismal font to the pouring rain outside."[67] The preacher came to know this because preachers who broadcast their services received thousands of letters and telephone calls expressing thanks and requesting copies of the sermons. That radio seemed to be bringing more Americans into the religious fold was significant indeed. Since the late nineteenth century, religious authority had been undermined by Darwinism, the ethics of industrial capitalism, and a reverence for science and technology. Radio, however, promised a reconciliation between religion and the corporate-industrial secular world, for it was the first technology that could bring religion into people's homes. Radio, the product of monopolistic capitalism, would help reassert precapitalist, Christian values in America.

Contemporary writers, whatever their hopes or biases, were all aware that they were witnessing a social transformation of monumental importance. Radio listeners constituted "the greatest audience ever assembled by any means for any purpose in the history of the world." This audience was "remarkable" and "totally different in several ways from anything before known."[68] First of all, it was huge; conceiving of an audience as hundreds of thousands or millions of people required a major imaginative leap. Second, the audience was invisible and unknown. The speaker or performer could not see facial responses or hear laughter, booing, or silence; nor was there applause. At the same time that the size of the speaker's audience had multiplied beyond anyone's calculation, his visual relationship with that audience was severed. Bruce Bliven thought that "so much listening without seeing" had "upset one of nature's subtle biological balances" and had created "what might be called 'a hunger of the eyes.' "[69]

Because this audience was invisible, scattered, and unknown, commentators were unsure about its character. Was this audience just like a mob, only dispersed, but equally capable of being excited and manipulated by an ambitious speaker who was newly empowered by radio? Or was the audience compromised more of people like the magazine writers themselves: discriminating, thoughtful, with values and ideas of their own, certainly impervious to the wiles of a disembodied voice? The answers to these questions were critical, for they were directly related to radio's potential as a tool for social control. The magazine articles on the topic revealed an uneasy ambivalence about the audience. What did these invisible listeners want? How pliant were they? Would radio be a "tremendous civilizer," increasing public demand for "the daily news of events, the opinions of leaders, the counsel of the wise, the comments of observers, [and] the hopes of the prophets," as one magazine suggested?[70] If so, how would such broadcasts be received? Might one of these leaders or prophets be able to "tell men what to think and say and how to act"? Would he be able to "shape them to a common, uniform, subservient mediocrity"?[71] Would this audience be content to hear primarily "outrageous rubbish, verbal and musical," and would it be swayed by the "appalling mass of solemn bunk and some really vicious propaganda" that was already flowing through the ether?[72]

Implicit in virtually all of the magazine articles written in the early 1920s about radio's promise was a set of basic, class-bound assumptions about who should be allowed to exert cultural authority in the ether. The *New Republic* stated the position baldly. Radio, asserted the editors, "is mainly under control of men unfitted by training and personality for posts of such importance." These were businessmen, ignorant of radio's "proper use" and "indifferent as to whether it is used properly or not." Such men were not unlike those who first controlled motion pictures: "fly-by-nights, adventurers and reformed pushcart peddlers, not one in a hundred of whom had reached the social level where one takes one's hat off indoors." The proper use of radio, according to the *New Republic,* was educational: radio should be "an intellectual force." Radio could never fulfill this mission as long as it was managed by those whose interest was music or entertainment. Such men, the magazine asserted, "are admirably fitted to assemble orchestras, pianists and singers; but when it comes to lectures and addresses they are about as competent as Florenz Ziegfeld is to run Columbia University."[73]

It was not that members of the educated bourgeoisie objected to radio being used to influence those millions of invisible listeners; the bourgeoisie's major concern was that those exerting the influence embrace genteel values about what culture should be. The subtext of these magazine articles maintained that radio should be edifying—should appeal to the intellect rather than the emotions—should elevate musical tastes, and should promote contemplation and the ability to discriminate between the worthy and the base. Radio, by providing the perfect instrument for delivering high culture to the masses, could produce a new mid-culture that combined the content of high culture with the techniques of commercialized entertainment. As one writer put it, "The man who directs a broadcasting station must combine the astuteness of P. T. Barnum and the good taste of a Gatti Cassaza."[74]

Here was a captive audience of millions. It was true they could turn their dial, but the hope running through these articles was that, given a choice between quality and banality, the audience would prefer quality. Thus did radio present the educated bourgeoisie with an opportunity to exert social control through culture. All the talk of cultural unity, of bringing the culturally dispossessed into the radio family, of leveling class distinctions based on money or education, revealed an intense desire to have this technology affirm and extend the cultural tastes and norms of the upper tiers of the middle class. The great fear was that, if it was not properly managed, radio would extend the cultural authority of the P. T. Barnums and "reformed pushcart peddlers" of America who had so debased, in genteel eyes, the leisure time of the masses.

Radio, then, meant progress for all. The technology would bring improvement to many areas of American life and thus benefit everyone, the ignorant and the well read, the poor and the rich, the individual and the institution. In these press accounts there was no tension between corporate ambitions and individuals desires: they were really the same thing.

EPILOGUE

IN THE SPRING OF 1922, the radio boom was just beginning. To many people at the time, the entire enterprise seemed filled with uncertainty: Who would be allowed to broadcast? How would interference between competing stations be controlled? Where would the money come from to support broadcasting? Such questions dominated magazine articles, executive board meetings, and the annual Washington Radio Conferences organized by Secretary of Commerce Herbert Hoover from 1922 to 1925. Congestion in the airwaves became intolerable as department stores, newspapers, universities, and churches across the country, as well as AT&T, Westinghouse, and GE, established their own radio stations. In many areas, competition over access to the spectrum required that stations "time-share," taking turns using the same wavelength. When such informal arrangements broke down, cacophony ensued. The cost of broadcasting rose as radio technology became more expensive and as licensing fees to use that technology, and the wages charged by performers, increased dramatically. A range of financing schemes was debated in the press, the industry, and the government, from a licensing fee on sets, to municipal funding, to advertising. At the same time, government officials complained that the Radio Act of 1912 did not provide them with adequate guidelines or power to make broadcasting more orderly. Between 1922 and 1927, broadcasting appeared to be in a state of economic, regulatory, and ethereal chaos.

Despite this chaos and uncertainty, however, critical precedents had been set prior to 1922 that guided how broadcasting would be managed. Most importantly, because radio technology was in the hands of corporations, the ether would have to turn a profit. Members of the radio trust—

RCA, GE, Westinghouse, and AT&T—had gained control of radio technology and, like their predecessor Marconi, sought to establish a technological and organizational system impervious to competition. Marconi's model was instructive: create a communications network, seek to monopolize message handling, and sell temporary access to the ether to interested clients. This is precisely what the radio trust did. The trust manufactured millions of dollars worth of apparatus and also set up stations around the country. Through patent suits and the imposition of licensing fees for using trust-controlled technology, the trust was able to reduce competition or make competitors pay for the privilege of broadcasting. The trust had the technical, financial, and organizational resources to shape programming content, to influence public policy, and to determine how broadcasting would maximize profits. AT&T, through its flagship station WEAF, introduced advertising over the airwaves in 1922. AT&T described itself as a communications firm that did not produce its own messages, but that sold access to the airwaves to people who had messages to send. Marconi had established the precedent for this model; now AT&T elaborated and extended it to produce staggering profits.

Regulatory guidelines for resolving disputes in the ether existed, as well. The state would intervene, ostensibly on behalf of "the people," and decide which wavelengths would be allocated to the various competing claimants to the spectrum. In 1912, the state gave the preferred portions of the spectrum to the commercial wireless companies and the military, and relegated individuals unaffiliated with corporations or the government to the least desirable wavelengths. This pattern of regulation was repeated in the 1920s. Herbert Hoover, whose trademark was industry-government cooperation, sought to alleviate etheral congestion in 1923 by dividing radio stations into three classes—high power, medium power, and low power—and assigning the most preferred and least congested wavelengths to the high-power stations, while consigning the low-power stations to the one wavelength (360 meters) that was already overcrowded. It will come as no surprise that the high-power stations were owned by AT&T, GE, and Westinghouse while the low-power stations belonged to universities, churches, and labor unions. These stations were still required to time-share, and many were only allowed to broadcast during the day. This preferential treatment toward the technologically most powerful (and richest) commercial stations, and the regulatory marginalization of smaller, noncommercial stations, persisted through the Radio Act of 1927 and the Communications Act of 1934. As in 1912, the state remained an important ally of corporate interests, legitimating their often preemptive claims to the spectrum, and constraining the transmitting activities of those with less power and money. Certainly the federal government, especially the justice department, was not always friendly to corporate interests, and since 1930 RCA and other members of the radio trust have been subject to antitrust suits and consent decrees. But the government's role in these cases has been to determine when oligopoly went too far, not to challenge its basic legitimacy.

Also in place by 1922 was the dominant conception about what the ether was and who had a legitimate claim on how it was used. As a result of the previous debates among commercial wireless companies, government officials, the amateurs, and the press, the ether was now considered a common property resource in which all Americans had an interest. To protect that interest, however, and to save the resource from being overrun and having its value destroyed, the ether needed custodians. Through the ongoing public discourses about managing "the air," the military had been rejected as appropriate caretaker, and the amateurs cast as agents of etheral anarchy. The badge of legitimacy went to the communications corporations, who burnished its authority by presenting themselves as acting out of benevolent, farsighted paternalism. There were dissenters from this conception of spectrum management, especially among amateurs, educators, and religious groups, and there was some resentment in the 1920s about a potential corporate monopoly of the air. But there was no major break in this ideological frame concerning who was best qualified to serve as warden of the ether.

Technically, economically, legislatively, and ideologically, the elements of America's broadcasting system were, thus, in place by 1922. The constellation of these factors, and how they interacted, had been shaped by the larger historical forces that were redefining American society at the turn of the century. Had the technical developments, the corporate strategies, or the journalistic frameworks been different, or had that period in history not been marked by consolidation and centralization in the public and private sectors, and by the marginalization of diversity in the ideological sphere, the use of radio in America may have been quite different: after all, national networks and radio advertising were not inevitable. There were other alternatives, as demonstrated by the way radio was managed in other countries. That those alternative courses were not taken tells us a great deal about how American society in the early twentieth century rationalized the connections among technology, ideology, and power.

. . .

SO MUCH HAD CHANGED in the United States since that fall day in 1899 when Marconi first demonstrated his new invention, the wireless telegraph, before awed Americans. The wireless, which Marconi meant to send Morse code messages between a specific sender and a specific receiver, which he developed for institutional clients, and which he struggled to make secret and private, had become, by 1922, radio, a device marketed to consumers so that they could hear programs broadcast to a vast audience of nonpaying listeners. The technology was under corporate control, and it would be corporations that would decide, for the most part, what was transmitted in the ether and what was not.

The nearly twenty-five-year process that produced this transformation involved the dynamic interplay of individual insights and oversights, organizational ambition and recalcitrance, and technological breakthroughs and errors. Sharply competing ideas about how the invention should be used, and by whom, informed the process from the start. In this interchange among men, machines, and ideas, which affected the social construction of radio broadcasting, the role played by conflict is striking. In this case, although the concept of social construction has been quite valuable, the word *construction* itself is misleading. It suggests a more cooperative process than occurred with radio. Radio broadcasting, despite the references to cultural unity in the press, was the result of battles over technological control and corporate hegemony, and of visions about who should have access to America's newly discovered frontier environment, the electromagnetic spectrum.

The corporations forming the radio trust won these battles in the end. This is not surprising, given that monopolistic capitalism was, by the 1920s, the established way of managing the American economy. But we must remember how they succeeded, because as late as 1920 they were not planning on radio broadcasting. One key to their success was the way they were able to control, and interlink, the three arenas of technology, business strategy, and the press. Here the role that bureaucracy plays in technological change becomes very clear. By the turn of the century, and especially after the panic of 1907, these corporations had what the inventors did not: well-staffed, well-funded, and separate yet related departments to deal specifically with each of these arenas. AT&T, G.E., and Westinghouse had industrial research laboratories, and while many engineers in the labs, particularly in the early years, felt the pressure of market considerations, others were more free to work on an emerging technology such as radio. These companies employed carefully selected engineers, who also appreciated market considerations, to serve as the liaisons between the research labs and the company's top brass. Such companies were also establishing their own public relations departments, staffed by men whose specialty was understanding and manipulating the rhetoric surrounding technology and business. Thus, their success was not due solely to the fact that the large electrical firms had more money than the individual inventors; more important was that their financial power was expressed through and reinforced by a carefully articulated bureaucratic structure that addressed, specifically, these three arenas of technology, business strategy, and the press.

That success in all three arenas was critical to controlling this emerging technology was demonstrated by the navy's failure to retain control of radio at the end of World War I. By April 1917, the navy's internal organization had changed to accommodate radio's presence at sea and on shore; the navy was now capable of coordinating technical development and bureaucratic implementation. The war was used to justify the external strategy of navy takeover, suppression of the amateurs, and expansion of military hegemony in the ether. Many of the ideological shifts that occurred during the war, the rhetorical exaggerations that legitimated a range of social changes and excesses, shifted back to prewar frameworks after the armistice, however. The navy's success in the ideological arena between 1915 and 1919 deluded Daniels and his supporters into thinking that this success would be lasting. They did not appreciate that the influence the military enjoys over public discourse during wars can, and usually does, dissipate along with the smoke of the last battle. The press supported private, capitalist control over radio: it had done so prior to 1915 and did so after November 1918. The navy could not overcome the bias against government control of public utilities, which had reached new heights by war's end. Nor did the navy control the technology. Thus, while in 1919 the navy enjoyed titular power over America's radio networks, it was no match for the communications corporations, which had carefully cultivated their turf in all three arenas.

Bureaucratic control over radio had its advantages. Certainly continuous wave technology advanced much more rapidly under the auspices of G.E., AT&T, and the U.S. Navy. The patent moratorium during the war, and the subsequent cross-licensing agreements, made possible the coordination of a complete technological system. These organizations simply had the financial, legal, and human resources to achieve these ends, resources unavailable to individuals.

Technological progress and systems building came at a cost, however; the price was individual initiative and freedom in the ether. Control over radio technology put these corporations in an extremely powerful position, not just economically, but culturally, as well. The way the state promoted and protected corporate interests in the airwaves over those of individuals added to corporate power. The radio trust was thus able to co-opt the amateur vision of how radio should be used, and to use the airwaves for commercial ends, to try to promote cultural homogeneity, to mute or screen out diversity and idiosyncracy, and to advance values consonant with consumer capitalism.

The broadcasting boom marks a critical turning point in consumer culture and the corporate role in shaping that culture. Certainly the fact that corporations were sending music and other forms of entertainment into people's homes was revolutionary, as was the eventual sponsorship of these shows by increasingly brazen advertisements. Another major turning point was reached, too; it concerned corporate insight into and sensitivity to the marketplace. While the amateur audience increased every year after 1907, it was catered to only by small companies such as De Forest's or Gernsback's Electro-Importing Company. The amateurs were too inconsequential a subculture in the eyes of G.E. and AT&T to be considered as anything but a quirky fringe group or a nuisance. Even as late as 1919, when RCA was formed, the oversight persisted. RCA would provide communications to institutional clients, not individual consumers. RCA had the resources to rectify its shortsightedness and to capitalize on the boom, so the myopia was hardly fatal. One cannot help but think that since they were surprised in this way, RCA executives and the many other corporate leaders who witnessed this ambush by the audience cultivated a more calculated, opportunistic outlook toward other American activities seemingly removed from the profit potential. The concept of who might be considered a potential consumer expanded. This sense of untapped domestic markets, this corporate conviction that there were millions who would willingly reorient their activities, values, and dreams around consumerism, took root and flowered as executives observed, and then managed, the broadcasting boom.

Legitimation of this revolutionary transformation was provided by the mainstream, popular press, which, by the 1920s, had itself invested in the assumptions underlying consumer capitalism. The press was no more objective then than it is now, and its biases caused certain aspects of radio's development to be ignored while other aspects were reviewed and celebrated repeatedly.

What was emphasized and what was ignored? Whether saving lives at sea or bringing lectures to the farmer, radio was consistently cast as the agent of American democracy and altruism. Wireless in 1900 would allow individuals to communicate with whomever they wanted whenever they wanted. Thus, through wireless, Americans could circumvent hated monopolies such as Western Union; the benefits of modern communication would be made available to all. In the 1920s, radio again was portrayed as a democratic agent, leveling class differences, making politicians more accountable to the people, and spreading education "for free." Radio, then, would do nothing less than resurrect the values of the early Republic and, through the power of technology, restore their primacy in an era of monopolistic capitalism.

In the pages of the press, those who stood to benefit financially from radio were able to downplay potential remuneration and to emphasize their humanitarian goals and their commitment to "give the people what they want." Marconi was a modest, selfless hero, and AT&T, in 1915, was a progressive-looking corporation primarily interested in bringing the benefits of modern technology to everyone. Neither Marconi nor Vail was described as a single-minded organization builder determined to establish his firm's corporate hegemony; nor was it suggested that it was profits, not altruism, which fueled such a quest. The theme of altruism, insisting as it did that radio work was guided first and foremost by what "the people" needed and wanted, led to one of the basic myths supporting broadcasting programming since the 1920s. According to the myth, broadcasters are servants of the people, giving the people what they want to hear or see. Because the people can turn their dial, or shut their radio (or television) off entirely, it is the audience, the myth asserts, which ultimately has control over programming content. This illusion of power residing with the audience rather than with the broadcasters was perpetuated in countless articles in the early 1920s and emerged out of the journalistic conventions that cast radio as an agent of altruism, restored democracy, and individual control. References to democracy and to audience participation equated consumption with power. The early 1920s may have been the historical period when the myth of audience power, which rested on the myth of consumer choice, became reified and held up as evidence that Americans possessed unprecedented political and economic freedom. This myth masked the corporate acquisition of control over the content and patterns of mass communication in America, and thus veiled the less romantic and less appealing realities of industrial capitalism.

·   ·   ·

THE EMERGENCE AND the tenacity of such myths, and the economic and political systems that sustain them, are best understood by linking behind-the-scenes developments in laboratories and offices with the public portrayal of those developments in the popular press. This is why a historical approach that regards technology as socially constructed is so useful, for it requires one to consider how individuals and institutions shaped the design of machines, and to analyze how the uses to which machines are put have been legitimated in the culture at large. The social construction approach demands that we integrate institutional and economic history, individual biography, and the history of technology with a critical perspective on how certain ideas and belief systems became dominant. It requires that the history of technology be construed as cultural history in the broadest possible sense. My aim has been to use this sort of interdisciplinary approach and to show that it is the best way to reconstruct and analyze the connections between technology and ideology, and, thus, between the rise and maintenance of power.

Although a great deal has changed in American broadcasting over the past sixty years, much that was established between 1899 and 1922 remains the same. Major corporations control both broadcasting technology and access to the spectrum, and they shape what kinds of messages we get and the range of ideas to which we are exposed in the public sphere. Corporate-military cooperation in the development of communications technology has reached unprecedented and some would say unsettling proportions. And the myths and heroes through which the mass media justify this state of affairs are strikingly similar to those devised at the beginning of the century. We are still told how much control we have over media content, that what we get is what we demand and want, and that the media are our servants. We are told that because there are at least three networks, plus cable, we have access to a wide diversity of information and perspectives. News stories about emerging technologies, from computers to SDI, are presented as if the stories are completely objective, free of value judgments about who should control technology and why. Just as the press helped shape the early history of radio, so do the mass media today define and delimit the public discourses surrounding how technology is, and should be, embedded in work and leisure, in the existing power structure, and in our very thoughts.

NOTES

Collections frequently cited have been identified by the following abbreviations:

Clark Collection. Also referred to as "Radioana," the name George Clark gave the collection of correspondence, photographs, reminiscences, press releases, and articles pertaining to the history of radio which he spent thirty years amassing and organizing. Located in the Archives Center, National Museum of American History, Smithsonian Institution.

NC PAPERS. Reginald Fessenden Papers, North Carolina State Archives, Raleigh, North Carolina.

Chelmsford Collection. Company archives, Marconi Company Limited, Chelmsford, England.

GMB Private Collection. Letters in the possession of Guglielmo Marconi's daughter, Gioia Marconi Braga.

NA. National Archives. All material cited is from Record Group 19, File 18301.

INTRODUCTION

1. "Radio—The New Social Force," *Outlook,* Mar. 19, 1924, 465.
2. The term and the concept are from Peter L. Berger and Thomas Luckmann, *The Social Construction of Reality* (Garden City, N.Y.: Doubleday, 1966).
3. See Herbert J. Gans, *Deciding What's News* (New York: Vintage Books, 1980); Gaye Tuchman, *Making News: A Study in the Construction of Reality* (New York: Free Press, 1978); and Todd Gitlin, *The Whole World Is Watching* (Berkeley and Los Angeles: University of California Press, 1980).
4. Jurgen Habermas, *Legitimation Crisis* (Boston: Beacon Press, 1975).
5. Exceptions include Daniel J. Czitrom, *Media and the American Mind* (Chapel Hill: University of North Carolina Press, 1982); and Wyn Wachhorst, *Thomas Alva Edison: An American Myth* (Cambridge: MIT Press, 1981).
6. John F. Kasson, *Civilizing the Machine* (New York: Grossman, 1976), 187–90; Ray Ginger, *Age of Excess* (New York: Macmillan, 1965), 152–53.
7. Alfred D. Chandler, Jr., *Strategy and Structure: Chapters in the History of the*

American Industrial Enterprise (Cambridge: MIT Press, 1962); and idem, *The Visible Hand: The Managerial Revolution in American Business* (Cambridge: Belknap Press, Harvard University Press, 1977).

8. Chandler, *Visible Hand*, 332; and Thomas Cochran, *Business in American Life: A History* (New York: McGraw-Hill, 1972), 152–59.

9. David F. Noble, *America by Design: Science, Technology, and the Rise of Corporate Capitalism* (New York: Oxford University Press, 1979).

10. Louis Galambos, *The Public Image of Big Business in America, 1880–1940* (Baltimore: Johns Hopkins University Press, 1975), 5.

11. Robert H. Weibe, *The Search for Order* (New York: Hill & Wang, 1967), chap. 5, "A New Middle Class"; Neil Harris, "The Lamp of Learning: Popular Lights and Shadows," in *The Organization of Knowledge in Modern America, 1860–1920*, ed. Alexandra Oleson and John Voss (Baltimore: Johns Hopkins University Press, 1979), 431.

12. Michael Schudson, *Discovering the News: A Social History of American Newspapers* (New York: Basic Books, 1978), 67.

13. Edwin and Michael Emery, *The Press and America*, 5th ed. (Englewood Cliffs, N.J.: Prentice-Hall, 1984), 231.

14. Ibid., 253–67.

15. Paul Avrich, *The Haymarket Tragedy* (Princeton: Princeton University Press, 1984), chap. 3.

16. Ibid., 256. For an elegant discussion of the relationship between the rise of the middle class and the rise of the press, see James W. Carey, "Criticism and the Press," *Review of Politics* 36 (April 1974): 228–29.

17. The evolution of and characteristics associated with the self-made man are described in Irvin G. Wylie, *The Self-Made Man in America* (New Brunswick, N.J.: Rutgers University Press, 1954); for an excellent discussion of the magazine hero, see Theodore P. Greene, *America's Heroes* (New York: Oxford University Press, 1970).

18. Kasson, *Civilizing the Machine*, 39–42.

19. Galambos found that "in the 1880s and early 1890s, Jay Gould and the Vanderbilts had been the major symbols of corporate business, and both names bore a heavy legacy of negative connotations" (*Public Image*, 181).

20. Czitrom, *Media*, 8–9.

21. Cited in Frank Luther Mott, *A History of American Magazines*, vol. 4; *1885–1905* (Cambridge: Harvard University Press, 1957), 11. Mott estimates that after the Civil War, the number of magazines nearly doubled every ten years until the total reached approximately 5,500 by 1900.

ONE: MARCONI AND THE AMERICA'S CUP

1. Quoted in *Webster's Guide to American History* (Springfield, Mass.: G. & C. Merriam, 1971), 428.

2. Wyn Wachhorst, *Thomas Alva Edison: An American Myth* (Cambridge: MIT Press, 1981), 15.

3. James Oliver Robertson, *American Myth, American Reality* (New York: Hill &

Wang, 1980), 174–80; Alan Trachtenberg, *The Incorporation of America* (New York: Hill & Wang, 1982), 5.

4. For more information on Dewey see Mark Sullivan, *Our Times*, vol. 1, *The Turn of the Century, 1900–1904* (New York: Charles Scribner's Sons, 1926), 309–43; and Ronald Spector, *Admiral of the New Empire* (Baton Rouge: Louisiana State University Press, 1974).

5. *New York Tribune*, Sept. 30, 1899, 6, and Oct. 1, 1899, 1.

6. *New York Herald*, Oct. 1, 1899, 1.

7. *Harper's Weekly*, Sept. 30, 1899, 954; James Barnes, "The Story of Dewey's Welcome Home," *Outlook*, Oct. 7, 1899, 299; "Admiral Dewey as a National Hero," *Century Magazine*, Oct. 1899, 928.

8. Ernest Knaufft, "Dewey Day Decorations in New York," *Review of Reviews*, Oct. 1899, 458–62.

9. John Barrett, "Admiral George Dewey," *Harper's Monthly*, Oct. 1899, 812.

10. Ibid.; "Dewey's Home-Coming," *Outlook*, Oct. 7, 1899, 291; Winston Churchill, "Admiral Dewey: A Character Sketch," *Review of Reviews*, June 1898, 686; Barnes, "Story of Dewey's Welcome Home," 299.

11. *New York Herald*, Oct. 1, 1899, 4.

12. Barrett, "Admiral George Dewey," 800.

13. Ibid., 799.

14. *Harper's Weekly*, Sept. 30, 1899, 954.

15. There were striking similarities between the celebrations for Dewey and for Lindbergh. See John W. Ward, "The Meaning of Lindbergh's Flight," *American Quarterly* 10 (Spring 1958): 7.

16. See Robert Nisbet, *History of the Idea of Progress* (New York: Basic Books, 1980).

17. *Popular Science Monthly* 56, no. 1 (1899): 26.

18. "The New Century," *Nature* 63, no. 1627 (1901): 221.

19. "A Century of Progress in the United States," *Scientific American* 83, no. 26 (1900): 400–402.

20. *Popular Science Monthly* 52 (Dec. 1897): 263.

21. Don C. Seitz, *The James Gordon Bennetts* (Indianapolis: Bobbs-Merrill, 1928), 372.

22. John Tebbel, *The Media in America* (New York: Mentor, 1974), 184; Michael Schudson, *Discovering the News: A Social History of American Newspapers* (New York: Basic Books, 1978), 31–35.

23. *New York Herald*, Oct. 1, 1899, sec. 4, 3.

24. Ibid., 1.

25. William Bissing, "Telegraphy without Line Wires," *Electrical World* 33, no. 2 (1899): 55.

26. For more detailed information on Henry, Faraday, and Maxwell, see *Electrical World* 33, no. 2 (1899): 56; "The Wireless Telegraph," *New York Times*, Apr. 9, 1899, 22; David Keith Chalmers MacDonald, *Faraday, Maxwell, and Kelvin* (Garden City, N.Y.: Anchor Books, 1964); Rollo Appleyard, *Pioneers of Electrical Communications* (London: Macmillan, 1930); Alvin F. Harlow, *Old Wires and New Waves* (New York: D. Appleton-Century, 1936); W. Rupert Maclaurin, *Invention and Innovation in the Radio Industry* (New York: Macmillan, 1949),

9–15; R.A.R. Tricker, *The Contributions of Faraday and Maxwell to Electrical Science* (New York: Pergamon Press, 1966); Hugh G. J. Aitken, *Syntony and Spark: The Origins of Radio* (New York: John Wiley & Sons, 1976).

27. For a more complete discussion of these experiments on induction, see J. J. Fahie, *A History of Wireless Telegraphy* (New York: Dodd, Mead, 1900). See also Elliot N. Sivowitch, "A Technological Survey of Broadcasting's 'Pre-History,' 1876–1920," *Journal of Broadcasting* 15, no. 1 (1970–71).

28. This information is from *The Dictionary of Scientific Biography* (New York: Charles Scribner's Sons, 1974), 9:204.

29. See the article on von Helmholtz in ibid., 6:242–43; see also Friedrich Kurylo and Charles Susskind, *Ferdinand Braun: A Life of the Nobel Prizewinner and Inventor of the Cathode-Ray Oscilloscope* (Cambridge: MIT Press, 1981).

30. For an elegant and more complete discussion of Hertz, see Aitken's chapter on Hertz in *Syntony and Spark*, 48–74; also H. Hertz, "On the Mechanical Actions of Electrical Waves Propagated in Conductors," *Electrical World* 58, no. 4 (1891): 66; "Hermann von Helmholtz," *Electrical World* 20, no. 10 (1892): 144; *Electrical World* 23, no. 4 (1894): 107; W. J. Baker, *A History of the Marconi Company* (London: Methuen, 1970), 18–20; Maclaurin, *Invention and Innovation*, 15–17.

31. *New York Times,* May 7, 1898, 12.

32. Announcement of Lodge's coherer appears in *Electrical World* 24, no. 1 (1894); see Aitken, *Syntony and Spark,* 103–8.

33. Arthur V. Abbott, "Electrical Radiation—III," *Electrical World* 33, no. 21 (1899): 701.

34. Information on Marconi is from Degna Marconi, *My Father, Marconi* (New York: McGraw-Hill, 1962), chaps. 1–5; Keith Geddes, *Guglielmo Marconi, 1874–1937* (London: Her Majesty's Stationery Office, 1974).

35. Marconi, *My Father,* 22–23.

36. Ibid., 13, 21–22.

37. For contemporary descriptions of Marconi's apparatus, see *Electrical World* 30, no. 5 (1897): 138; 33, no. 13 (1899): 417; and 33, no. 4 (1899): 110. Also, *New York Times,* May 26, 1897, 6, and May 7, 1898, 12.

38. Geddes, *Guglielmo Marconi,* 5; Baker, *History of the Marconi Company,* 26.

39. *New York Times,* May 26, 1897, 6.

40. Aitken, *Syntony and Spark,* 224; Geddes, *Guglielmo Marconi,* 9.

41. Geddes, *Guglielmo Marconi,* 11.

42. *Electrical World* 33, no. 12 (1899): 377.

43. *New York Times,* Mar. 29, 1899, 1.

44. *New York Herald,* Oct. 1, 1899, sec. 4, 1.

45. Ibid.

46. Ibid., 7.

47. Marconi's Reminiscences of the American Tests, Chelmsford Collection, HIS 72.

48. *New York Herald,* Oct. 6, 1899, 7.

49. *New York Tribune,* Mar. 30, 1899, 8.

50. *New York Herald,* Oct. 1, 1899, sec. 5, 4.

51. *New York Times,* Jan. 14, 1902, 1; and Jan. 15, 1902, 8.

52. *New York Herald,* Oct. 4, 1899, 7.

53. See *Electrical World*, 39, no. 9 (1902): 387; *New York Times*, Jan. 25, 1903, 10. Marconi's conviction that his inventions were not derivative, but represented original technological breakthroughs, comes across in company and personal correspondence, and in patent applications.
54. Marconi's Reminiscences.
55. *New York Tribune*, Sept. 22, 1899, 6.
56. Marconi, *My Father*, 34, 49–50.
57. *Electrical World* 40, no. 13 (1902): 485.
58. Ibid., 33, no. 20 (1899): 643.
59. The term *technological display* comes from Michael L. Smith's excellent article "Selling the Moon: The U.S. Manned Space Program and the Triumph of Commodity Scientism," in *The Culture of Consumption*, ed. Richard Wightman Fox and T. J. Jackson Lears (New York: Pantheon, 1983).
60. *New York Times*, Nov. 5, 1897, 6.
61. *Popular Science Monthly* 56, no. 1 (1899): 72.
62. *New York Herald*, Oct. 4, 1899, 7.
63. Daniel J. Czitrom, *Media and the American Mind* (Chapel Hill: University of North Carolina Press, 1982), 3.
64. *Popular Science Monthly* 56, no. 1 (1899): 72.
65. *New York Times*, May 7, 1899, 20.
66. P. T. McGrath, "Authoritative Account of Marconi's Work in Wireless Telegraphy," *Century Magazine*, Mar. 1902, 782.
67. Czitrom, *Media*, 21–29.
68. *New York Times*, Aug. 15, 1899, 6.
69. *New York Herald*, Oct. 2, 1899, 8.
70. *New York Times*, Mar. 9, 1902, 6.
71. Ibid., May 1, 1902, 8.
72. Ibid., Aug. 15, 1899, 6; *New York Herald*, Oct. 2, 1899, 8.
73. *New York Times*, May 7, 1899, 20.
74. *Electrical World* 34, no. 20 (1899): 730.

TWO: COMPETITION OVER WIRELESS TECHNOLOGY

1. See Sally Gregory Kohlstedt, *The Formation of the American Scientific Community: The American Association for the Advancement of Science, 1848–1860* (Urbana: University of Illinois Press, 1976), 17–22. See also David F. Noble, *America by Design: Science, Technology, and the Rise of Corporate Capitalism* (New York: Oxford University Press, 1979), 20–49.
2. Daniel J. Kevles, *The Physicists* (New York: Alfred A. Knopf, 1978), 17–19.
3. Wyn Wachhorst, *Thomas Alva Edison: An American Myth* (Cambridge: MIT Press, 1981), 35.
4. *Electrical World* 30, no. 7 (1897): 181.
5. Ibid., 31, no. 2 (1898): 75; 30, no. 18 (1897).
6. Ibid., 32, no. 7 (1898): 166.
7. Ibid., 30, no. 7 (1897): 182.
8. *New York Times*, Dec. 16, 1901, 9.

9. *Electrical World* 33, no. 14 (1899): 444.
10. *New York Times,* Jan. 19, 1903, 1.
11. Ibid., May 7, 1899, 20; *Electrical World* 33, no. 14 (1899): 444.
12. *New York Times,* Mar. 23, 1902, 6.
13. For a more elegant and detailed discussion of this process, see Hugh G. J. Aitken, *Syntony and Spark: The Origins of Radio* (New York: John Wiley & Sons, 1976), 198–99.
14. Marconi to H. Cuthbert Hall, Aug. 22, 1902, GMB Private Collection.
15. Ibid., Oct. 1, 1902, GMB Private Collection.
16. For a discussion of the impact of personalities within organizational settings and the dynamic relationships between individual entrepreneurs and business bureaucracies, see Harold C. Livesay, "Entrepreneurial Persistence through the Bureaucratic Age," *Business History Review* 51, no. 4 (1977): 415–43.
17. Marconi's daughter observes that some people meeting Marconi for the first time found him to be taciturn and difficult to know. See Degna Marconi, *My Father, Marconi* (New York: McGraw-Hill, 1962), 50.
18. Ibid., 49.
19. See, for example, Marconi's letters to Hall, in which he praises his manager's diligence and compliments his skill in drafting and explaining contracts. Marconi to H. Cuthbert Hall, Aug. 20, 1901, and June 19, 1906, GMB Private Collection.
20. Ibid., Oct. 1, 1902, GMB Private Collection.
21. A much more detailed and sophisticated discussion of Lodge's work appears in Aitken, *Syntony and Spark,* 80–168.
22. W. J. Baker, *A History of the Marconi Company* (London: Methuen, 1970), 54; *Electrical World* 37, no. 11 (1901); 453; Aitken, *Syntony and Spark,* 250.
23. *Electrical World* 37, no. 25 (1901): 1082.
24. A. Frederick Collins, "The Lodge-Muirhead System of Wireless Telegraphy," *Electrical World* 42, no. 5 (1903): 174; idem, "Coherers: The Development, Construction, Operation, and Function of Electric Wave Detectors, *Electrical World* 38, no. 7 (1901): 252; B. T. Judkins cited in George H. Clark, "The Life and Creations of John Stone Stone" (1946), 70, Clark Collection.
25. *Electrical World* 41, no. 15 (1903): 613.
26. E. Rutherford, "A Magnetic Detector of Electrical Waves and Some of Its Applications," *Phil. Trans. Roy. Soc.* (London) 189A (1897); also *Proc. Roy. Soc.* (London) 60 (1896); George W. Pierce, *Principles of Wireless Telegraphy* (New York: McGraw-Hill, 1910), 145–46.
27. Elmer E. Bucher, *Practical Wireless Telegraphy* (New York: Wireless Press, 1917), 148; Rupert Stanley, *Text Book on Wireless Telegraphy* (New York: Longmans, Green, 1914), 220.
28. Marconi to H. Cuthbert Hall, June 9, 1902, GMB Private Collection.
29. Noble, *America by Design,* 95–97.
30. Recent scholarship is focusing more on this type of man, the entrepreneur within the organizational structure. See Livesay, "Entrepreneurial Persistence," and also Leonard S. Reich, "Irving Langmuir and the Pursuit of Science and Technology in the Corporate Environment," *Technology and Culture* 24, no. 2 (1983): 199–221.
31. There is no scholarly biography of Fessenden. The following material is based on

the bitter and defensive biography written by his wife, Helen Fessenden, *Fessenden: Builder of Tomorrows* (New York: Coward-McCann, 1940).

32. Ibid., 22. Hugh G. J. Aitken first brought my attention to the probability that Fessenden never received a college degree.

33. James E. Brittain, author of a forthcoming biography of Ernst Alexanderson, brought this connection to my attention. James Brittain to the author, Sept. 10, 1979.

34. James E. Brittain, "The Alexanderson Alternator: An Encounter between Radio Physics and Electrical Power Engineering" (Paper delivered before the Society for the History of Technology, Oct. 31, 1982), 3.

35. Fessenden, *Fessenden*, 74.

36. See Report of the Chief of the Weather Bureau in the *Annual Reports* of the Department of Agriculture (Washington, D.C.: Government Printing Office), 1898–1900.

37. Fessenden, *Fessenden*, 83.

38. A. Frederick Collins, "Fessenden's Work in Wireless Telegraphy," *Electrical World* 42, no. 12 (1903): 475; Gleason Archer, *The History of Radio to 1926* (New York: American Historical Society, 1938), 68; A. P. Morgan, *Wireless Telegraphy and Telephony* (New York: Norman W. Henley, 1912), 58–59.

39. Stanley, *Text Book*, 66–70; James Brittain to the author, Sept. 10, 1979.

40. Fessenden to Charles Steinmetz, June 1, 1900, NC Papers.

41. Ibid., Feb. 10, 1901, NC Papers.

42. Charles Steinmetz to Reginald Fessenden, July 16, 1901, Clark Collection.

43. WFH at General Electric to Fessenden, Aug. 30, 1905, Clark Collection.

44. Fessenden to Charles Steinmetz, July 3, 1901, NC Papers.

45. The two biographies of De Forest, the first virtually ghostwritten by the inventor, and the second an autobiography, are characterized by their fervid, melodramatic prose and their historical unreliability. These are Georgette Carneal, *Conqueror of Space: The Life of Lee De Forest* (New York: Horace Liveright, 1930), and Lee De Forest, *Father of Radio* (Chicago: Wilcox & Follett, 1950). The best source on De Forest remains Samuel Lubell's "Magnificent Failure," published in three successive issues of the *Saturday Evening Post* in January 1942. De Forest's diaries are available at the Library of Congress, Manuscript Division, and through Yale University. This discussion is based on Lubell and the diaries.

46. See Wachhorst, *Thomas Alva Edison*, 23–25.

47. Lubell, "Magnificent Failure," Jan. 17, 1942, 78.

48. De Forest to Marconi, Sept. 22, 1899, Chelmsford Collection, HIS 1.

49. *Electrical World* 40, no. 6 (1902): 227; *New York Times*, July 25, 1902, 8; Carneal, *Conqueror of Space*, 110; Donald McNicol, *Radio's Conquest of Space* (New York: Murray Hill Books, 1946), 115; Ellison Hawks, *Pioneers of Wireless* (London: Methuen, 1927), 261–74; George G. Blake, *History of Radio Telegraphy and Telephony* (London: Chapman & Hall, 1928), 82–83.

50. Material on Stone is from Clark's "Life and Creations of John Stone Stone"; Clark's manuscript was published by Frye & Smith, San Diego, Calif., in 1946.

51. Ibid., 23–25.

52. Ibid., 55–67, 78, 103.

53. *Electrical World* 38, no. 25 (1901): 1024.

54. *Scientific American* 86, no. 1 (1902): 4.

55. *Electrical World* 36, no. 16 (1900): 618.

56. G. H. Barbour, "Recent Practices in Wireless Station Construction," *Electrical World* 49, no. 9 (1907): 438.

57. *New York Times,* Dec. 16, 1901, 9; *Electrical World* 46, no. 8 (1905): 294.

58. H. Cuthbert Hall to Marconi, Nov. 27, 1901, Chelmsford Collection, HIS 80.

59. Baker, *History of the Marconi Company,* 66.

60. Henry Herbert McClure, "Messages to Mid-Ocean; Marconi's Own Story of His Latest Triumph," *McClure's Magazine* 23 (April 1902): 526.

61. H. Cuthbert Hall to Marconi, July 12, 1901, Chelmsford Collection, HIS 77.

62. *Electrical World* 38, no. 15 (1901): 596–97.

63. Scientists doubt whether Marconi received this signal. See J. A. Ratcliffe, "Scientists' Reactions to Marconi's Transatlantic Signal," *Proc. Inst. Elect. Engineers* (England) 121, no. 9 (1974): 1033–38.

64. *Scientific American* 86, no. 1 (1902): 2.

65. Carl Snyder, "Wireless Telegraphy and Signor Marconi's Triumph," *Review of Reviews* 25 (Feb. 1902): 173.

66. Ibid., 174.

67. P. T. McGrath, "Marconi and His Transatlantic Signal," *Century Magazine* 63, (Mar. 1902): 775.

68. Ray Stannard Baker, "Marconi's Achievement," *McClure's Magazine* 18 (Feb. 1902).

69. Snyder, "Wireless Telegraphy," 173.

70. Lee De Forest, Diary, Dec. 22, 1901, and Jan. 13, 1902, Library of Congress.

71. *Electrical World* 39, no. 1 (1902): 31.

72. Ibid., 38, no. 25 (1901): 1023, 1011.

THREE: VISIONS AND BUSINESS REALITIES OF THE INVENTORS

1. Thomas R. Navin and Marian V. Sears, "The Rise of a Market for Industrial Securities, 1887–1902," *Business History Review* 29 (June 1955): 106. This thorough and extremely clear article describes the relationship between the merger movement and the expansion of the industrial securities market in much more detail.

2. Ray Ginger, *Age of Excess* (New York: Macmillan, 1965), 218–19; Robert Sobel, *The Big Board: A History of the New York Stock Market* (New York: Free Press, 1965), 149–56.

3. Mark Sullivan, *Our Times,* vol. 2, *America Finding Herself* (New York: Charles Scribner's Sons, 1927), 2:358.

4. Ibid., 359. Alexander Dana Noyes, *Forty Years of American Finance* (New York: G. P. Putnam's Sons, 1909), 294–301; idem, *The Market Place* (Boston: Little, Brown, 1938), 195.

5. Carl Snyder, "Wireless Telegraphy and Signor Marconi's Triumph," *Review of Reviews* 25 (Feb. 1902): 173.

6. Alfred D. Chandler, Jr., *Strategy and Structure: Chapters in the History of the*

American Industrial Enterprise (Cambridge: MIT Press, 1962), 1–17.

7. C. H. Taylor to Degna Marconi, Aug. 7, 1952, GMB Prvate Collection.
8. *New York Times,* Nov. 23, 1899, 1.
9. For a more detailed study of the British company see W. J. Baker, *A History of the Marconi Company* (London: Methuen, 1970).
10. This comes across especially clearly in his letters to H. Cuthbert Hall. See Marconi to Hall, Nov. 26, 1902, June 25, 1906, Aug. 6, 1906, all in GMB Private Collection.
11. Marconi to H. Cuthbert Hall, Mar. 15, 1903, GMB Private Collection.
12. John Bottomley, Minutes of Regular Quarterly Meeting of the Board of Directors, Marconi Wireless Telegraph Company of America, June 9, 1903, Chelmsford Collection, HIS 237.
13. H. Cuthbert Hall to Marconi, July 1, 1901, Chelmsford Collection, HIS 80.
14. Baker, *History of the Marconi Company,* 42.
15. Ibid., 59. For the best discussion of this subject, see the chapter on Marconi in Hugh G. J. Aitken, *Syntony and Spark: The Origins of Radio* (New York: John Wiley & Sons, 1976), esp. 233.
16. Baker, *History of the Marconi Company,* 59.
17. Marconi to H. Cuthbert Hall, Aug. 20, 1901, GMB Private Collection.
18. Aitken, *Syntony and Spark,* 239.
19. *New York Times,* Mar. 30, 1902, 2.
20. W. W. Bradfield to C. A. Butlin, Director of General Posts and Telegraphs, Santo Domingo, Feb. 23, 1902, Clark Collection.
21. *New York Times,* Oct. 8, 1901, 3.
22. H. Cuthbert Hall to Marconi, Apr. 25, 1904, Chelmsford Collection, HIS 81.
23. Marconi to H. Cuthbert Hall, July 11, 1904, GMB Private Collection.
24. John Bottomley to H. Cuthbert Hall, Jan. 19, 1906, Chelmsford Collection, HIS 235.
25. Undated, unsigned memo on the Babylon, Long Island, Station, Chelmsford Collection, HIS 79.
26. Bottomley, Minutes of Regular Quarterly Meeting of the Board of Directors, June 9, 1903.
27. J. D. Jerrold Kelley, Chairman, Executive Committee of the Marconi Wireless Telegraph Company of America, to Marconi, May 21, 1902, Chelmsford Collection, HIS 235.
28. "Marconi Wireless Telegraphy: A Record of Its Achievements, Complements of Munroe and Munroe" (1903), Clark Collection.
29. E. H. Moeran to Florence S. Hoyt, Jan. 25, 1902, quoted in George Clark's abstracts of Marconi Company letters, Clark Collection.
30. E. H. Moeran to the Marconi Wireless Telegraph Company, Jan. 7 and Feb. 7, 1902, quoted in Clark abstracts, Clark Collection.
31. Bottomley, Minutes of Regular Quarterly Meeting of the Board of Directors, June 9, 1903.
32. E. H. Moeran to G. Kemp, Sept. 27, 1901, Clark abstracts, Clark Collection.
33. E. H. Moeran to Marconi Wireless Telegraph Company, Oct. 1, 1901, Clark abstracts, Clark Collection.

34. Bottomley, Minutes of Regular Quarterly Meeting of the Board of Directors, June 9, 1903.
35. Minutes of Meeting of Board of Directors of Marconi Wireless Telegraph Company of America, undated but probably 1910, Chelmsford Collection, HIS 237.
36. Marconi to H. Cuthbert Hall, Sept. 4, 1903, GMB Private Collection.
37. W. H. Bentley, Marconi Wireless Telegraph Company of America, to G. Branchi, Italian Consul General, Jan. 13, 1904, Chelmsford Collection, HIS 235.
38. *New York Times,* Dec. 21, 1902, 1, and Jan. 20, 1903, 1.
39. Construction began in 1902. See *New York Times,* Apr. 30, 1902, 1.
40. John Bottomley to Marconi, Aug. 31, 1905, Chelmsford Collection, HIS 235.
41. W. H. Bentley to Board of Directors, Marconi Wireless Telegraph Company of America, Apr. 22, 1904, Chelmsford Collection, HIS 235.
42. John Bottomley to H. Cuthbert Hall, Feb. 20, 1906, Chelmsford Collection, HIS 235.
43. Ibid., Jan. 19, 1906, Chelmsford Collection, HIS 235.
44. Marconi to H. Cuthbert Hall, Oct. 27, 1904, GMB Private Collection.
45. Cleveland Abbe (on behalf of Willis Moore) to Fessenden, Dec. 21, 1899, Clark Collection.
46. Fessenden to President Theodore Roosevelt, June 16, 1902, NC Papers.
47. Telegram from Willis Moore to Fessenden, July 31, 1902, NC Papers.
48. Fessenden to Hay Walker, Jr., Nov. 16, 1903, NC Papers.
49. Ibid., July 6, 1904, NC Papers.
50. Ibid., Oct. 19, 1904, NC Papers.
51. George H. Clark, "The Life and Creations of John Stone Stone" (1946), 91, Clark Collection.
52. Fessenden to Ernst Berg, Jan. 2, 1904; Berg to Fessenden, Jan. 5, 1904; Fessenden to Berg, Jan. 14, 1904; Clark Collection.
53. Fessenden to E. W. Rice, Mar. 17, 1904; Rice to H. W. Young, NESCO Sales Manager, June 27, 1904; Clark Collection.
54. Hay Walker, Jr., to Fessenden, July 21, 1905, NC Papers.
55. A. A. Isbell to W. J. Glaubitz, Jan. 27, Feb. 1, and Apr. 25, 1905; Isbell to NESCO headquarters, Washington, D.C., June 22, 1905; Clark Collection.
56. M. F. Westover to Fessenden, Dec. 1, 1905, Clark Collection.
57. Fessenden to Hay Walker, Jr., Oct. 15, 1903, NC Papers.
58. Ibid., and Nov. 16, 1903, NC Papers.
59. Ibid., Jan. 19, 1904, NC Papers.
60. Sam Scoggins to Fessenden, May 3, 1904, NC Papers.
61. Fessenden to Hay Walker, Jr., Oct. 17, 1903, NC Papers.
62. Ibid., Oct. 19, 1904, NC Papers.
63. Ibid., Jan. 4, 1905.
64. Mr. Dean to Hay Walker, Jr., Dec. 9, 1903, NC Papers.
65. Fessenden to Hay Walker, Jr., Nov. 4, 1904, NC Papers.
66. Hay Walker, Jr., to Fessenden, July 7, 1904, NC Papers.
67. Fessenden to Hay Walker, Jr., Oct. 4, 1903, NC Papers; Charles Bright to Fessenden, Jan. 24, 1907, and Memorandum, NESCO to Lloyd's, May 1, 1906, Clark Collection.

68. Hay Walker, Jr., to Fessenden, June 10, 1904, NC Papers.
69. Fessenden to Hay Walker, Jr., June 13, 1904, NC Papers.
70. Thomas Given to Fessenden, Mar. 11, 1905, NC Papers.
71. Fessenden *vs.* NESCO, Affidavits for Complainant in Reply (1911), 60, NC Papers.
72. Hay Walker, Jr., to Fessenden, June 17, 1905, NC Papers.
73. Memorandum, Expenses of NESCO, May 10, 1906, NC Papers.
74. Hay Walker, Jr., to Fessenden, Feb. 23, 1905, NC Papers.
75. Ibid., May 16, 1904, NC Papers.
76. See, for example, W. Rupert Maclaurin, *Invention and Innovation in the Radio Industry* (New York: Macmillan, 1949), 59, 62.
77. Fessenden to Marconi, Oct. 6, 1899, Chelmsford Collection, HIS 2.
78. *New York Times,* Dec. 18, 1899, 1.
79. Fessenden to Hay Walker, Jr., July 6, 1904, NC Papers.
80. Lee De Forest, Diary, Dec. 1, 1901, Library of Congress.
81. Ibid., July 20, 1902.
82. Ibid., Jan. 13 and Feb. 9, 1902.
83. Samuel Lubell, "Magnificent Failure," *Saturday Evening Post,* Jan. 24, 1942, 21.
84. De Forest, Diary, Feb. 9, 1902.
85. *New York Times,* Feb. 8, 1903, 12.
86. Lubell, "Magnificent Failure," 35; Frank Fayant, "The Wireless Telegraph Bubble," *Success* 10, no. 157 (1907): 508.
87. *Current Literature* 53, no. 6 (1907): 677; John Firth Scrapbook on Wireless, Clark Collection.
88. Original De Forest stock promotion material in the Clark Collection; see also Frank Fayant, "Fools and Their Money," *Success* 10, no. 152 (1907): 9–11.
89. H. J. Brown, De Forest operator, to John Firth, Apr. 26, 1904; John Firth, "Story of My Life," 1937, Clark Collection; *New York Times,* July 3, 1904, 8.
90. *Wireless News* (New York) 1, no. 2 (1903), Clark Collection.
91. Roy Mason, "The History of the Development of the United Fruit Company's Radio Telegraph System," *Radio Broadcast,* Sept. 1922, 377–78.
92. Ibid., 378–80.
93. Notes by Abraham White to George Clark, Oct. 1941, Clark Collection.
94. Lubell, "Magnificent Failure," 35.
95. *New York Times,* May 1, 1902, 8.
96. Ibid., Sept. 6, 1904, 2, Sept. 11, 1902, 12, and Nov. 12, 1904, 1; *Electrical World* 44, no. 10 (1904).
97. Clark, "Life and Creations of John Stone Stone," 1.
98. Ibid., 65–71.

FOUR: WIRELESS TELEGRAPHY IN THE NEW NAVY

1. See, for example, Gleason Archer, *The History of Radio to 1926* (New York: American Historical Society, 1938), 76.
2. *Annual Report of the Secretary of the Navy,* 1915, 42–43.
3. Robert Seager II, "Ten Years before Mahan: The Unofficial Case for the New Navy, 1880–1890," *Mississippi Valley Historical Review* 40 (Dec. 1953): 503.

4. Ibid.

5. *Review of Reviews* 19 (May 1899): 554. For more information see also Captain Alfred Thayer Mahan, *The Influence of Sea Power upon History, 1660–1783* (Boston: Little, Brown, 1890), and the many contemporary reviews printed in such magazines as *Blackwood's Magazine* 148 (Oct. 1890): 576; *Atlantic* (Oct. 1890): 563; *Living Age* 187 (Nov. 1890): 401.

6. Ray Ginger, *Age of Excess* (New York: Macmillan, 1965), 159.

7. Cited in G. S. Clark, "Captain Mahan's Counsels to the United States," *Nineteenth Century* 43 (Feb. 1898): 293.

8. Mahan, *Influence of Sea Power*, 83.

9. Seager, "Ten Years before Mahan," 497.

10. "The Rebuilt Navy of the United States: A Letter from Mr. Roosevelt," *Review of Reviews* 17 (Jan. 1898): 68.

11. *Electrical World* 49, no. 2 (1907): 83.

12. Ibid., 38, no. 25 (1901): 1012.

13. Elting Morison, *Men, Machines, and Modern Times* (Cambridge: MIT Press, 1966), 98–122; Lance C. Buhl, "Mariners and Machines: Resistance to Technological Change in the American Navy, 1865–1869," *Journal of American History* 61, no. 4 (1974): 704; and Harold and Margaret Sprout, *The Rise of American Naval Power, 1776–1918* (Princeton: Princeton University Press, 1939), 165–82.

14. Sprout, *Rise of American Naval Power*, 270–80, and illus. facing 218.

15. At the turn of the century, these were: Bureau of Navigation, Bureau of Ordnance, Bureau of Equipment, Bureau of Construction and Repairs, Bureau of Steam Engineering, Bureau of Yards and Docks, Bureau of Medicine and Surgery, Bureau of Supplies and Accounts.

16. Charles Oscar Paullin, *Paullin's History of Naval Administration, 1775–1911* (Annapolis: U.S. Naval Institute Press, 1968), 438; Sprout, *Rise of American Naval Power*, 274; Robert Greenhalgh Albion, *Makers of Naval Policy, 1798–1947* (Annapolis: U.S. Naval Institute Press, 1980), 7, 12.

17. Sprout, *Rise of American Naval Power*, 193.

18. *Annual Report of the Secretary of the Navy*, 1905, 3; Sprout, *Rise of American Naval Power*, 193.

19. *Annual Report of the Secretary of the Navy*, 1900, 1905.

20. Albion, *Makers of Naval Policy*, 212–13; *Annual Report of the Secretary of the Navy*, 1909, 6.

21. Sprout, *Rise of American Naval Power*, 274, 271.

22. Ibid., 168; Alfred Thayer Mahan, *From Sail to Steam: Recollections on Naval Life* (New York: Harper, 1908), 270.

23. Prior to 1907, squadrons were organized into fleets on a temporary basis, as during the Spanish-American War.

24. L. S. Howeth, *History of Communications-Electronics in the United States Navy* (Washington, D.C.: Government Printing Office, 1963).

25. Paullin, *Paullin's History*, 447.

26. Albion, *Makers of Naval Policy*, 212–17.

27. Howeth, *History of Communications,* 26.

28. Lt. J. B. Blish, Report to Bureau of Equipment, Nov. 13, 1899, NA, box 83.

29. Marconi to Wireless Board, Oct. 29, 1899, NA, box 83.

30. George H. Clark, "Radio in War and Peace" (1940), 14, Clark Collection.

31. Lt. Comdr. J. T. Newton, Report on the Marconi System, Nov. 13, 1899, NA, box 83.

32. Ibid.

33. Lt. G. W. Denfeld to Chief, Bureau of Equipment, Nov. 1, 1899, NA, box 83.

34. Adm. R. B. Bradford, Chief, Bureau of Equipment, to Secretary of the Navy, Dec. 1, 1899, NA, box 83.

35. Ibid.

36. William Moody, Secretary of the Navy, to the American Marconi Wireless Telegraph Company, Sept. 25, 1903, NA, box 85.

37. Comdr. F. M. Barber to Adm. R. B. Bradford, Chief, Bureau of Equipment, Dec. 6, 1901, NA, box 83.

38. Ibid., June 22, 1906, NA, box 90.

39. John Bottomley, Minutes of Regular Quarterly Meeting of the Board of Directors, Marconi Wireless Telegraph Company of America, June 9, 1903, Chelmsford Collection, HIS 237.

40. Naturally, there are exceptions to this characterization. Bradley A. Fiske and William S. Sims both developed various inventions during their naval careers. The difficulty Sims encountered in trying to get the navy to adopt continuous-aim firing is described in Morison, *Men, Machines, and Modern Times,* 17–44. Even Fiske "was inclined to the opinion that radio had no military usefulness whatever" (Howeth, *History of Communications,* 65).

41. Morison, *Men, Machines, and Modern Times,* 9.

42. Hugh G. J. Aitken, *Syntony and Spark: The Origins of Radio* (New York: John Wiley & Sons, 1976), 322.

43. Comdr. F. M. Barber to Adm. R. B. Bradford, Chief, Bureau of Equipment, Apr. 2, 1902, NA, box 85.

44. Ibid., Dec. 31, 1901, NA, box 84.

45. Ibid., Jan. 30, 1902, NA, box 84.

46. Ibid., Apr. 15, 1902, NA, box 85.

47. Ibid., Feb. 19, 1902, NA, box 84.

48. Ibid., Dec. 31, 1901, NA, box 84; July 29, 1902, NA, box 85; June 17, 1902, NA, box 85.

49. Ibid., May 22, 1908, NA, box 89.

50. Ibid., Nov. 28, 1901, NA, box 83.

51. Ibid., Jan. 15, 1902, NA, box 84.

52. Ibid., Apr. 22, 1902, NA, box 85.

53. Ibid., Feb. 11, 1907, NA, box 89.

54. Ibid., Jan. 20, 1902, NA, box 84.

55. Adm. R. B. Bradford, Chief, Bureau of Equipment, to Comdr. F. M. Barber, Jan. 13, 1902, NA, box 84; Dec. 6, 1901, NA, box 83; Dec. 14, 1901, NA, box 83.

56. Howeth, *History of Communications,* 52, 43.

57. Clark, "Radio in War and Peace," 33.

58. Adm. R. B. Bradford, Chief, Bureau of Equipment, to the Secretary of the Navy, Dec. 13, 1902, NA, box 85.

59. Adm. R. B. Bradford, Chief, Bureau of Equipment, to Comdr. F. M. Barber, Apr. 19, 1902, NA, box 84.

60. Ibid.

61. Comdr. F. M. Barber to Adm. R. B. Bradford, Chief, Bureau of Equipment, Apr. 4, 1902, NA, box 86.

62. Ibid., Dec. 6, 1902, NA, box 86.

63. Ibid., Jan. 30, 1902, NA, box 84, and May 14, 1902, NA, box 88; Adm. R. B. Bradford Chief, Bureau of Equipment, to Comdr. F. M. Barber, Feb. 11, 1902, NA, box 84.

64. Slaby-Arco to Comdr. F. M. Barber, July 19, 1902, NA, box 84.

65. Report, Wireless Telegraph Board, to Adm. R. B. Bradford Chief, Bureau of Equipment, Dec. 3, 1902, NA, box 85.

66. Comdr. F. M. Barber to Adm. R. B. Bradford, Chief, Bureau of Equipment, Nov. 7, 1902, NA, box 83.

67. Clark, "Radio in War and Peace," 35.

68. *Electrical World* 43, no. 24 (1902).

69. Fessenden to Chief, Bureau of Equipment, May 8, 1903, NA, box 85.

70. Adm. R. B. Bradford, Chief, Bureau of Equipment, to Fessenden, May 21, 1903, NA, box 85.

71. Comdr. F. M. Barber to Adm. R. B. Bradford, Chief, Bureau of Equipment, Feb. 18, 1908, NA, box 89.

72. Adm. R. B. Bradford, Chief, Bureau of Equipment, to Fessenden, May 21, 1903, SI.

73. Fessenden to Lt. Hudgins, USS *Topeka*, May 26, 1903; Fessenden to Adm. Henry N. Manney, June 13, 1903, Clark Collection.

74. Capt. C. H. Arnold, Wireless Telegraphy Board (on behalf of Bradford), to Fessenden, June 7, 1903, Clark Collection.

75. Charles Darling, Acting Secretary of the Navy, to NESCO, Dec. 15, 1903, Clark Collection.

76. *Electrical World* 36, no. 5 (1900): 157.

77. Ibid., 36, no. 8 (1900): 273.

78. Ibid., 40, no. 10 (1902): 354.

79. Ibid., 39, no. 15 (1902): 656.

80. *New York Times*, Mar. 4, 1902, 3, and Mar. 30, 1902, 2.

81. *Electrician* 48, no. 22 (1902): 848.

82. *Electrical World* 40, no. 16 (1902): 629.

83. "The Politics of Radio Telegraphy," *Edinburgh Review* 207 (Apr. 1908): 466.

84. Marconi to H. Cuthbert Hall, Oct. 1, 1902, GMB Private Collection.

85. *Electrical World* 42, no. 21 (1903): 836; John Waterbury, "Wireless Telegraphy Conference," *North American Review* 177 (Nov. 1903): 655–66. For further information on the conferences see John D. Tomlinson, *The International Control of Radio Communications* (Ann Arbor, Mich.: J. W. Edwards, 1945), 11–45.

86. *Electrical World* 42, no. 11 (1903): 423.

87. Delegates to the International Wireless Conference to John Hay, Secretary of State, Aug. 14, 1903, NA, box 88.
88. *Electrical World* 46, no. 3 (1905): 86, and 43, no. 17 (1904): 759; *New York Times,* July 10, 1904, pt. 5, 26.
89. *Electrical World* 46, no. 3 (1905): 86.
90. Archer, *History of Radio,* 75.
91. *Electrical World* 44, no. 1 (1904): 25; *New York Times,* June 26, 1904, 1.
92. *New York Tribune,* May 29, 1904, 6; *New York Times,* May 29, 1904, 3.
93. Chief, Bureau of Equipment, to Wireless Telegraph Board, Apr. 6, 1904, NA, box 84; W.H.G. Bullard, "The Naval Radio Service," *Proc. Inst. Radio Engineers* 3, no. 1 (1915): 9–10.
94. *Electrical World* 44, no. 8 (1904): 284, and 44, no. 7 (1904): 241.
95. *New York Times,* Sept. 10, 1906, 6.
96. *New York Tribune,* July 6, 1904, 6.
97. *Electrical World* 43, no. 23 (1904): 1068; 44, no. 9 (1904): 319.
98. Ibid., 43, no. 24 (1904); 48, no. 10 (1906): 470.
99. *New York Times,* Sept. 10, 1906, 6.
100. NESCO press release, Oct. 17, 1904, Clark Collection.
101. H. J. Glaubitz, letter to the editor, *Electrical World* 43, no. 26 (1904): 1201.
102. Ibid.
103. Marconi Wireless Telegraph Company of America to Chief, Bureau of Equipment, May 9, 1904, NA, box 89.
104. Ibid., May 2, 1904, NA, box 89.
105. Chief, Bureau of Equipment, to Marconi Wireless Telegraph Company of America, May 10, 1904, NA, box 89.
106. Comdr. F. M. Barber to Chief, Bureau of Equipment, July 25, 1903, NA, box 90. Barber was able to get Slaby-Arco down to $1,077.50 a set.
107. Ibid., July 8, 1904, NA, box 12.
108. Chief, Bureau of Equipment, to Commandant, Navy Yard, New York, Nov. 25, 1903, NA, box 88; Comdr. F. M. Barber to Chief, Bureau of Equipment, July 10, 1904, NA, box 12.
109. James Boyle to Fessenden, Sept. 17, 1904, Clark Collection.
110. Comdr. F. M. Barber to Chief, Bureau of Equipment, Feb. 20, 1906, NA, box 89.
111. George H. Clark, "The Life and Creations of John Stone Stone" (1946), 92–94, Clark Collection.
112. Fessenden to Charles Bonaparte, Secretary of the Navy, May 5, 1906; Fessenden to President Theodore Roosevelt, May 14, 1906; NESCO to Adm. Henry N. Manney, June 14, 1905, Clark Collection.
113. Fessenden to Charles Bonaparte, Secretary of the Navy, May 5, 1906, Clark Collection.
114. Secretary of the Navy to Secretary of Agriculture, Sept. 15, 1903, NA, box 83.
115. Charles Bonaparte, Secretary of the Navy, to Fessenden, Apr. 19, 1906, Clark Collection.
116. NESCO *vs.* De Forest Wireless Telegraph Company et al., 1906, Clark Collection.

117. Fessenden to President Theodore Roosevelt, May 14, 1906, Clark Collection.

118. Comdr. F. M. Barber to Chief, Bureau of Equipment, June 23, 1904, NA, box 12.

119. A. W. Greely, Chief, Signal Officer, U.S. Army, to Hay Walker, Jr., and Thomas Given, Nov. 9, 1905, Clark Collection.

120. Comdr. F. M. Barber to Chief, Bureau of Equipment, Sept. 14, 1903, NA, box 89.

121. Hay Walker, Jr., to Fessenden, Mar. 29, 1905, NC Papers.

122. Fessenden to Lt. Comdr. Cleland Davis, Nov. 30, 1906, Clark Collection.

123. Chief, Bureau of Equipment, to Comdr. F. M. Barber, Aug. 15, 1902, NA, box 85.

124. Hay Walker, Jr., to James Hayden, Nov. 15, 1905, NC Papers.

125. Information on the contract for these stations is from the *New York Times,* July 10, 1904, pt. 5, 26; H. W. Young to Hay Walker, Jr., June 29, 1904; NESCO memo, "In Regard to the West India Wireless Contract," May 7, 1906, Clark Collection.

126. Chief, Bureau of Equipment, to Comdr. F. M. Barber, July 9, 1904, NA, box 12.

127. Comdr. F. M. Barber to Chief, Bureau of Equipment, Nov. 29, 1904, NA, box 89.

128. Lee De Forest, *Father of Radio* (Chicago: Wilcox & Follett, 1950), 178–81; Frank E. Butler, "How Wireless Came to Cuba," *Radio Broadcast,* Nov. 1924– Apr. 1925, 916–20.

129. De Forest to Frank Butler, Apr. 20, 1906; De Forest to Francis X. Butler, Oct. 14, 1905, Clark Collection.

130. Lee De Forest, Diary, Aug. 1905, Library of Congress.

131. Clark, "Life and Creations of John Stone Stone," 116.

132. Capt. C. H. Arnold, President, Wireless Telegraph Board, to Chief, Bureau of Equipment, July 10, 1903, NA, box 85.

133. Howeth, *History of Communications,* 65.

134. George H. Clark, "Radio in the U.S. Navy," Clark Collection.

135. Howeth, *History of Communications,* 65.

136. Lt. J. M. Hudgins to Secretary of the Navy, Feb. 15, 1904, NA, box 83.

137. Chief, Bureau of Equipment, to Commander in Chief, North Atlantic Fleet, Mar. 22, 1905, NA, box 32; F. S. Doane, Master of Light-Vessel no. 85, to Capt. W. G. Cutler, Inspector, July 20, 1909, NA, box 79; W. G. Cutler to Lighthouse Board, July 21, 1909, NA, box 79; Susan J. Douglas, "Exploring Pathways in the Ether" (Ph.D. diss., Brown University, 1979), 306–12.

138. Paullin, *Paullin's History,* 406. *Annual Report of the Secretary of the Navy,* 1883, 17, 107; 1884, 16–19.

139. Commandant, Navy Yard, Washington, D.C., to Chief, Bureau of Equipment, Mar. 9, 1909, NA, box 76.

140. Clark, "Radio in War and Peace," 316–17.

141. Ibid., 317.

142. Chief, Bureau of Equipment, to Commandant, Navy Yard, Washington, D.C., June 5, 1907, NA, box 13.

143. S. M. Kintner, General Manager, NESCO, to General Storekeeper, Navy Yard, Brooklyn, Oct. 5, 1911, NA, box 13.

144. John Firth, Wireless Specialty Apparatus Company, to Chief, Bureau of Equipment, Nov. 9, 1909, NA, box 82; see also complaints from William Walker, Massachusetts Wireless Equipment Company, to Chief, Bureau of Equipment,

Dec. 31, 1908, NA, box 75; and Douglas, "Exploring Pathways in the Ether," 159–60.

145. Commandant, Navy Yard, Mare Island, to Chief, Bureau of Equipment, Apr. 16, 1904, NA, box 83.

146. Lt. E. H. Dodd, Mare Island, to Chief, Bureau of Equipment, Apr. 6, 1910, NA, box 81.

147. W. L. Howard, Inspection Officer, to Commandant, U.S. Navy Yard, Philadelphia, June 16, 1910, NA, box 80; C. D. Mills, Chief Electrician, Tatoosh Island, Wash., to Inspector, Navy Yard, Puget Sound, Jan. 19, 1910, NA, box 80; Inspector of Equipment to Commandant, Navy Yard, Norfolk, June 29, 1909, NA, box 80; Chief Electrician, Navy Yard, to Inspector of Equipment, Navy Yard, Charleston, S.C., June 10, 1909, NA, box 78.

148. *New York Times,* Sept. 27, 1906, 7.

149. *Electrical World* 46, no. 5 (1905): 165; *New York Times,* Sept. 10, 1906, 6.

150. *New York Times,* Nov. 19, 1904, 1; *New York Herald,* July 14, 1904.

151. James R. Sheffield, Attorney for Marconi Wireless Telegraph Company of America, to President Theodore Roosevelt, July 19, 1904, Hooper Papers, Library of Congress, Washington, D.C.

152. Comdr. F. M. Barber to Chief, Bureau of Equipment, June 1, 1906, NA, box 90.

153. *Electrical World* 48, no. 19 (1906): 904.

154. *New York Times,* Nov. 4, 1906, 7.

155. Ibid., Nov. 3, 1907, pt. 5, 1.

156. *Electrical World* 49, no. 2 (1907): 83. Details on the conference are in *International Wireless Convention* (Washington, D.C.: Government Printing Office, 1907), Clark Collection.

157. Comdr. F. M. Barber to Chief, Bureau of Equipment, Nov. 9, 1906, NA, box 89.

158. *International Wireless Convention.*

159. "The Politics of Radio-Telegraphy," *Edinburgh Review* 207, no. 424 (1908): 466.

160. *New York Times,* Jan. 16, 1906, 8.

161. "Politics of Radio-Telegraphy," 473.

162. "President Roosevelt and Emperor William at Odds," *Current Literature* 42 (Apr. 1907): 372; "Dictator of Europe," *Harper's Weekly* 49 (July 8, 1905): 974; "Chief of Police of Europe," *Review of Reviews* 32 (July 1905): 93.

163. W. Van Schierbrand, "Is Kaiser Wilhelm II of Normal Mind?" *Lippincott's* 78 (Nov. 1906): 619–25.

FIVE: INVENTORS AS ENTREPRENEURS

1. Wyn Wachhorst, *Thomas Alva Edison: An American Myth* (Cambridge: MIT Press, 1981), 40–42.

2. Ray Stannard Baker, "Marconi's Achievement," *McClure's Magazine* 18 (Feb. 1902): 291.

3. *Electrical World* 42, no. 11 (1903): 468.

4. *Electrician* 48, no. 13 (1902): 501.

5. *New York Times,* Oct. 17, 1907, 1; *Electrical World* 50, no. 17 (1907): 794.

6. *New York Times,* Sept. 26, 1907, 8, and Oct. 18, 1907, 10.

7. Ibid., Oct. 17, 1907, 1.
8. *New York Tribune,* Oct. 18, 1907, 6.
9. *Outlook* 87 (Oct. 26, 1907): 372.
10. "Transatlantic Marconigrams Now and Hereafter," *World's Work* 15 (Dec. 1907): 9624.
11. Reginald Fessenden, "Regular Wireless Service between America and Europe," *Scientific American Supplement* 64, no. 1663 (1907): 319. *Electrician* 60, no. 6 (1907): 200; 60, no. 3 (1907): 77, 100; 60, no. 8 (1907): 273.
12. *Electrical World* 49, no. 26 (1907): 1315.
13. *Review of Reviews* 36 (Dec. 1907): 647.
14. Ibid., 37 (Feb. 1908): 131.
15. "The Lessons of the Recent Wall Street Panic," *World's Work* 14 (May 1907): 8829.
16. *New York Times,* Oct. 23, 1907, 1; William Justus Boies, "Trust Companies and the Panic," *Review of Reviews* 36 (Dec. 1907): 680.
17. Boies, "Trust Companies," 680.
18. *Review of Reviews* 37 (Feb. 1908): 131.
19. Robert Sobel, *Panics on Wall Street* (New York: Macmillan, 1968), 305–21.
20. Frank Fayant, "The Wireless Telegraph Bubble," *Success Magazine* 10 (June 1907): 387.
21. "Transatlantic Marconigrams," 9626.
22. George H. Clark, "The Life and Creations of John Stone Stone" (1946), 145–46, Clark Collection.
23. H. W. Southworth to Lawrence Sherman, Trustee for Stone Telegraph and Telephone, Nov. 29, 1909, Clark Collection.
24. Helen Fessenden, *Fessenden: Builder of Tomorrows* (New York: Coward-McCann, 1940), 126.
25. Fessenden to Hay Walker, Jr., Jan. 24 and Feb. 11, 1905, NC Papers.
26. Ibid., July 7, 1905, NC Papers.
27. Fessenden, *Fessenden,* 126.
28. Fessenden to Hay Walker, Jr., Dec. 14, 1904, NC Papers.
29. Fessenden, *Fessenden,* 141.
30. Hay Walker, Jr., to Fessenden, Aug. 10, 1905, NC Papers.
31. Fessenden, *Fessenden,* 141.
32. Fessenden to Hay Walker, Jr., Dec. 22, 1905, NC Papers.
33. Fessenden to H. W. Fisher, Apr. 5, 1906; Fessenden to Charles Bright, June 11, 1906, Clark Collection.
34. S. M. Kintner, "Pittsburgh's Contribution to Radio," *Proc. Inst. Radio Engineers* 20, no. 12 (1932): 1853; Fessenden to Clarence Feldman, Delft University, Holland, 1906, Clark Collection.
35. Fessenden to Clarence Feldman, Jan. 11, 1906, Clark Collection.
36. Fessenden to *Scientific American,* Jan. 8. 1907.
37. Ernst Alexanderson, Reminiscences, 16, 18, Columbia Oral History Library; Alexanderson to Fessenden, June 18, 1906, Clark Collection.
38. Alexanderson, Reminiscences, 17–18; Gleason Archer, *The History of Radio to 1926* (New York: American Historical Society, 1938), 69, 115–18.

39. Fessenden to A. Dempster at G.E., Sept. 28, 1906, Clark Collection.
40. Fessenden to H. G. Reist at G.E., Nov. 6, 1906, Clark Collection; Alexanderson, Reminiscences, 18–19.
41. Transcript of telephone conversation between A. E. Kennelly and Fessenden, Oct. 18, 1906, Clark Collection.
42. Fessenden to J. N. Aylesworth, East Orange, N.J., Aug. 28, 1906, Clark Collection.
43. Fessenden, *Fessenden,* 153–54.
44. Fessenden to Clarence Feldman, Sept. 21, 1906, and June 11, 1907, Clark Collection.
45. Fessenden to Comdr. Cleland Davis, Nov. 30, 1906, Clark Collection.
46. Hay Walker, Jr., to Fessenden, June 17, 1905, NC Papers.
47. Fessenden to Bell Telephone, Dec. 12, 1906, Clark Collection.
48. Memo, Apr. 2, 1907, cited in W. Rupert Maclaurin, *Invention and Innovation in the Radio Industry* (New York: Macmillan, 1949), 65.
49. John Brooks, *Telephone: The First Hundred Years* (New York: Harper & Row, 1976), 122–24.
50. Leonard S. Reich, *The Making of American Industrial Research: Science and Business at G.E. and Bell, 1876–1926* (Cambridge: Cambridge University Press, 1985), chap. 7.
51. Maclaurin, *Invention and Innovation,* 66.
52. Fessenden to Thomas Given, June 26, 1907, NC Papers.
53. Thomas Given to Fessenden, June 28, 1907, NC Papers.
54. Hay Walker, Jr., to Fessenden, Oct. 25, 1907, NC Papers.
55. Fessenden to Franklin Reed, Jan. 2, 1908; Fessenden to Cleland Davis, July 16, 1908, Clark Collection.
56. For example, Charles D. Guthrie, Chief Operator, USS *Kentucky,* Guantanamo Bay, Cuba, to Fessenden, Jan. 18, 1907, Clark Collection.
57. R. E. Thompson to George Clark, Recollections of John Firth, 1941; Frank A. Hinners to George Clark, Recollections of John Firth, June 9, 1941, Clark Collection.
58. Fessenden, *Fessenden,* 162.
59. Frank A. Hinners to George Clark, June 9, 1941, in Firth, Reminiscences.
60. Fessenden, *Fessenden,* 163.
61. Hay Walker, Jr., and Fessenden to Col. John Firth, Sept. 12, 1908, NC Papers.
62. Hay Walker, Jr., to Fessenden, Aug. 25, 1908, NC Papers.
63. Fessenden to Hay Walker, Jr., Sept. 24, 1908, NC Papers.
64. From Stokes's obituary, *New York Times,* May 20, 1926, 25.
65. Fessenden to Hay Walker, Jr., Oct. 19, 1908, NC Papers.
66. Fessenden to Clarence Feldman, Feb. 9, 1909; B. A. MacNab to Fessenden, May 12, 1909; Fessenden to MacNab, June 2, 1909, all in Clark Collection. P. Schubert, *The Electric Word* (New York: Macmillan, 1928), 79–80.
67. Fessenden to T. N. Vail, Feb. 25, 1910, Clark Collection.
68. Cited in Maclaurin, *Invention and Innovation,* 64, 66.
69. Fessenden, *Fessenden,* 171.
70. Bent to Fessenden, Aug. 22, 1910, NC Papers.

71. Alexanderson, Reminiscences, 16–17.
72. George Clark to D. W. Todd, Nov. 24, 1910, Clark Collection.
73. Unsigned letter to Hay Walker, Jr., May 5, 1911, NC Papers.
74. Hay Walker, Jr., to Fessenden, Dec. 10, 1910, NC Papers.
75. Francis Clay to Fessenden, Jan. 7, 1911, NC Papers.
76. Fessenden, *Fessenden*, 183–87; and Brant Rock staff to Hay Walker, Jr., Dec. 30, 1910, NC Papers.
77. Fessenden to Francis Clay, Nov. 29, 1910, NC Papers.
78. Fessenden, *Fessenden*, 191.
79. Lee De Forest, Diary, Sept. 30, 1906, Library of Congress.
80. Marconi to H. Cuthbert Hall, July 11, 1904, GMB Private Collection.
81. Samuel Lubell, "Magnificent Failure," *Saturday Evening Post*, Jan. 24, 1942, 21; De Forest to Frank Butler, Oct. 23, 1906, Clark Collection.
82. De Forest, Diary, Sept. 30, 1906.
83. Lubell, "Magnificent Failure," 36.
84. Maclaurin, *Invention and Innovation*, 46–48.
85. J. A. Fleming, letter to the editor, *Electrical World* 48, no. 23 (1906): 1117.
86. *Electrical World* 48, no. 18 (1906): 836.
87. Lubell, "Magnificent Failure," 36.
88. Lloyd Espenschied et al., "Discussion of 'A History of Some Foundations of Modern Radio-Electronic Technology,'" *Proc. Inst. Radio Engineers* 47, no. 7 (1959): 1263.
89. Donald G. Little, Reminiscences, 6, Columbia Oral History Library.
90. Lloyd Espenschied, Reminiscences, 6–8, Columbia Oral History Library.
91. De Forest interviewed by Gordon Greb (1959), 6, Columbia Oral History Library.
92. George G. Blake, *History of Radio Telegraphy and Telephony* (London: Chapman & Hall, 1928), 159; Arthur H. Morse, *Radio: Beam and Broadcast* (London: Ernest Benn, 1925), 30–31, 75 ff.
93. Morse, *Radio*, 75; Blake, *History of Radio*, 185–86; George W. Pierce, *Principles of Wireless Telegraphy* (New York: McGraw-Hill, 1910).
94. De Forest to Frank Butler, Apr. 22, 1907, Clark Collection.
95. *New York Times*, Feb. 14, 1909, 1.
96. Herbert T. Wade, "Wireless Telegraphy by the De Forest System," *Review of Reviews* 35, no. 16 (1907): 685.
97. *Electrical World* 57, no. 15 (1911): 923.
98. *New York Times*, Dec. 17, 1906, 10.
99. Ibid., Jan. 23, 1910, sec. 5, 6.
100. *Electrical World* 55, no. 2 (1910): 86.
101. Erik Barnouw, *A Tower in Babel* (New York: Oxford University Press, 1966), 26.
102. Lubell, "Magnificent Failure," 36.
103. Ibid., 38; De Forest to Frank Butler, Mar. 2, 1908, Clark Collection; *Electrical World* 51, no. 18 (1908): 898; Lee De Forest, "Progress in Radio Telephony," *Electrical World* 53, no. 1 (1909): 13.
104. Lubell, "Magnificent Failure," 38; De Forest to George Clark, Jan. 17, 1933, Clark Collection.

105. De Forest to Frank Butler, July 2, 1907, Clark Collection; Lubell, "Magnificent Failure," 38.

106. *New York Times,* Jan. 14, 1910, 2; *Electrical World* 55, no. 2 (1910): 98.

107. Lubell, "Magnificent Failure," 38.

108. "Warns Wives of Careers," *New York Times,* July 28, 1911, 18.

109. Lubell, "Magnificent Failure," 21.

110. "The Ownership of Wireless Equipment," *Wireless Age* 2, no. 10 (1915): 719; Thomas E. Clark, Reminiscences, 13, 33, Columbia Oral History Library.

111. James L. Charlton, "Wireless Telegraphy on the Atlantic Coast of the United States," *Electrical World* 59, no. 4 (1912): 197.

112. *Annual Report of the Commissioner of Navigation* (Washington, D.C.: Government Printing Office, 1911), app. M, 202; John Firth to S. M. Kintner, May 8, 1911, Clark Collection.

113. Clark, Reminiscences, 13.

114. *Annual Report of the Commissioner of Navigation* (1911), app. M, 206; (1912), 38. C. A. Butlin to Capt. Simpson, Marconi Wireless, Oct. 25, 1911, Clark Collection.

115. *Collier's,* May 7, 1910; *Electrical World* 56, no. 10 (1910): 547.

116. Marconi as cited in *Electrical World* 57, no. 25 (1911): 1592.

117. See Hugh Aitken's excellent description of the development of the disc discharger in *Syntony and Spark: The Origins of Radio* (New York: John Wiley & Sons, 1976), 276–80.

118. Ibid., 277.

119. *Electrical World* 57, no. 25 (1911): 1592; Donald McNicol, *Radio's Conquest of Space* (New York: Murray Hill Books, 1946), 102–3; W. J. Baker, *A History of the Marconi Company* (London: Methuen, 1970), 117–20.

120. Robert Merriam has a working disc discharger on display at his New England Wireless Museum in East Greenwich, R.I.

121. W. J. Baker, *History of the Marconi Company,* 112; *New York Times,* Mar. 29, 1908, pt. 3, 2.

122. Admiralty to the Marconi Company, cited in W. P. Jolly, *Marconi* (New York: Stein & Day, 1972), 157–58.

123. Aitken, *Syntony and Spark,* 261–82; Baker, *History of the Marconi Company,* 120; *New York Times,* May 9, 1910, 8.

124. Cited in Keith Geddes, *Guglielmo Marconi: 1874–1937* (London: Her Majesty's Stationery Office, 1974), 27.

125. Jolly, *Marconi,* 164–67.

126. John Bottomley to H. Cuthbert Hall, Jan. 10 and Jan. 17, 1908, Chelmsford Collection.

127. Ibid., Jan. 10, 1908; H. Cuthbert Hall to John Bottomley, Jan. 19, 1908, Chelmsford Collection.

128. John Bottomley to Marconi, Jan. 17, 1908, Chelmsford Collection.

129. Thorn Mayes, "A Brief History of the Marconi Wireless Telegraph Company of America, 1899–1919," pamphlet in the Chelmsford Collection.

130. *New York Times,* Feb. 2, 1912, 1; Baker, *History of the Marconi Company,* 137.

131. Marconi to Beatrice O'Brien Marconi, Dec. 13, 1908, GMB Private Collection.

132. *New York Times,* Jan. 15, 1911; Baker, *History of the Marconi Company,* 133.

133. Geddes, *Guglielmo Marconi*, 28; *New York Times*, Feb. 22, 1911, 4, and Nov. 4, 1911, 7.
134. *New York Times*, Feb. 22, 1911, 4.
135. Ibid., June 16, 1910, 1.
136. Ibid.
137. Ibid., May 30, 1911, 1.
138. John L. Hogan to S. M. Kintner of NESCO, May 10, 1911, Clark Collection.
139. John Bottomley, *Annual Report*, Marconi Wireless Telegraph Company of America, 1912, Chelmsford Collection.
140. Marconi to Beatrice O'Brien Marconi, Mar. 22, 1912, GMB Private Collection.
141. Ibid., Apr. 29, 1912, GMB Private Collection.

SIX: POPULAR CULTURE AND POPULIST TECHNOLOGY

1. *New York Times*, Nov. 3, 1907, pt. 5, 1.
2. Charles Barnard, "A Young Expert in Wireless Telegraphy," *St. Nicholas* 35 (April 1, 1908): 530–32.
3. John Higham, "The Reorientation of American Culture in the 1890s," in *The Origins of Modern Consciousness*, ed. John Weiss (Detroit: Wayne State University Press, 1965), 26–28.
4. See Roderick Nash's introduction in *The Call of the Wild*, ed. Nash (New York: George Braziller, 1970), 2–4.
5. Ernest Thompson Seton, *Boy Scouts of America: A Handbook of Woodcraft, Scouting, and Life-craft*, excerpted ibid., 20.
6. E. Anthony Rotundo, "Learning about Manhood: Gender Ideals and the Middle-Class Family in Nineteenth-Century America" (Paper delivered at the Smithsonian-Smith Conference on the Conventions of Gender, Feb. 16–17, 1984), 11.
7. Theodore P. Greene, *America's Heroes* (New York: Oxford University Press, 1970), 70.
8. Seton, *Boy Scouts of America*, 20.
9. E. F. Bleiler, ed., *Eight Dime Novels* (New York: Dover, 1974).
10. Ibid., 67, 65, 126.
11. Greene, *America's Heroes*, 113, 128.
12. Russel Nye, *The Unembarrassed Muse* (New York: Dial, 1950), 169.
13. See John Kasson, *Amusing the Million* (New York: Hill & Wang, 1978).
14. Robert A. Morton, "The Amateur Wireless Operator," *Outlook* 94 (Jan. 15, 1910): 131.
15. *Electrical World* 51, no. 9 (1908): 423; 54, no. 24 (1909): 1401.
16. For information on crystal receivers, see "History of the Wireless Specialty Apparatus Company," dictated by G. W. Pickard to G. H. Clark, 1931, Clark Collection; *Electrical World* 48, no. 23 (1906): 1100; A. P. Morgan, *Wireless Telegraphy and Telephony* (New York: Norman W. Henley, 1912), 52, 57; Elmer E. Bucher, *Practical Wireless Telegraphy* (New York: Wireless Press, 1917), 132; De Forest to *Electrical World* 48, no. 10 (1906): 491; WSA to Director of Naval Intelligence, London, July 10, 1908, Clark Collection, *Electrical World* 48, no. 21 (1906): 994.
17. Allen Chapman, *The Radio Boys' First Wireless* (New York: Grosset & Dunlap, 1922), 63.

18. Edgar S. Love, Reminiscences, 2, Columbia Oral History Library.

19. W. V. Albert, Chief Electrician, USN, to Commandant, Navy Yard, Boston, Feb. 26, 1908, NA, box 76. Hugh G. J. Aitken provided the information on the Model T. Coils.

20. Donald G. Little, Reminiscences, 5, Columbia Oral History Library, *New York Times,* May 31, 1909, 1; E. L. Bragdon, Reminiscences, 4, Columbia Oral History Library; Love, Reminiscences, 2.

21. Clinton B. De Soto, *Two Hundred Meters and Down: The Story of Amateur Radio* (West Hartford, Conn.: American Radio Relay League, 1936), 3.

22. *Electrical World* 57, no. 13 (1911): 760; W. B. English, Jr., United Wireless, Boston, to the Bureau of Equipment, Jan. 5, 1910, NA, box 80; *New York Times,* Mar. 29, 1912, 12.

23. Francis A. Collins, *The Wireless Man* (New York: Century, 1912), 29.

24. Francis A. Collins, "An Evening at the Wireless Station," *St. Nicholas* 39 (Oct. 1912): 1110.

25. Ibid., 1111.

26. Victor Appleton, *Tom Swift and His Wireless Message* (New York: Grosset & Dunlap, 1911), 179–96.

27. A. R. Carman, "In Marconiland," *Canadian Magazine* 32 (Mar. 1909): 426–33; George A. England, "Wooed by Wireless," *Cosmopolitan* 44 (Apr. 1908): 497–501; J. F. Wilson, "Sparks," *McClure's Magazine* 37 (June 1911): 149–54.

28. *New York Times,* Aug. 20, 1967, 88.

29. Edgar Felix, Reminiscences, Columbia Oral History Library.

30. Hugo Gernsback, letter to the editor, *New York Times,* Mar. 29, 1912, 12.

31. *Harper's Weekly* 53 (Jan. 30, 1909): 1.

32. Ibid.

33. *New York Times,* Jan. 26, 1909, 1.

34. "A Wireless Victory," *Harper's Weekly* 53 (Jan. 30, 1909): 7.

35. Arthur D. Howden Smith, "Men of the Wireless," *Putnam's* 6 (Apr. 1909): 73–74.

36. "How Binns Flashed His Calls for Help," *New York Times,* Jan. 26, 1909, 4.

37. Ibid.

38. *Current Literature* 46 (March 1909): 248.

39. *New York Times,* Feb. 4, 1909, 8.

40. Smith, "Men of the Wireless," 76.

41. *New York Times,* Jan. 27, 1909, 1.

42. A. Henry Savage Landor, "The Latest Drama of the Sea," *Harper's Weekly* 53 (Feb. 6, 1909): 7.

43. *New York Times,* Feb. 6, 1909, 8.

44. Smith, "Men of the Wireless," 78.

45. *New York Times,* Feb. 8, 1909, 8.

46. Smith, "Men of the Wireless," 78.

47. Ibid., 77.

48. Robert Sloss, "Binns and 'The Men of the Broken Ear,'" *Harper's Weekly* 53 (Feb. 13, 1909): 15.

49. Lloyd Jacquet, "The Heritage of the Radio Club of America," *Fiftieth Anniversary Golden Yearbook* (New York: RCA, 1959), 4.

50. Felix, Reminiscences.
51. Collins, *Wireless Man,* 29–31.
52. Felix, Reminiscences.
53. Collins, "Evening at the Wireless Station," 1113.
54. Love, Reminiscences, 5; Bragdon, Reminiscences.
55. Jacquet, "Heritage of the Radio Club."
56. Collins, *Wireless Man,* 26.
57. De Soto, *Two Hundred Meters,* 23–24; Felix, Reminiscences; "The Birth of ARRL," *Fifty Years of ARRL,* Hartford, Conn., 9 (copy at the New England Wireless and Steam Museum, East Greenwich, R.I.); *New York Times,* Jan. 31, 1909, 18, and Mar. 29, 1912, 12.
58. *New York Times,* Jan. 31, 1909, 18, and Mar. 29, 1912, 12.
59. *Electrical World* 56, no. 3 (1910): 139; Collins, *Wireless Man,* 42–47.
60. De Soto, *Two Hundred Meters,* 37–41.
61. Ibid., 40.
62. Ibid., 3, 28; *New York Times,* Jan. 31, 1909, 18.
63. *New York Times,* Jan. 28, 1910, 8; Morton, "Amateur Wireless Operator," 131.
64. De Soto, *Two Hundred Meters,* 23–24.
65. Collins, "An Evening at the Wireless Station," 1113.
66. Bragdon, Reminiscences, 6.
67. Francis Hart, logbook, entry for Nov. 23, 1907, Clark Collection.
68. *New York Times,* Jan. 28, 1910, 8, and Jan. 30, 1910, 4; *Electrical World* 57, no. 13 (1911): 760; Morton, "Amateur Wireless Operator."
69. *Electrical World* 55, no. 10 (1910): 610.
70. Benjamin Wolf, Chief Electrician, U.S. Navy Yard, New York, to B. F. Walling, Capt. of Yard, Jan. 10, 1910, NA, box 82.
71. Hart logbook, entries for Jan. 9, Sept. 18, and Sept. 21, 1909.
72. *Electrical World* 56, no. 3 (1910): 139.
73. *Scientific American* 106, no. 12 (1912): cover; *New York Times,* Mar 31, 1912, 14, and Apr. 21, 1912, pt. 5, 2.
74. *Annual Report of the Navy Department* (Washington, D.C.: Government Printing Office, 1909), 280.
75. Morton, "Amateur Wireless Operator," 132–33.
76. Incident cited in *New York Times,* Feb. 1, 1910, 8; Edwin L. Powell, letter to the editor, *Scientific American* 106, no. 25 (1912): 563.
77. John V. Purssell, letter to the editor, *Scientific American* 106, no. 23 (1912): 515.
78. Powell, letter to the editor.
79. F. L. Coombs, letter to the editor, *New York Times,* Apr. 21, 1912, 12.
80. Purssell, letter to the editor.
81. H. E. Rowson, letter to the editor, *Scientific American* 106, no. 23 (1912): 515.
82. *New York Times,* Apr. 29, 1910, 18.
83. *Electrical World* 67, no. 9 (1906): 437.
84. Lt. S. C. Hooper to Chief, Bureau of Steam Engineering, "Report on Radio-Telegraphy in the Atlantic Fleet, Spring Target Practice, 1912," Hooper Papers, Library of Congress.
85. S. C. Hooper, "Navy History: Radio, Radar, Sonar," transcript of recordings,

Office of Naval History. Washington, D.C., cited in L. S. Howeth, *History of Communications-Electronics in the United States Navy* (Washington, D.C.: Government Printing Office, 1963), 193.

SEVEN: THE *TITANIC* DISASTER AND THE FIRST RADIO REGULATION

1. Henry Cabot Lodge to Elihu Thomson, Feb. 3, 1908, Clark Collection.
2. *New York Times,* Jan. 30, 1909, 8.
3. Ibid., Feb. 4, 1912, 12.
4. "Hearings Before a Subcommittee of the Committee on Naval Affairs of the House of Representatives on H. J. Resolution 95," 1910, cited in L. S. Howeth, *History of Communications-Electronics in the United States Navy* (Washington, D.C.: Government Printing Office, 1963), 156.
5. Ibid.
6. For a general discussion of overexploitation of common property resources, see Garrett Hardin's "The Tragedy of the Commons" in *Economic Foundations of Property Law,* ed. Bruce A. Ackerman (Boston: Little, Brown, 1975), 4.
7. House Reports, 60th Cong. 2d sess., Dec. 7, 1908–March 4, 1909, vol. 1, report 2086, 2.
8. Senate Documents, 60th Cong., 2d sess., 1908–9, vol. 21, doc. 700.
9. Public Law 262, 61st Cong.; see Frank J. Kahn, *Documents of American Broadcasting* (New York: Appleton-Century-Crofts, 1968), 6–7; House Reports, 61st Cong., 1st and 2d sess., 1909–10, vol. 3, report 1373.
10. House Reports, 61st Cong., 2d sess., 1909–10, vol. 2, 1910, report 892, 6.
11. Ibid., 6, 8, letter from P. Schwarzhaupt.
12. House Reports, 61st Cong., 2d sess., 1909–10, vol. 2, 1910, report 924, 2; *New York Times,* Mar. 29, 1910, 6.
13. House Reports, 61st Cong., 1st and 2d sess., 1909–10, vol. 3, 1910, report 1373.
14. NESCO to Senator P. C. Knox, Feb. 1, 1908, Clark Collection.
15. House Reports, 61st Cong., 2d sess., 1909–10, vol. 2, 1910, report 924, 2.
16. *New York Times,* Mar. 31, 1910, 10.
17. Ibid., Apr. 29, 1910, 18; Clinton B. De Soto, *Two Hundred Meters and Down: The Story of Amateur Radio* (West Hartford, Conn.: American Radio Relay League, 1936), 32.
18. Hugo Gernsback, "The Old EIC Company Days," *Radio-Craft,* March 1938, 635.
19. George E. Burghead, "A History of the Radio Club of America, Inc.," *Fiftieth Anniversary Golden Yearbook* (New York: RCA, 1959), 17.
20. *Electrical World* 51, no. 12 (1908): 590.
21. *New York Times,* Mar. 23, 1912, 12.
22. Ibid., Mar. 31, 1910, 10. For use of the term *czar,* see House Reports, 61st Cong., 2d sess., 1909–10, vol. 2, 1910, report 924.
23. Robert A. Morton, Jr., "Curbing the Wireless Meddler," *Scientific American* 106, no. 12 (1912): 266.
24. *New York Times,* Mar. 23, 1912, 12.
25. *Electrical World* 51, no. 12 (1908): 594.
26. *Scientific American* 106, no. 13 (1912): 282.

27. *New York Times,* Mar. 8, 1910, 5.
28. House Reports, 61st Cong., 2d sess., 1909–10, vol. 2, 1910, report 924, 2.
29. Ibid., 7.
30. Ibid.
31. Senate Reports. 62d Cong., 2d sess., Dec. 4, 1911–Aug. 26, 1912, vol. 2, report 698.
32. House Reports, 61st Cong., 2d sess., 1909–10, Vol. 2, 1910, report 892; *Electrical World* 55, no. 13 (1910): 786.
33. T. J. Styles, letter to the editor, *New York Times,* Jan. 31, 1910, 8.
34. De Soto, *Two Hundred Meters,* 30.
35. Senate Reports, 62d Cong., 2d sess., Dec. 4, 1911–Aug. 26, 1912, vol. 2, report 698.
36. House Reports, 61st Cong., 2d sess., 1909–10, vol. 2, 1910, report 892.
37. Information on the *Titanic* from the *New York Times,* Apr. 10–Apr. 23, 1912; and "*Titanic* Disaster: Hearings Before a Subcommittee of the Committee on Commerce, United States Senate," Senate Reports, 62d Cong. 2d sess., Dec. 4, 1911–Aug. 26, 1912, vol. 28, doc. 726.
38. *New York Times,* Apr. 15, 1912, 2.
39. Ibid., Apr. 16, 1912, 1, and Apr. 19, 1912, 2.
40. Ibid., Apr. 24, 1912, 1; *Scientific American* 106, no. 23 (1912): 510.
41. Marconi to Beatrice O'Brien Marconi, Apr. 16, 1912, GMB Private Collection.
42. *New York Times,* Apr. 17, 1912, 4, and Apr. 18, 1912, 1.
43. *Electrical World* 59, no. 17 (1912): 880.
44. *Literary Digest* 49, no. 1 (1914): 16; *New York Times,* Apr. 18, 1912, 1.
45. *New York Times,* Apr. 21, 1912, 1.
46. *Scientific American* 106, no. 25 (1912): 563.
47. *New York Times,* Apr. 19, 1912, 1, 10.
48. Ibid., Apr. 20, 1912, 8.
49. Ibid., Apr. 18, 1912, 15.
50. Marconi to Beatrice O'Brien Marconi, Apr. 16, 1912, GMB Private Collection.
51. *New York Times,* Apr. 17, 1912, 12.
52. Quotation from the *Chicago Record-Herald* cited in "The Loss of the *Titanic:* A Poll of the Press," *Outlook* 101 (May 4, 1912): 22a.
53. *Electrical World* 59, no. 17 (1912): 880.
54. "The *Titanic* and Its Heroes," *World's Work* 24 (June 1912): 143.
55. For an excellent discussion of how both ideology and economic interest influenced the American regulatory process, see Morton Keller, "The Pluralist State: American Economic Regulation in Comparative Perspective, 1900–1930," and Samuel P. Hays, "Political Choice in Regulatory Administration," both in *Regulation in Perspective,* ed. Thomas K. McCraw (Cambridge: Harvard University Press, 1981), 56–94 and 124–54.
56. *Electrical World* 59, no. 17 (1912): 880; 59, no. 23, (1912): 1242.
57. H. E. Rawson, letter to the editor, *Scientific American* 106, no. 23 (1912): 515.
58. House Reports, 62d Cong., 2d sess., Dec. 4, 1911–Aug. 26, 1912, vol. 3, report 582.
59. *New York Times,* Apr. 20, 1912, 2; *Electrical World* 60, no. 4 (1912): 183; *New York Times,* May 2, 1912, 1, and June 4, 1912, 2.

60. Kahn, *Documents*, 8–16; *Literary Digest* 45, no. 17 (1912): 716.
61. John Bottomley, *Annual Report*, Marconi Wireless Telegraph Company of America, 1912, Chelmsford Collection.
62. "'Wireless' and the 'Titanic,' an Authorized Interview with Guglielmo Marconi," *World's Work* 24 (June 1912): 225.
63. *New York Times*, Oct. 7, 1912, 11, and Dec. 15, 1912, pt. 5, 3.
64. E. L. Bragdon, Reminiscences, 7, Columbia Oral History Library.
65. De Soto, *Two Hundred Meters*, 34–35.
66. Edgar S. Love, Reminiscences, 6, Columbia Oral History Library.
67. Gabriel Kolko, *The Triumph of Conservatism* (New York: Free Press, 1963), and James Weinstein, *The Corporate Ideal in the Liberal State, 1900–1918* (Boston: Beacon Press, 1968).
68. Harvey J. Levin, *The Invisible Resource* (Baltimore: Johns Hopkins Press for Resources for the Future, 1971), 9, 18.

EIGHT: THE RISE OF MILITARY AND CORPORATE CONTROL

1. *New York Times*, Oct. 22, 1915, 1.
2. Ibid., Jan. 26, 1915, 1.
3. Ibid., Sept. 30, 1915, 1.
4. Ibid., Oct. 1, 1915, 1.
5. Leonard S. Reich, *The Making of American Industrial Research: Science and Business at G.E. and Bell, 1876–1926* (Cambridge: Cambridge University Press, 1985), chap. 6.
6. Ibid.
7. For more details see Hugh G. J. Aitken, *The Continuous Wave* (Princeton: Princeton University Press, 1985), 234–49.
8. Georgette Carneal, *A Conqueror of Space: The Life of Lee De Forest* (New York: Horace Liveright, 1930), 257; Gleason Archer, *The History of Radio to 1926* (New York: American Historical Society, 1938), 106–9.
9. *New York Times*, Nov. 27, 1913, 15.
10. Ibid.
11. Ibid., Dec. 10, 1913, 9.
12. Ibid., Dec. 6, 1913, 4.
13. Samuel Lubell, "Magnificent Failure," *Saturday Evening Post*, Jan. 31, 1942, 27.
14. Ibid., Jan. 24, 1942, 43.
15. Carneal, *Conqueror of Space*, 265.
16. W. Rupert Maclaurin, *Invention and Innovation in the Radio Industry* (New York: Macmillan, 1949), 78.
17. Lubell, "Magnificent Failure," Jan. 31, 1942, 40.
18. Ibid.
19. Carneal, *Conqueror of Space*, 267.
20. *New York Times*, Oct. 22, 1915, 3.
21. Lubell, "Magnificent Failure," Jan. 31, 1942, 38.
22. Ibid., 41.
23. *New York Times*, Sept. 30, 1915, 1.
24. Ibid., Oct. 1, 1915, 3.

25. Edward Marshall, "What Transcontinental Wireless Phone Means," *New York Times Magazine,* Oct. 17, 1915, sec. 4, 9.
26. "A Wireless Telephone Message across the Sea," *Literary Digest* 51 (Oct. 16, 1915): 833–34.
27. "The Success of Wireless Telephony," *World's Work* 51 (Nov. 1915): 14.
28. "Hello! Aloha!" *Independent* 84 (Oct. 11, 1915): 43.
29. Cited in Theodore P. Greene, *America's Heroes* (New York: Oxford University Press, 1970), 285.
30. Ibid., 289. See also Richard Hofstadter, *The Age of Reform* (New York: Random House, 1955), 195–96.
31. Cited in Greene, *America's Heroes,* 294.
32. Ernst Alexanderson, Reminiscences, 21, Columbia Oral History Library.
33. Ibid., 23.
34. *New York Times,* Sept. 23, 1953, 31, and Oct. 1, 1913, 6.
35. Ibid., Oct. 1, 1913, 6.
36. *Review of Reviews* 49 (March 1914): 327–35.
37. W. J. Baker, *A History of the Marconi Company* (London: Methuen, 1970), 150–71.
38. For a more detailed discussion of Elwell and the Federal Company, see Aitken, *Continuous Wave,* chap. 3.
39. Helen Fessenden, *Fessenden: Builder of Tomorrows* (New York: Coward-McCann, 1940), 214.
40. Ibid., 325–26.
41. Ibid., 328.
42. The exact amount of the award remains a mystery. The $500,000 figure comes from Maclaurin, *Invention and Innovation,* 63.
43. *Annual Report of the Secretary of the Navy,* 1910.
44. Ibid., 1912, 38–39.
45. Ibid.
46. Aitken, *Continuous Wave,* 90–91.
47. L. S. Howeth, *History of Communications-Electronics in the United States Navy* (Washington, D.C.: Government Printing Office, 1963), 114. Material on Hooper is from George H. Clark, "Radio in War and Peace," 1940, Clark Collection. In 1940, Clark, who had worked as a radio inspector for the navy between 1907 and 1919, persuaded Hooper to dictate his memoirs to Clark. Some of the material is based on Clark's own recollections of this period and some on Hooper's reminiscences.
48. Clark, "Radio in War and Peace," 70.
49. Ibid.
50. Ensign C. H. Maddox, "Report of Battle Practice, Autumn 1911" (Washington, D.C.: Government Printing Office, 1911), 230, Clark Collection.
51. Clark, "Radio in War and Peace," 84–85.
52. Lt. S. C. Hooper to Chief, Bureau of Steam Engineering, "Report on Radio-Telegraphy in the Atlantic Fleet, Spring Target Practice, 1912," Hooper Papers, Library of Congress.
53. Howeth, *History of Communications,* 195.

54. Ibid.
55. Clark, "Radio in War and Peace," 115. This is one of many examples cited by Clark.
56. Ibid., 120.
57. Ibid., 122.
58. Howeth, *History of Communications*, 195.
59. Ibid., 196.
60. Clark, "Radio in War and Peace," 158.
61. Aitken, *Continuous Wave*, 94–95.
62. Robert Greenhalgh Albion, *Makers of Naval Policy, 1798–1947* (Annapolis: U.S. Naval Institute Press, 1980), 13 and 218–19.
63. *Annual Report of the Secretary of the Navy*, 1915, 10, 263; 1916, 27.
64. Cited in Arthur S. Link, *Wilson: The Struggle for Neutrality, 1914–1915* (Princeton: Princeton University Press, 1960), 66.
65. *New York Times*, Aug. 6, 1914, 4.
66. Aitken, *Continuous Wave*, 283.
67. *New York Times*, Jan. 29, 1914, 1.
68. Ibid., Aug. 6, 1914, 4.
69. Ibid., Aug. 7, 1914, 7.
70. Ibid., Aug. 6, 1914, 4.
71. Ibid., Aug. 23, 1914, pt. 2, 7.
72. Ibid., Aug. 22, 1914, 5.
73. Ibid., Sept. 6, 1914, 14, and Sept. 10, 1914, 6.
74. Ibid., Sept. 11, 1914, 6.
75. Ibid., Sept. 20, 1914, 8.
76. Ibid., Oct. 9, 1914, 5.
77. Ibid., Oct. 22, 1914, 10.
78. Ibid., Apr. 23, 1915, 1.
79. Casualty figures from Link, *Wilson*, 372; infant death figures from *Literary Digest* 50 (May 22, 1915): 1198.
80. New York Times, May 11, 1915, 8.
81. Ibid., May 10, 1915, 8.
82. Ibid., July 1, 1915, 1.
83. Ibid., 2.
84. Charles Bright, "Wireless in Wartime," *Nineteenth Century* 77 (April 1915): 874–78; see also "Clandestine Wireless Stations Here and Abroad," *Scientific American* 116 (April 7, 1917): 340.
85. "To Choke German Wireless," *Literary Digest* 51 (July 31, 1915): 210.
86. *New York Times*, July 21, 1915, 7.
87. "What Our Government Found at Sayville," *Literary Digest* 51 (Sept. 11, 1915): 525.
88. *Annual Report of the Secretary of the Navy*, 1916, 27.
89. *New York Times*, Oct. 14, 1916, 10.
90. Ibid., Feb. 4, 1917, 9, Feb. 8, 1917, 4, and Apr. 8, 1917, 1.
91. Ibid., Nov. 30, 1914, 2.
92. Ibid.

93. Bright, "Wireless in Wartime," 875.
94. *Annual Report of the Secretary of the Navy,* 1917, 44.
95. Clark, "Radio in War and Peace," 301–2.
96. Ibid.
97. For more details on negotiations surrounding the New Brunswick alternator, see Aitken, *Continuous Wave,* 304–14.
98. Howeth, *History of Communications,* 253, 322, 247–48.
99. *Annual Report of the Secretary of the Navy,* 1918, 22.
100. Aitken, *Continuous Wave,* 292–94.
101. *Annual Report of the Secretary of the Navy,* 1918, 23.
102. John Brooks, *The Telephone* (New York: Harper & Row, 1975), 157–59.
103. David Kennedy, *Over Here* (New York: Oxford University Press, 1980), 253–56.
104. *New York Times,* Nov. 25, 1918, 1, 5.
105. Ibid., Dec. 18, 1918, 17.
106. Ibid., Jan. 18, 1919, 18.
107. Ibid., Jan. 30, 1919, 3.
108. Ibid., Feb. 8, 1919, 10.
109. Ibid., July 24, 1919, 1.
110. Ibid., July 25, 1919, 10.
111. "Navy Control of Wireless," *Independent* 99 (Sept. 20, 1919): 386.
112. Aitken, *Continuous Wave,* 322–23.
113. Details on these provisions in ibid., 345–50.
114. *New York Times,* Jan. 9, 1918, 1–2.
115. Ibid., Jan. 5, 1920, 1.
116. Aitken, *Continuous Wave,* 248.
117. Ibid., 440.

NINE: THE SOCIAL CONSTRUCTION OF AMERICAN BROADCASTING

1. Department of Commerce, *Annual Report* (Washington, D.C.: Government Printing Office, 1913), 148.
2. Clinton B. De Soto, *Two Hundred Meters and Down: The Story of Amateur Radio* (Hartford, Conn.: American Radio Relay League, 1936), 34.
3. Hiram Percy Maxim, "The Amateur in Radio," *Annals of the American Academy of Political and Social Science* 142 (Mar. 1929): 32.
4. Department of Commerce, *Annual Report,* 1916, 968.
5. Ibid., 1918, 810.
6. De Soto, *Two Hundred Meters,* 48.
7. Department of Commerce, *Annual Report,* 1916, 970–71.
8. Joseph D. Cappa, interviewed by Gordon Greb, 4, Columbia Oral History Library.
9. Edgar Felix, Reminiscences, Columbia Oral History Library.
10. Lloyd Espenschied, Reminiscences, 3, Columbia Oral History Library.
11. Georgette Carneal, *A Conqueror of Space: The Life of Lee De Forest* (New York: Horace Liveright, 1930); Felix, Reminiscences.

12. *New York Times,* Oct. 27, 1916, 7.
13. Ibid., Nov. 8, 1916, 6; Samuel Lubell, "Magnificent Failure," *Saturday Evening Post,* Jan. 31, 1942, 42.
14. *New York Times,* Dec. 31, 1916, 4.
15. Espenschied, Reminiscences, 13.
16. De Soto, *Two Hundred Meters,* 45–46.
17. Ibid., 43.
18. *New York Times,* Feb. 22, 1916, 2.
19. Ibid., Mar. 8, 1917, 9.
20. "Seventy-five Thousand American Boys Have This Enthusiasm," *American Magazine* 81 (June 1916): 104.
21. T. A. Collins, "Almost a Soldier," *Woman's Home Companion* 43 (October 1916): 32.
22. *New York Times,* Apr. 4, 1917, 14, and Apr. 7, 1917, 7.
23. "Work for Wireless Amateurs," *Literary Digest* 55 (Aug. 18, 1917): 23.
24. Ibid., 24.
25. De Soto, *Two Hundred Meters,* 68.
26. Erik Barnouw, *A Tower In Babel* (New York: Oxford University Press, 1966), 66–71.
27. Gleason Archer, *The History of Radio to 1926* (New York: American Historical Society, 1938), 200–201.
28. "The Long Arm of Radio Is Reaching Everywhere," *Current Opinion* 72 (May 1922): 684.
29. "Astonishing Growth of the Radiotelephone," *Literary Digest,* April 15, 1922, 28.
30. "'Listening In,' Our New National Pastime," *Review of Reviews* 67 (Jan. 1923): 52; Waldemar Kaempffert, "Radio Broadcasting," *Review of Reviews* 65 (Apr. 1922): 399.
31. "Radio: The New Social Force," *Outlook,* Mar. 19, 1924, 465.
32. *New York Times,* Mar. 2, 1922, 20.
33. Archer, *History of Radio,* 241.
34. Barnouw, *Tower in Babel,* 125.
35. A. Leonard Smith, Jr., "Broadcasting to Millions," *New York Times,* Feb. 19, 1922, sec. 7, 6.
36. Joseph K. Hart, "Radiating Culture," *Survey,* Mar. 18, 1922, 949.
37. Waldemar Kaempffert, "The Social Destiny of Radio," *Forum* 71 (June 1924): 772.
38. "That Prospective Communication with Another Planet," *Current Opinion* 66 (Mar. 1919): 170; "Those Martian Radio Signals," *Scientific American* 122 (Feb. 14, 1920): 156.
39. Thomas Walker, "Can We Radio a Message to Mars?" *Illustrated World* 33 (Apr. 1920): 242.
40. Hart, "Radiating Culture," 948.
41. Kaempffert, "Social Destiny of Radio," 771–72.
42. Stanley Frost, "Radio Dreams That Can Come True," *Collier's* 69 (June 10, 1922): 18.

43. Hart, "Radiating Culture," 949.
44. Frost, "Radio Dreams," 9.
45. Bruce Bliven, "How Radio Is Remaking Our World," *Century Magazine* 108 (June 1924): 149.
46. Ibid., 152.
47. Waldemar Kaempffert, "Social Destiny of Radio," 771.
48. Howard Vincent O'Brien, "It's Great to Be a Radio Maniac," *Collier's,* 74 (Sept. 13, 1924): 16.
49. Orange Edward McMeans, "The Great Audience Invisible," *Scribner's Magazine* 73 (Apr. 1923): 413.
50. O'Brien, "It's Great to Be a Radio Maniac," 16.
51. McMeans, "Great Audience Invisible," 412.
52. O'Brien, "It's Great to Be a Radio Maniac," 16.
53. Frost, "Radio Dreams," 18.
54. Bliven, "Radio Is Remaking Our World," 148.
55. McMeans, "Great Audience Invisible," 411.
56. Ibid.
57. O'Brien, "It's Great to Be a Radio Maniac," 16.
58. Frost, "Radio Dreams," 18.
59. Kaempffert, "Social Destiny of Radio," 768.
60. Hart, "Radiating Culture," 949.
61. Kaempffert, "Social Destiny of Radio," 768.
62. Raymond Francis Yates, "What Will Happen to Broadcasting?" *Outlook* 136 (Apr. 9, 1924): 604.
63. Hart, "Radiating Culture," 949.
64. Kaempffert, "Social Destiny of Radio," 768, 772.
65. Bliven, "Radio Is Remaking Our World," 153.
66. L. H. Rosenberg, "A New Era in Wireless," *Scientific American* 124 (June 4, 1921): 449.
67. McMeans, "Great Audience Invisible," 415.
68. Ibid., 411.
69. Bliven, "Radio Is Remaking Our World," 147.
70. Frost, "Radio Dreams," 18.
71. Hart, "Radiating Culture," 949.
72. Bliven, "Radio Is Remaking Our World," 154, 151.
73. "The Future of Radio," *New Republic* 40 (Oct. 8, 1924): 136.
74. Waldemar Kaempffert, "The Progress of Radio Broadcasting," *Review of Reviews* 66 (Sept. 1922): 305.

INDEX

Abbe, Cleveland, 45

Aerials, 16, 33, 54–55, 153, 180–181, 279

Aitken, Hugh G. J., xx, 113, 284

Alexander, J. W., 225, 281, 287

Alexanderson, Ernst F. W., 155, 164, 252–53

Alger, Horatio, 189, 192

Alternator, high frequency, 246, 257–58; and Alexanderson, 155, 252–53; and Fessenden, 46–47, 89, 153, 154–56, 160, 185; and General Electric, 278, 285; and Marconi Company, 253, 278, 284, 285

Amateur operators, 170, 291, 317; and broadcasting, 293, 299–303; estimated numbers, 198, 205, 207, 293, 298, 299; lobby against regulation, 223, 283; and masculinity, 190–92; and national relays, 296–97; and popular culture, 192–95, 198–200, 206–7; press coverage of, 187–90, 191–92, 210–11, 212–13, 297–98; regulation of, 234–36, 292–93; and technical adaptation, 197–98; and *Titanic* disaster, 228–29, 233; and U.S. Navy, 207–15, 220–21, 222–23; and World War I, 297–98

American De Forest Wireless Telegraph Company, 92–98, 167–68; publicity stunts, 92–94, 97–98; and stock promotion, 92–94, 97–98; strategy and structure, 92–98; and United Fruit, 95–96

American Magazine, 297

American Marconi Company. *See* Marconi Wireless Telegraph Company of America

American Radio Relay League (ARRL), 206, 295–97, 298

American Telephone and Telegraph (AT&T), 278, 281, 315, 316, 318, 319, 320; acquires audion rights, 243–44, 246–47, 293; and Fessenden, 157, 159–60, 163–64; and RCA, 288–90; press coverage of, 240–41, 247–52; transatlantic wireless, 240–41, 247–49; transcontinental telephone, 241; transcontinental wireless, 241. *See also* Bell Telephone

American Wireless Telephone and Telegraph Company, 56–57

America's Cup Races: of 1899, 9–10, 19–20, 29; of 1901, 56; of 1903, 94

Apgar, Charles, 273–74

Arcs, oscillating, 171, 172, 246, 254–55, 259

BOOKS IN THE SERIES

INVENTING AMERICAN BROADCASTING, 1899–1922

Designed by Martha Farlow.

Composed by the Composing Room of Michigan, Inc., in Rotation.

Printed by the Maple Press Company on 50-lb. Eggshell Cream Offset and bound in Holliston Roxite A cloth with Mohawk Ticonderoga endsheets.